Preparation, Characterization and Application of the Delivery System for Food Products

Preparation, Characterization and Application of the Delivery System for Food Products

Editors

Qing Guo
Shuai Chen

MDPI

Basel • Beijing • Wuhan • Barcelona • Belgrade • Novi Sad • Cluj • Manchester

Editors
Qing Guo
Tianjin University of Science and Technology
Tianjin
China

Shuai Chen
Wuhan University
Wuhan
China

Editorial Office
MDPI AG
Grosspeteranlage 5
4052 Basel, Switzerland

This is a reprint of articles from the Special Issue published online in the open access journal *Foods* (ISSN 2304-8158) (available at: https://www.mdpi.com/journal/foods/special_issues/QY182A1RS4).

For citation purposes, cite each article independently as indicated on the article page online and as indicated below:

Lastname, A.A.; Lastname, B.B. Article Title. *Journal Name* **Year**, *Volume Number*, Page Range.

ISBN 978-3-7258-2431-1 (Hbk)
ISBN 978-3-7258-2432-8 (PDF)
doi.org/10.3390/books978-3-7258-2432-8

Contents

About the Editors vii

Shuai Chen and Qing Guo
Preparation, Characterization and Application of the Delivery System for Food Products
Reprinted from: *Foods* **2023**, *12*, 4187, doi:10.3390/foods12234187 1

Wendi Teng, Zixiao Zhou, Jinxuan Cao and Qing Guo
Recent Advances of Natural Pentacyclic Triterpenoids as Bioactive Delivery System for Synergetic Biological Applications
Reprinted from: *Foods* **2024**, *13*, 2226, doi:10.3390/foods13142226 4

Yan Dai, Lei Wang, Xingyi Chen, Angxin Song, Laping He, Lingyuan Wang and Diandian Huang
Lentinula edodes Sing Polysaccharide: Extraction, Characterization, Bioactivities, and Emulsifying Applications
Reprinted from: *Foods* **2023**, *12*, 3289, doi:10.3390/foods12173289 20

Chen Meng, Yashu Chen, Xintian Wang, Hongjian Chen and Qianchun Deng
Effect of Different Temperatures on the Storage Stability of Flaxseed Milk
Reprinted from: *Foods* **2023**, *12*, 3571, doi:10.3390/foods12193571 37

Meishan Wu, Xiaoye He, Duo Feng, Hu Li, Di Han, Qingye Li, et al.
The Emulsifying Properties, In Vitro Digestion Characteristics and Storage Stability of High-Pressure-Homogenization- Modified Dual-Protein-Based Emulsions
Reprinted from: *Foods* **2023**, *12*, 4141, doi:10.3390/foods12224141 51

Di Zhang, Xianrui Chen, Nitong Bu, Liying Huang, Huanglong Lin, Lizhen Zhou, et al.
Biosynthesis of Quercetin-Loaded Melanin Nanoparticles for Improved Antioxidant Activity, Photothermal Antimicrobial, and NIR/pH Dual-Responsive Drug Release
Reprinted from: *Foods* **2023**, *12*, 4232, doi:10.3390/foods12234232 73

Siyu Wang, Yuan Cheng, Jingyi Wang, Miao Ding and Ziluan Fan
Antioxidant Activity, Formulation, Optimization and Characterization of an Oil-in-Water Nanoemulsion Loaded with Lingonberry (*Vaccinium vitis-idaea* L.) Leaves Polyphenol Extract
Reprinted from: *Foods* **2023**, *12*, 4256, doi:10.3390/foods12234256 87

Ziyan Wang, Jingwen Li, Chao Peng, Bin Li, Qian Shen and Yijie Chen
Physicochemical Quantitative Analysis of the Oil–Water Interface as Affected by the Mutual Interactions between Pea Protein Isolate and Mono- and Diglycerides
Reprinted from: *Foods* **2024**, *13*, 176, doi:10.3390/foods13010176 102

Fan Su, Yongqiang Wu, Yanping Cao and Shaojia Wang
Differences in the Chromogenic Effect of Corn Starch and Potato Starch on Paprika Red Pigment and Structural Characterisation
Reprinted from: *Foods* **2024**, *13*, 191, doi:10.3390/foods13020191 115

Nasrin Beigmohammadi, Seyed Hadi Peighambardoust, Asad Mohammad Amini and Kazem Alirezalu
Enhancing Encapsulation Efficiency of Chavir Essential Oil via Enzymatic Hydrolysis and Ultrasonication of Whey Protein Concentrate–Maltodextrin
Reprinted from: *Foods* **2024**, *13*, 1407, doi:10.3390/foods13091407 130

Liping Wang and Ziyun Wu
A Simple High-Throughput Technology for Microorganism Detection and Quantitative Analysis
Reprinted from: *Foods* **2024**, *13*, 2954, doi:10.3390/foods13182954 . . . . **152**

About the Editors

Qing Guo

Dr. Qing Guo carried out his doctoral research work at the China Agricultural University (Beijing, China) and the University of Massachusetts, Amherst (USA), and his Ph.D. was awarded by China Agricultural University. His research is focused on the development of novel emulsion-based delivery systems, especially for the development of starch-based emulsifiers and the effects of their interfacial behavior on the stability and digestion behavior of emulsions. He has published a total of 19 SCI papers, including 12 SCI papers as the first author or corresponding author in journals such as *Food Hydrocolloids* and *Food Chemistry*, with a cumulative Impact Factor of 90.1. Additionally, he has contributed as a co-author to the publication of two books. He is the Associate Editor of *Foods* (SCI, JCR Q1, IF:4.7) and an editorial board member of the *World Journal of Food Science and Technology*. He also serves as a reviewer for many high-quality journals in the field of food.

Shuai Chen

Dr. Shuai Chen carried out his doctoral research work at the China Agricultural University (Beijing, China) and the University of Massachusetts, Amherst (USA), and his Ph.D. was awarded by China Agricultural University. His research is focused on food hydrocolloids and protein–polysaccharide nanocomposites and delivery systems for bioactive compounds. He has published more than 60 papers, several patents, and several books. He is the Associate Editor of several journals, such as *eFood* (SCI, JCR Q1, IF:4.0), *Foods* (SCI, JCR Q1, IF:4.7), and *Frontiers in Nutrition* (SCI, JCR Q1, IF:4.0), and an editorial board member of the *Journal of Future Foods*, *Food Innovation and Advances*, *Food & Medicine Homology*, etc. He also serves as a reviewer for more than twenty high-quality journals in the field of food, such as *Trends in Food Science & Technology*, *Comprehensive Reviews in Food Science and Food Safety*, *Critical Reviews in Food Science and Nutrition*, *Food Chemistry*, and *Journal of Agricultural and Food Chemistry*.

foods MDPI

Editorial

Preparation, Characterization and Application of the Delivery System for Food Products

Shuai Chen [1,*] and Qing Guo [2,3]

1 School of Public Health, Wuhan University, Wuhan 430071, China
2 School of Food Science and Technology, Tianjin University of Science & Technology, Tianjin 300457, China; qingguo@tust.edu.cn
3 State Key Laboratory of Food Nutrition and Safety, Tianjin University of Science & Technology, Tianjin 300457, China
* Correspondence: chenshuai2016@hotmail.com; Tel.: +86-18811356812

1. Introduction

In the dynamic and evolving landscape of food science and technology, the quest to develop innovative and effective delivery systems for bioactive compounds remains a focal point of research and development [1]. This pursuit is fueled by the ever-increasing demand for functional foods that offer not only enhanced nutritional value, but also targeted and controlled release of bioactive components [2], thereby imparting health benefits to consumers [3]. The Special Issue of the journal *Foods* represents a significant milestone in the ongoing exploration of delivery systems designed to revolutionize the way we consume and benefit from bioactive compounds.

2. Delivery Systems for Food Products

Bioactive compounds, such as polysaccharides, peptides, polyphenols, alkaloids, sterols, saponins, etc., have gained recognition for their potential to promote human health and well-being [4]. However, realizing the full therapeutic potential of these compounds often requires overcoming several challenges, including their stability during food processing and storage [5], controlled release in the gastrointestinal tract, and interaction with the complex milieu of the intestinal microbiota [6].

3. Controlled Release of Bioactive Compounds

Controlled release technologies have emerged as a key focus area within food science [7]. The ability to modulate the release of bioactive compounds in the digestive system not only enhances their bioavailability, but also ensures that they reach their intended target sites, where their effects are most pronounced [8,9]. In this Special Issue, we delve into the latest advancements in controlled release mechanisms, from encapsulation techniques to nanostructured delivery systems. We explore how these innovative approaches can safeguard bioactive compounds from degradation, improve their solubility, and enable their sustained release, ultimately enhancing their efficacy.

4. Enhancing Stability and Bioavailability

Stability and bioavailability are critical factors governing the efficacy of bioactive compounds in functional foods [10,11]. The articles in this Special Issue explore various strategies for enhancing the stability of these compounds during food processing and storage. Furthermore, they investigate how delivery systems can facilitate the release of bioactives at specific points along the digestive tract, optimizing their absorption and utilization by the body.

Citation: Chen, S.; Guo, Q. Preparation, Characterization and Application of the Delivery System for Food Products. *Foods* **2023**, *12*, 4187. https://doi.org/10.3390/foods12234187

Received: 27 October 2023
Revised: 14 November 2023
Accepted: 15 November 2023
Published: 21 November 2023

5. Resistant Starch: A Resilient Carrier

Resistant starch, a unique dietary fiber, has garnered attention for its potential as a carrier for bioactive compounds [12]. Its resistance to digestion in the small intestine allows it to serve as a selective delivery vehicle for bioactives to the colon, where they can interact with the diverse and influential community of intestinal flora. This Special Issue also explores the utilization of resistant starch as a carrier system, shedding light on its potential to modulate gut health and microbiota composition.

6. Gut Health

The impact of bioactive compounds on gut health is a burgeoning area of research, with implications far beyond digestion [13]. This Special Issue also includes articles that provide insights into how bioactive delivery systems can influence the gut microbiota, potentially leading to improved metabolic outcomes, enhanced immune function, and reduced risk of chronic diseases.

7. Functional Food: From Concept to Reality

Ultimately, the culmination of research relating delivery systems for bioactive compounds is the creation of functional foods that offer tangible health benefits [14,15]. This Special Issue bridges the gap between scientific inquiry and practical application, with a focus on translating research findings into food product development. From fortified beverages to novel snacks, the potential for functional foods is vast, and this Special Issue offers a glimpse into their exciting future.

8. Conclusions

As we embark on this journey through the pages of the Special Issue, we invite readers to explore the cutting-edge research, innovative technologies, and transformative insights that promise to shape the future of food science. This collection of articles represents a collaborative effort by researchers, scientists, and industry experts dedicated to harnessing the power of delivery systems to unlock the full potential of bioactive compounds in food products. Together, we embark on a path toward healthier, more nutritious, and more enjoyable foods that have the potential to revolutionize our approach to nutrition and well-being.

Author Contributions: Conceptualization, S.C. and Q.G.; writing—original draft preparation, S.C.; writing—review and editing, S.C. and Q.G.; funding acquisition, S.C. All authors have read and agreed to the published version of the manuscript.

Funding: This research was funded by the National Natural Science Foundation of China (32202234); China Postdoctoral Science Foundation (2022M712473); the Fundamental Research Funds for the Central Universities (2042023kf1064); The Postdoctoral Innovative Research Position of Hubei Province; Hubei Provincial Administration of Traditional Chinese Medicine, Traditional Chinese Medicine Research Fund (ZY2023Q033); and the Shuguang project of Wuhan Science and Technology Bureau (2023KZ00593).

Conflicts of Interest: The authors declare no conflict of interest.

References

1. Zou, L.; Gao, H.; Liu, W.; McClements, D.J. Chapter 6–Hydrocolloids for the encapsulation and delivery of active compounds. In *Handbook of Hydrocolloids*, 3rd ed.; Phillips, G.O., Williams, P.A., Eds.; Woodhead Publishing: Cambridge, CA, USA, 2021; pp. 157–194.
2. Fathi, M.; Donsi, F.; McClements, D.J. Protein-based delivery systems for the nanoencapsulation of food ingredients. *Compr. Rev. Food Sci. Food Saf.* **2018**, *17*, 920–936. [CrossRef] [PubMed]
3. Peng, S.; Li, Z.; Zou, L.; Liu, W.; Liu, C.; McClements, D.J. Improving curcumin solubility and bioavailability by encapsulation in saponin-coated curcumin nanoparticles prepared using a simple pH-driven loading method. *Food Funct.* **2018**, *9*, 1829–1839. [CrossRef] [PubMed]

4. Zhang, R.; Han, Y.; Xie, W.; Liu, F.; Chen, S. Advances in protein-based nanocarriers of bioactive compounds: From microscopic molecular principles to macroscopical structural and functional attributes. *J. Agric. Food Chem.* **2022**, *70*, 6354–6367. [CrossRef] [PubMed]
5. Zhang, L.; McClements, D.J.; Wei, Z.; Wang, G.; Liu, X.; Liu, F. Delivery of synergistic polyphenol combinations using biopolymer-based systems: Advances in physicochemical properties, stability and bioavailability. *Crit. Rev. Food Sci. Nutr.* **2020**, *60*, 2083–2097. [CrossRef] [PubMed]
6. Ma, G.; Chen, Y. Polyphenol supplementation benefits human health via gut microbiota: A systematic review via meta-analysis. *J. Funct. Foods* **2020**, *66*, 103829. [CrossRef]
7. Liu, Z.; Liu, C.; Sun, X.; Zhang, S.; Yuan, Y.; Wang, D.; Xu, Y. Fabrication and characterization of cold-gelation whey protein-chitosan complex hydrogels for the controlled release of curcumin. *Food Hydrocoll.* **2020**, *103*, 105619–105629. [CrossRef]
8. Liu, L.; Yang, S.; Chen, F.; Cheng, K. Polysaccharide-zein composite nanoparticles for enhancing cellular uptake and oral bioavailability of curcumin: Characterization, anti-colorectal cancer effect, and pharmacokinetics. *Front. Nutr.* **2022**, *9*, 846282. [CrossRef] [PubMed]
9. Li, X.; Chen, W.; Hao, J.; Xu, D. Construction of different properties single and double cross-linked binary emulsion filled gels based on rice bran oil body emulsion. *Colloids Surf. A Physicochem. Eng. Asp.* **2023**, *665*, 131238. [CrossRef]
10. Gan, C.; Liu, Q.; Zhang, Y.; Shi, T.; He, W.; Jia, C. A novel phytosterols delivery system based on sodium caseinate-pectin soluble complexes: Improving stability and bioaccessibility. *Food Hydrocoll.* **2022**, *124*, 107295. [CrossRef]
11. Ban, C.; Jo, M.; Park, Y.H.; Kim, J.H.; Han, J.Y.; Lee, K.W.; Kweon, D.; Choi, Y.J. Enhancing the oral bioavailability of curcumin using solid lipid nanoparticles. *Food Chem.* **2020**, *302*, 125328. [CrossRef] [PubMed]
12. Dong, D.; Geng, T.; Cui, B.; Yuan, C.; Guo, L.; Zhao, M.; Zou, F.; Liu, P.; Zhang, H. Effect of octenyl succinic anhydride modified starch on soy protein-polyphenol binary covalently linked complexes. *Front. Nutr.* **2023**, *10*, 1093250. [CrossRef] [PubMed]
13. Mithul Aravind, S.; Wichienchot, S.; Tsao, R.; Ramakrishnan, S.; Chakkaravarthi, S. Role of dietary polyphenols on gut microbiota, their metabolites and health benefits. *Food Res. Int.* **2021**, *142*, 110189–110207. [CrossRef] [PubMed]
14. Gupta, E.; Mishra, P. Functional food with some health benefits, so called as superfood: A review. *Curr. Nutr. Food Sci.* **2020**, *17*, 144–166.
15. McClements, D.J. Delivery by design (DbD): A standardized approach to the development of efficacious nanoparticle–and microparticle–based delivery systems. *Compr. Rev. Food Sci. Food Saf.* **2018**, *17*, 200–219. [CrossRef] [PubMed]

foods

MDPI

Review

Recent Advances of Natural Pentacyclic Triterpenoids as Bioactive Delivery System for Synergetic Biological Applications

Wendi Teng [2], Zixiao Zhou [2], Jinxuan Cao [2] and Qing Guo [1,*]

[1] State Key Laboratory of Food Nutrition and Safety, School of Food Science and Technology, Tianjin University of Science and Technology, Tianjin 300457, China
[2] Key Laboratory of Geriatric Nutrition and Health, School of Food and Health, Beijing Technology and Business University, Beijing 100048, China; wenditeng@btbu.edu.cn (W.T.); zzxiao525@hotmail.com (Z.Z.); caojinxuan@btbu.edu.cn (J.C.)
* Correspondence: qingguo@tust.edu.cn

Abstract: Bioactive compounds have drawn much attention according to their various health benefits. However, poor dissolvability, low stability and limited bioavailability largely narrow their applications. Although a variety of nontoxic ingredients have been rapidly developed as vehicles to deliver bioactive compounds in the last few years, most of them are non-bioactive. Pentacyclic triterpenoids, owing to their unique self-assembly and co-assembly behaviors and different physiological functions, can construct bioactive carriers due to their higher biodegradability, biocompatibility and lower toxicity. In this paper, the basic classification, biological activities and physicochemical properties of pentacyclic triterpenoids were summarized. Additionally, applications of self-assembled and co-assembled pentacyclic triterpenoids as bioactive delivery systems to load bioactive components and future research directions were discussed. This study emphasizes the potential of pentacyclic triterpenoids as bioactive delivery systems, offering a new perspective for constructing self- or co-assemblies for further synergetic biological applications.

Keywords: pentacyclic triterpenoids; natural small molecules; biological activity; self-assembly; bioactive nanocarriers

Citation: Teng, W.; Zhou, Z.; Cao, J.; Guo, Q. Recent Advances of Natural Pentacyclic Triterpenoids as Bioactive Delivery System for Synergetic Biological Applications. *Foods* **2024**, *13*, 2226. https://doi.org/10.3390/foods13142226

Academic Editor: Lubomir Lapcik

Received: 17 June 2024
Revised: 11 July 2024
Accepted: 13 July 2024
Published: 16 July 2024

1. Introduction

Natural active components attract much attention due to their bioactivities in enhancing the nutritional values and preventing the risks of various diseases like cancers, obesity and diabetes [1,2]. However, bioactive compounds often cannot be completely absorbed by human body [3]. Most of them are not dissolved in water and are very sensitive to environmental stimuli like heat and oxygen, resulting in the decline of stability and limited bioavailability [4]. Moreover, after the bioactive components passed through the gastrointestinal tract, various enzymes and gastric acids would largely decrease the bioactivity [5]. For example, poor solubility of hydrophobic compounds such as resveratrol greatly prevent their dissolution in water and absorption under gastrointestinal conditions in the human body [6,7]. Liposoluble vitamins are often not stable at the stage of processing and storage, or under gastrointestinal conditions [8]. Meanwhile, limited diffusion of the components crossing the intestinal mucus and low permeability through intestinal epithelium largely impeded their bioavailability [9].

In order to overcome these shortcomings, some food matrices were used to embed bioactive compounds to increase their stability and bioavailability [10,11]. Many scientists are focusing on developing delivery platforms including emulsions, nanoparticles (NPs), liposomes and hydrogels, which increase their water solubility, protect from hostile external conditions, and improve intestinal cell permeability [12–14]. Nevertheless, these delivery platforms are principally macromolecules or amphiphilic compounds, showing insufficient bioactivities along with easy aggregation, low encapsulation efficiency (EE), loading capacity (LC) and potential safety concerns [15–17]. Thus, it is essential to develop new delivery carriers, especially with good bioactivities or biological activities. Recently, some natural small molecules (NSMs) have shown their potential to function as vehicles for their self-assembled supramolecular nanostructures [18–20]. The NSMs with bioactivities exhibited superb effects for bioactive delivery and remarkable synergistic bioactivities [21,22]. Among those NSMs, pentacyclic triterpenoids, as a kind of phytochemical, are secondary plant metabolites with chemical structures consisting of five-membered carbon rings [23]. Pentacyclic triterpenoids attracted great attention owing to their remarkable bioactivities but without toxicity [24]. Additionally, avoiding complicated chemical modification, they could form nanostructures via self-assembly or co-assembly, which would greatly enhance their water solubility, bioactivity and oral bioavailability, with no extra components involved [20]. Further, the self/co-assembled nanostructures have great potential as carriers to embed hydrophobic foods and drugs [4]. Pentacyclic triterpenoids showed synergistic biological activities compared with conventional drug delivery systems. Besides the advantages of higher water solubility, improved stability and controlled release, pentacyclic triterpenoids are capable of targeted delivery, offering promising roles for them except for purely biologically active components.

Many reviews have been published on the bioactivities or physicochemical properties of pentacyclic triterpenoids [25–27]. However, there are no comprehensive reports on the applications of pentacyclic triterpenoids as nanocarriers for the encapsulation of bioactive components. In this review, the basic classification, biological activities and physicochemical properties of pentacyclic triterpenoids were first summarized. Additionally, applications of self-assembled and co-assembled pentacyclic triterpenoids as delivery systems to load bioactive components were compared and discussed. Last, challenges and future research directions of pentacyclic triterpenoids as delivery systems were proposed and analyzed. This work offers comprehensive knowledge of pentacyclic triterpenoids and highlights their potential as bioactive nanocarriers, offering a novel perspective for constructing pentacyclic triterpenoids-based self/co-assemblies for synergetic effects.

2. Distribution of Pentacyclic Triterpenoids in Nature

Triterpenes are composed of six isoprene units with 30 carbon atoms, are secondary metabolites in nature [23]. Pentacyclic triterpenoids consist of 6-6-6-6-5 or 6-6-6-6-6 as skeletal units, which are one kind of active triterpenes commonly found in plants, monocotyledons, dicotyledons, fungi, marine organisms and pteridophytes [25]. They often exist in Leguminosae, Araliaceae, Cucurbitaceae, Compositae, Poaceae and other plants [28]. According to the skeletal structures, the main types of pentacyclic triterpenes were summarized in Figure 1, including oleanane-type, ursane-type, lupane-type and friedelane-type [29].

Oleanane-type are also called β-amyrin-derived pentacyclic triterpenoids. They are always presented in the form of free or combined saponins [30]. Oleanolic acid, glycyrrhetinic acid, arjunolic acid, maslinic acid, esculentoside and polygalacic acid belong to oleanane-type pentacyclic triterpenoids [31]. Ursane-type are as well known as α-amyrin-derived pentacyclic triterpenoids, which have 6-6-6-6-6 as skeletal units. They are often presented in the free form or glycosides from Ilex chinensis, Sanguisorba officinalis and others. Ursolic acid, corosolic acid and asiatic acid are the typical representatives [32]. There are 6-6-6-6-5 as skeletal units for the lupane-type pentacyclic triterpenoids. Among them, betulinic acid, betulonic acid, betulinol and lupeol are widely studied [33]. Friedelane originates from oleanane via methyl translocation. Friedelin and friedelanol, extracted from tripterygium wilfordii, belong to friedelane-type pentacyclic triterpenoids [34]. Furthermore, tingenone exists in Euonymus, and pristimerin is presented in Celastrus [35].

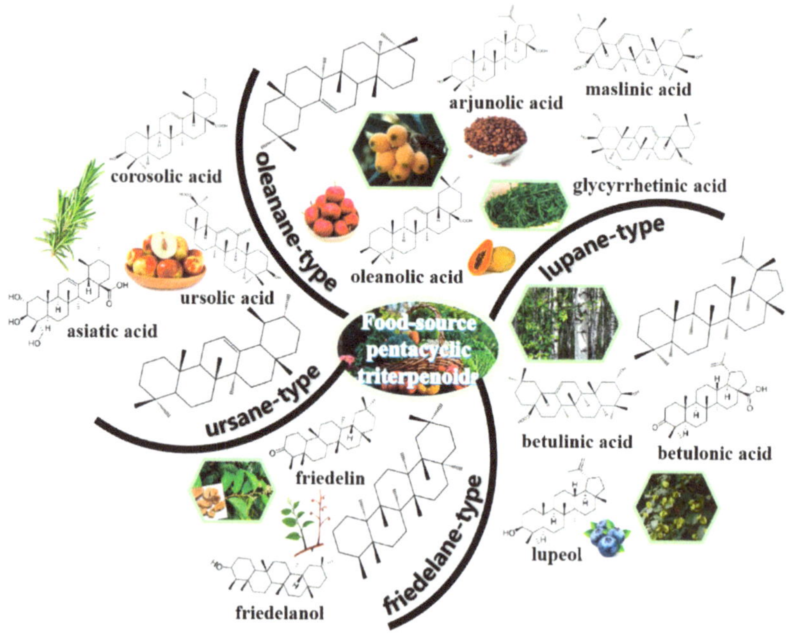

Figure 1. Overview of food-source pentacyclic triterpenoids. Four vital types are summarized, including oleanane-type (e.g., oleanolic acid, glycyrrhetinic acid, arjunolic acid, maslinic acid), ursane-type (e.g., ursolic acid, corosolic acid, asiatic acid), lupane-type (e.g., betulinic acid, betulonic acid, lupeol) and friedelane-type (e.g., friedelin, friedelanol).

Furthermore, a few other types of pentacyclic triterpenoids, like fernane and hopane, could also be found in nature [25]. Fernane and isofernane are the isomers of lupane. Fernenol, mainly extracted from Citrullus colocynthis leaves, is the typical representative of fernane-type pentacyclic triterpenoids [36]. Cylindrin, originating from Imperata cylindrica, is the typical representative of the isofernane type [37]. Hopane is the isomer of fernane. Hopane-type pentacyclic triterpenoids could be extracted from Aquilaria sinensis [38]. 2-hydroxydiplopterol was isolated from the metabolites produced by the halotolerant fungal strain Aspergillus variecolor B-17 [39].

3. Biological Activities of Pentacyclic Triterpenoids

Regularly intaking plant-based foods, such as pawpaw, Chinese ziziphus jujuba and strawberries could lower the incidence of various diseases [40]. The diverse structures of pentacyclic triterpenes determine their varied bioactivities. Pentacyclic triterpenes, with a perceptible improvement in the pharmacological activity, have been shown to have significant therapeutic and tissue-protective effects, including anti-tumor, hepatoprotective, glucose-regulating effects, among others (Figure 2), which are important for the improvement of body functions [25].

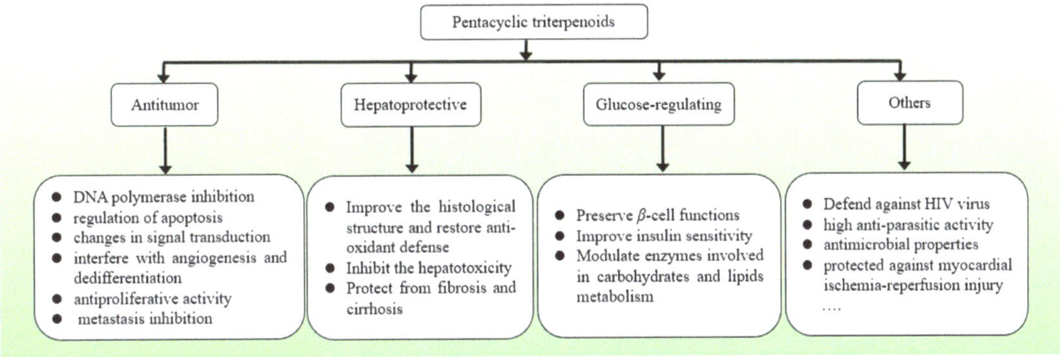

Figure 2. The main biological activities of pentacyclic triterpenoids include anti-tumor, hepatoprotective, glucose regulating and others.

3.1. Anti-Tumor Activity

In recent research, pentacyclic triterpenoids derived from plants have shown great potential antitumor activity [41]. They could function as molecular targets and cell proliferation regulatory, including inhibiting DNA polymerase, regulating apoptosis, modulating autophagy-related protein, altering the cell cycle in its early phases, changing signal transduction, interfering with angiogenesis and differentiation, inhibiting proliferation and metastasis [42]. Oleanolic acid showed potential antitumor activity in breast cancer, lung cancer and liver cancer, as well as no side effects [43]. The involved pathways included activation of AMP-activated protein kinase pathway, inhibition of PI3K Akt/mTOR/NF-κB signaling pathway, upregulation of p53 mediated activation of mitochondrial apoptotic pathway, and blocking the release of anti-apoptotic inhibitor of apoptotic protein (IAP) family proteins [44]. Ursolic acid exerted anticancer efficacy by inhibiting proliferation, increasing apoptosis, modulating growth receptors, decreasing cellular oxygen consumption, and enhancing NO levels [45]. Betulinic acid demonstrated a selective antitumor activity on human melanoma cell lines both in vitro and in vivo, but minimal toxicity against normal cells was observed [46]. Pentacyclic triterpenoids, naturally isolated active ingredients, could be used as a safer alternative in preventing cancers due to their cytotoxicity to cancer cells but less toxicity to normal cells.

3.2. Hepatoprotective Activity

The hepatoprotective activity of pentacyclic triterpenoids has a long history and mature application in China, from the hepatic protection of licorice in the traditional Chinese "Shen Nong Ben Cao Jing" to modern application of over-the-counter hepatoprotective drugs [47,48]. Glycyrrhetinic acid, as one of the most important functional components in licorice, was proven to have a hepatoprotective effect. Glycyrrhetinic acid demonstrated anti-inflammatory effects on propionibacterium acne-induced liver injury by inhibiting macrophage inflammatory protein (MIP)-1 in Kupffer cells by downregulating MyD88 and NF-κB. Glycyrrhetinic acid also regulated the proliferation of liver-infiltrating CD4+T cells and suppressed proinflammatory cytokines [49]. Oleanolic acid was used as a hepatic drug for more than 20 years due to hepatoprotective effects. Researchers concluded oleanolic acid could increase the activity of alanine, reduce liver inflammation, promote the regeneration of liver cells and enable rapid recovery of necrotic liver tissue. It not only prevented liver fibrosis and cirrhosis, but also resisted liver damage caused by various chemicals or active substances [50]. Ursolic acid was shown to protect tetrachloride-induced oxidative damage and greatly increase the survival rate of hepatocytes when compared to tetrachloride-exposed cells, suggesting ursolic acid had strong hepatic protective activity by maintaining the glutathione as well as restraining the malondialdehyde formation according to its radical scavenging functionalities [51]. Additionally, ursolic acid has been shown to protect against ethanol-mediated chronic liver injury through an antioxidant pathway [52].

3.3. Glucose-Regulating Activity

The antidiabetic activities of pentacyclic triterpenoids have been explored both in vivo and in vitro. Pentacyclic triterpenoids decreased plasma glucose, HbA1c, enhanced plasma insulin, muscle and liver glycogen, and showed protective effects on the liver, eye and heart of diabetic rats, suggesting that they had the potential to prevent diabetes and associated complications [53]. Oleanolic acid alleviates diabetes and metabolic syndrome by improving insulin sensitivity [54]. In rodent studies, oleanolic acid also attenuated fructose-induced hyperglycemia by regulating enzymes involved in carbohydrate digestion, insulin secretion and insulin signaling [55]. Ursolic acid was reported to inhibit pancreatic α-amylase activity and decrease the blood glucose level both in vivo and in vitro [56,57]. 2.5–10 μM of ursolic acid raised the expression of the transcription factors PPARγ, sterol regulatory factor binding protein 1c (SREBP-1c), fatty acid-synthase and fatty acid binding protein 4 (FABP4) [58]. Betulinic acid could enhance insulin sensitivity, and alleviate the blood glucose, inflammation, dyslipidemia and oxidative stress in high-fructose diet-induced metabolic syndrome by activating PI3K/Akt signal pathways [59]. Betulinic acid was also effective at ameliorating TNF-α-induced insulin resistance by preventing the negative regulator of insulin signaling and inflammation-activated protein kinase in adipocytes, thus potentially relieving insulin resistance [29,60].

3.4. Others

Pentacyclic triterpenoids have been reported in some other biological activities in the last few years. They were mainly used to inhibit viral replication and reduce reverse transcriptase activity and protease activity as a way to defend against self-replication of HIV [61]. Among 191 pentacyclic triterpenoids reported, 85 have shown high anti-parasitic activity against various species belonging to the genera of plasmodium, leishmania, trypanosoma, as well as various genera of nematoda [62]. Oleanolic acid protects against myocardial ischemia-reperfusion injury by enhancing glutathione and α-tocopherol-mediated mitochondrial antioxidation and ameliorates high glucose-induced endothelial dysfunction by activating PPARδ [63]. Ursolic acid has antimicrobial properties against several harmful pathogens including bacteria, viruses, fungi and parasites [64]. Ursolic acid could kill bacteria and inhibit bacterial growth by altering bacterial cell membranes and adhesion proteins [65]. In addition, cell morphology and expression levels of virulence factors, such as pili and fritillary were regulated by ursolic acid [66]. These highly biological activities with low toxicity made it possible for them to be used as alternatives to traditional chemotherapeutics.

4. Self-Assembly and Co-Assembly Properties of Pentacyclic Triterpenoids

Intermolecular noncovalent interactions, including electrostatic interactions, hydrogen bonding, hydrophobic interactions, π-π stacking interactions and van der Waals interactions could drive molecular assembly [67]. Spontaneous self-assembly forming supramolecular architectures have vital potentials in the fields of delivery vehicles, generation of hybrid materials, nanobiodiagnostics, and so on [68]. In the past few years, the self-assembly of natural compounds such as sugars, amino acids and fatty acids has been reported [69]. In 2011, Bag first found that several natural triterpenoids were of nano-meter dimensions with various rigid or flexible lengths, which could self-assemble in different solutions yielding nanoarchitectures like vesicles, fibers, flowers or spheres [70]. Among triterpenoids, pentacyclic triterpenoids are of great significance, which diversified structures lead to self-assembly properties: (1) Their fused 6-6-6-6-6 or 6-6-6-6-5 skeletal units provide a nano-sized rigid backbone; (2) The hydroxyl and carboxyl groups, with the amphiphilic nature and hydrogen bond, provide promising self-assembly properties; (3) The in-built centers of chirality make them useful as chiral building blocks. These different numbers and types of carboxyl, hydroxyl or other groups at different positions and spatial orientations have great influences on the self-assembly properties, leading to various sizes and morphologies [71]. Betulinic acid could self-assemble in 19 organic liquids and alcohol-water mixtures forming gels with a fibrillar network [72]. Betulin and lupeol enabled them to self-assemble in many liquids with various morphologies and similar rheological properties [73]. Likewise, both self-assembled glycyrrhetinic acid and oleanolic acid generated spheres, vesicles or fibrils to form supramolecular gels [74]. Additionally, ursolic acid was shown to stably self-assemble to supramolecular gel enhancing solubility and anticancer effects [75,76].

Co-assembly is an effective and convenient strategy to construct "multi-bioactive" delivery vehicles. Supramolecular co-assembly has been widely used to construct carrier platforms because of their practicality and simplicity [77]. With the increased self-assembled pentacyclic triterpenoids, constructing co-assembled vehicles along with multi-bioactivities has become possible. During the process of co-assembly, NPs could not only maintain the biological activity of each compound but also modulate the self-assembled morphologies and sizes to achieve combination therapy [22].

5. Applications of Pentacyclic Triterpenoids as Bioactive Delivery System

5.1. Directed Self-Assemblies

Pentacyclic triterpenoids are often widely distributed and could be derived from diversified plants. Their various biological activities help treat or prevent many diseases. However, low physicochemical stability and poor bioavailability limit the applications of bioactive compounds in various fields. For example, the bioavailabilities of curcumin, β-carotene and α-tocopherol were proved to be lower than 5%, which resulted in a waste of cost and time spent to produce them and showed inefficient health-promoting results [78]. In recent decades, encapsulating bioactive compounds to protect from multiple food processing and increase delivery efficiency has attracted great interest from food scientists, but nano-vehicles might go along with high toxicity as well as possible immunogenicity, therefore enhancing product costs and bringing unanticipated risks. Though various nontoxic ingredients have rapidly developed as vehicles to deliver bioactive compounds in the last few years, they are all non-bioactive [79]. Pentacyclic triterpenoids, owing to their unique self-assembly/co-assembly behaviors and different physiological functions, can construct carriers for further synergetic biological applications, which have brought great potential in many fields due to their more biodegradability, biocompatibility and lower toxicity. Until now, only a few pentacyclic triterpenoids have been developed for bioactive delivery, since not all pentacyclic triterpenoids can self-assemble to vehicles with a proper morphology and size, or good biological functions. All self-assembled pentacyclic terpenoids have been explored and compared in how to release active compounds and function as bioactive delivery carriers, and the specific information is listed in Table 1.

Table 1. Nanoparticle carriers for transporting bioactive components.

Carrier	Loaded Compounds	LC&EE	Interaction Forces	Physicochemical Properties	Functional Activities	References
Oleanolic acid	Car	32.6 ± 0.0%; 80.7 ± 0.1%.	Hydrogen bonding; hydrophobic interactions.	Increased aqueous solubility and protection against UV radiation ionic strength, and heat. Delayed release in simulated gastric conditions and controlled release in simulated intestinal conditions.	Enhanced hepatoprotective and antioxidant effects.	[80]
Oleanolic acid	PTX	17.1 ± 0.0%; 62.6 ± 0.0%.	Hydrophobic interactions; hydrogen bonding.	Increased stability, slowed and sustained release under acidic conditions.	Tumor inhibition rate was 76.54 ± 0.66%, about 18% higher than that of PTX group. Reduced liver injury.	[81]
Ursolic acid	PTX	23.12 ± 1.07%; 94.41 ± 4.28%,	Hydrophobic interactions; hydrogen bonding.	More stable in PBS for at least 15 days, in serum for 24 h and also under acidic conditions. Controlled release after entering the cells.	Prolonged plasma half-life. Tumor inhibition rate was 90.2%, 3.3 times higher than that of PTX group. Reduced liver injury.	[82]

Table 1. *Cont.*

Carrier	Loaded Compounds	LC&EE	Interaction Forces	Physicochemical Properties	Functional Activities	References
Ursolic acid	Rho B/CF/DOX		Hydrogen bonding; Van der Waals interactions.	Controlled release of Rho B at physiological conditions. Slow release of DOX at physiological conditions and pH 6.6.	–	[77]
Betulinic acid	Glyburide	–	–	Increase the delivery to the brain. Controllably released over three days.	Synergetic effects for ischemic stroke by antioxidant and anti-edema.	[23]
Betulonic acid	PTX	–	Hydrogen bonding; hydrophobic interaction.	Improved water solubility. More stable under acidic conditions.	Synergistic anti-tumor efficacy and minimize the side effects.	[83]
Betulonic acid	Ce6	–	π-π stacking; hydrophobic interactions.	Excellent water dispersity. Keep stable in water, cell cultural medium and PBS buffer (pH 7.4). Have better photostability.	prolonged blood circulation. Synergistic anticancer efficacy.	[84]
Arjunolic acid	Rho B/CF/DOX		Hydrogen bonding	Slow release of PTX at physiological pH (7.2).	–	[85,86]
Corosolic acid	Rho B/CF/DOX		Hydrogen bonding; Van der Waals interactions.	Triton X-100-triggered release of Rho-B.	–	[87,88]
Maslinic acid	Rho B/ CF/ DOX		Hydrogen bonding; lipophilic interactions.	Triton X-100-triggered release of DOX.	–	[71]

Car; carotene; PTX: paclitaxel; Rho B: Rhodamine B; CF: 5,6-carboxfluorescein; DOX: doxorubicin; Ce6: chlorin e6.

5.1.1. Oleanolic Acid

Limited water dispersibility and stability of β-carotene (Car) largely prevent its functions and uses in the field of the food industry. Self-assembled oleanolic acid was used as a delivery system to load hydrophobicCar. After the optimization, LC and EE reached 32.6% and 80.7%, respectively. Self-assembled oleanolic acid NPs could embed up to 300 mg/g of Car. Hydrogen bonding and hydrophobic interactions between oleanolic acid and Car were the main forces promoting self-assembly and capsulation. Compared to free Car, Car/oleanolic acid NPs greatly increased aqueous solubility and protected against UV radiation, ionic strength, and heat. Moreover, Car/oleanolic acid NPs offered delayed release in simulated gastric conditions and controlled release in intestinal conditions. In addition, Car/oleanolic acid NPs showed better hepatoprotective and antioxidant effects [80].

Paclitaxel (PTX) is an anti-cancer chemotherapeutic agent but with certain toxic side effects on normal cells. Self-assembled oleanolic acid was constructed to encapsulate PTX, with 17.1% of the maximal LC and 62.6% of maximal EE [81]. An amount of 280 mg/g of PTX could be encapsulated by oleanolic acid supramolecular NPs. Molecular simulation showed that the assembly process is mainly through hydrophobic interaction and hydrogen bonding interaction. Oleanolic acid could form an assembled nanostructure with the carboxyl group facing the continuous phase and the hydroxyl group dispersed phase. Along with the organic phase volatilized, oleanolic acid molecules gradually aggregate, forming a compactly packed bilayer shell structure through the hydrophobic interaction. Oleanolic acid went on to internally extend loosely to form the intermediate layer of oleanolic acid NPs. When adding the PTX, intermediate layers and the internal hydrophobic area were filled by aggregates via PTX-OA interaction, forming core-shell structures. PTX/oleanolic acid NPs showed slow and sustained release of PTX and were more stable under acidic conditions. In-vitro and in-vivo research demonstrated that PTX/ oleanolic acid NPs had synergistic antitumor effects due to different mechanisms, with an approximately 18% higher rate of tumor inhibition than the PTX group. Liver injury was largely decreased owing to the hepatoprotective activity of oleanolic acid. These studies showed that oleanolic acid could be used as bioactive nanocarriers to deliver hydrophobic nutrients, bioactive components and chemotherapeutic agents, so as to show synergetic effects and lowered side effects.

5.1.2. Ursolic Acid

Self-assembled ursolic acid NPs were developed as a new "self-contained bioactive nanocarrier" to load PTX by hydrogen bonding and hydrophobic interaction. The resulting PTX/ursolic acid NPs showed a uniformly dispersed nanosphere structure with 23.12 ± 1.07% of LC and 94.41 ± 4.28% of EE after optimization. The highest loading capacity of ursolic acid supramolecular NPs for PTX was up to 230 mg/g. They had excellent stability in PBS for at least 15 days, and in serum for 24 h. Likewise, they dissociated gradually and released active components after entering the cells. PTX/ursolic acid NPs significantly prolonged the plasma half-life and showed highly efficient synergistic antitumor efficacy. The inhibition rate in mice was up to 90%, 3.3 times higher than that in the PTX group. Further, PTX/ursolic acid NPs significantly alleviated liver injury, showing that ursolic acid, as a "self-contained bioactive nanocarrier", was biosafe and enabled reduced liver damage induced by chemotherapeutics via key antioxidant pathways [82].

At the same time, Braja Gopal Bag et al. demonstrated that the vesicular self-assembled ursolic acid gel was capable of entrapping cationic dye Rhodamine B (Rho-B), anionic dye 5,6-carboxfluorescein (CF) and chemotherapeutic agent doxorubicin (DOX) by comparative fluorescence intensity, respectively. Hydrogen bonding involving hydroxyl and carboxyl groups and the dispersion interactions by the triterpenoid backbones might play an important part in the self-assembly. An amount of 87% of Rho B was controllably released after 23 h 30 min at physiological conditions and 75% of DOX was released after 30 min at physiological conditions. Surprisingly, 88% of DOX was released after 60 min at pH 6.6, imitating the low pH environment of hypoxic tumors or endosomes [77]. All these works make ursolic acid possible as a prospective delivery vehicle for scientific and industrial fields.

5.1.3. Betulinic Acid

Glyburide is a promising tested agent for stroke by mediating cerebral edema, but has a limited ability to pass the blood-brain barrier [23]. Betulinic acid NPs were employed as a bioactive multifunctional vehicle to markedly increase the delivery of glyburide to the brain, leading to synergetic effects for ischemic stroke by nuclear factor erythroid 2-related factor 2 (Nrf2)/heme oxygenase (HO-1) antioxidant pathway, compared with either glyburide or betulinic acid NPs. 91% of glyburide was controllably released and retained in the brain over three days [23]. Currently, in clinical practice, a low dose (3 mg/d) of glyburide has been used, thus limiting its efficacy. However, a higher dose of glyburide would lead to hypoglycemia. Betulinic acid NPs decrease glyburide to be exposed to the circulatory system, lowering the risk of hypoglycemia. It not only enhanced the delivery of glyburide to brain tissue, but also decreased the side effects. Furthermore, glyburide-encapsulated betulinic acid NPs showed the simplest and high-efficiency way to intervene in both cerebral edema and oxidation, two critical targets for stroke treatment [51]. Additionally, glyburide/betulinic acid NPs improved patient convenience. Owing to the low brain retention and short blood half-life, continuous infusion for 72 h by osmatic pumps was needed for glyburide administration in vivo [12]. Glyburide/betulinic acid NPs would be sufficient to bring remarkable therapeutic benefits.

5.1.4. Betulonic Acid

Supramolecular self-assembled betulonic acid NPs were formed to load PTX mainly by hydrogen bond and hydrophobic interaction. The contact angle of PTX-loaded betulonic acid NPs was 77.5°, showing improved water dispersibility, and it helped keep stability under acidic conditions. In-vitro and in-vivo data found that PTX/betulonic acid NPs enabled precisive targeting, increased PTX bioavailability by increasing the permeability and retention effect, and showed superb synergistic anti-tumor efficacy by different antitumor mechanisms of betulonic acid and PTX. These self-assembled NPs showed various advantages of nanovehicles, like excellent water dispersibility, proper morphologies and sizes, prolonged plasma half-life and enhanced cell uptake and targeting. In addition, it increased biosafety and reduced the toxicity to normal tissues and cells [83].

In photodynamic therapy, chlorin e6 (Ce6) is often used as a photosensitizer because of its efficient antitumor effect. Assembly strategies were used to construct betulonic acid-mediated photosensitizers with no structural modifications. Intermolecular π-π stacking and hydrophobic interactions might be the primary driving forces. Ce6/betulonic acid NPs kept stable in water, cell cultural medium and PBS buffer, and also had better photostability and anti-photobleaching ability than free Ce6. The assembled Ce6/betulonic acid NPs prolonged blood circulation and exhibited a significantly increased, synergistic anticancer efficacy by promoting ROS generation [84].

5.1.5. Arjunolic Acid, Corosolic Acid and Maslinic Acid

Gel could often be utilized as smart scaffolds for extensive delivery, whose fibrillary structures enabled stable release and protected from premature degradation, extending the circulation time and lowering the cytotoxicity. Additionally, the fibrillary networks facilitated to be taken up easily by cells, thus increasing the therapeutic effects. Arjunolic acid, corosolic acid and maslinic acid can also respectively self-assemble to vesicular gels to entrap Rho B, CF or DOX in aqueous solvent mixtures respectively and controlled release of the entrapped molecules at physiological pH or mediated by surfactant, indicating their usefulness as delivery vehicles [85–87]. Researchers need to further explore their potential for use in health benefits.

5.2. Multi-Components Co-Assemblies

Different from multi-component-directed self-assemblies, supramolecular co-assemblies were novel chemical entities based upon various non-covalent interactions to combine with different components, which could function as co-delivery systems. All co-assembled pentacyclic terpenoids have been explored and the related information is listed in Table 2.

Table 2. Nanoparticle carriers for transporting bioactive components.

Carrier	Loaded Compounds	LC&EE	Interaction forces	Physicochemical Properties	Functional Activities	References
Oleanolic acid -glycyrrhetinic acid	PTX	15.1 ± 0.4%, 98.8 ± 1.0%	Hydrogen bonding; hydrophobic interactions.	Stable under acidic conditions. Have good dispersion and chemical stability.	Tumor inhibition rate was 82.6%, 23.7% higher than the PTX group. Reduced liver damage and nanotoxicity.	[22]
Glycyrrhetinic acid—oleanolic acid		-	Hydrogen bonding; hydrophobic interactions.	-	Tumor inhibition rate was Enhanced from 50.5% to 69.5%. Reliable biosafety.	[89]
Glycyrrhetinic acid-Liquidambaric acid		-	Hydrogen bonding; hydrophobic interactions.	-	Tumor inhibition rate was Enhanced from 37.3% to 82.9%. Reliable biosafety.	[89]
Ursolic acid -EGCG-aptamer		73.6%	Hydrophobic interactions; hydrogen bonding.	pH-responsive, released rapidly in acid conditions.	Synergistic anticancer effect. Enhanced tumor immune infiltration.	[90]

EGCG: epigallocatechin gallate.

5.2.1. Oleanolic Acid-Glycyrrhetinic Acid/PTX

A pure natural co-assembled delivery system was formed through directly co-assembling oleanolic acid and glycyrrhetinic acid through hydrogen bonding and hydrophobic interactions. PTX could be entrapped in the co-assembled vehicles with 15.1 ± 0.4% of maximum LC and 98.8 ± 1.0% of EE. The maximal loading of glycyrrhetinic acid-oleanolic acid co-assembled NPs for PTX was up to 150 mg/g. PTX-loaded oleanolic acid+ glycyrrhetinic acid NPs were relatively stable under acidic conditions, and also had good dispersion and chemical stability at room temperature over 2 months as well as in serum for at least 48 h. 84 ± 9.9% of PTX could gradually dissociate and continuously released after being engulfed by the cells. In-vitro and in-vivo studies have demonstrated synergistic anticancer activities and the tumor inhibition rate increased from 58.9% to 82.6%. In addition, compared with free PTX, they could alleviate liver damage by regulating key antioxidant pathways and reduce the risk of nanotoxicity, due to the unique anti-inflammatory and hepatoprotective effects of oleanolic acid and glycyrrhetinic acid [22].

5.2.2. Carrier-Free Nanodelivery System

A "core-shell" co-assembly carrier-free system was constructed based on ursolic acid self-assembled as core and epigallocatechin gallate (EGCG) self-polymerized to form a uniform layer as a shell with aptamer modification. The carrier-free system was pH-responsive and kept stable in a normal extracellular environment, but released rapidly in weak acid condition where tumor existed. It increased tumor cellular uptake and long circulation. Ursolic acid combined with EGCG activated the innate immunity resulting in a significant synergistic anti-hepatocellular carcinoma effect. Moreover, the carrier-free system showed low cytotoxicity in normal cells, good biosafety of organic tissues and enlarged tumor accumulation in vivo [90]. Glycyrrhetinic acid/oleanolic acid and Glycyrrhetinic acid/liquidambaric acid NPs were combined via a supramolecular assembly mechanism of non-covalent interactions to form a carrier-free system, which not only enhanced cellular uptake and passive targeting tumor ability, but also arrested cell cycle at different stages of cancer cells and induced apoptosis, demonstrating synergistic anticancer effects with reliable biosafety when comparing to every single component [89]. However, not all the co-assemblies would be effective. For example, co-assembled morphologies were not observed in ursolic acid/betulinic acid, glycyrrhetinic acid/lupeol and ursolic acid/liquidambaric acid [89]. It might be due to hydrophobic forces not driving to form supramolecular nanostructures based on their spatial orientation of carboxyl, hydroxyl or other groups.

6. Conclusions and Further Perspectives

Pentacyclic triterpenoids are pivotal natural active components with various biological activities, but poor water dispersibility, poor absorption and rapid metabolism, which is largely influenced their applications. Nowadays, a variety of vehicles have been developed to deliver pentacyclic triterpenoids [91]. Despite these merits, the drawback of these delivery systems is their low loading capacity. For example, the loading capacity of asiatic acid in glyceryltristearate solid lipid nanoparticles is just 2.1% to 5.0% [92]. Merely 3.35% to 12.05% ursolic acid was embedded by a nanostructured lipid carrier coated with chitosan oligosaccharides [93] (Das et al., 2017). Furthermore, large amounts of inactive vehicle materials not only expanded the cost but also brought safety concerns [94].

According to the current research, pentacyclic triterpenoids could be a good matrix that acts as biocompatible nanocarriers. Their spontaneous, reversible self-assembly results from the generation of intermolecular forces, mainly by hydrogen bonds of carboxyl and hydroxyl groups, as well as the dispersing forces of the non-polar triterpenoid skeleton. Discovering the self-assembly properties was recorded as a breakthrough in natural products. Unique nanostructures with various physical characteristics enabled them to form NPs in different sizes and shapes. Surprisingly, the loading capacity of self-assembled/co-assembled pentacyclic triterpenoids was quite higher than that of other delivery systems. Meanwhile, the spontaneous assembly of pentacyclic triterpenoids offered biological activities, which eliminated cytotoxicity and complicated synthesis; it not only provided effective strategies to develop novel therapies, but also helped establish new usage of old drugs.

Although the self-assembled NSMs have attracted much attention, various theoretical and technological problems still need to be overcome. We have proposed several research perspectives in both basic research and applications (Figure 3): (1) Though assembled pentacyclic triterpenoids were used for drug encapsulation, there have been few studies on entrapping food nutritional components for food applications. Nano-delivery systems based upon pentacyclic triterpenoids can be used in the food industry to develop products such as special medical foods, healthy nano-foods and nano-nutrient supplements to meet special needs for people with limited access to food, digestive and absorption disorders or specific disease states for nutrients or diets. They can also be used to prevent the occurrence of certain health problems and improve health status. It provides a novel strategy for nutraceuticals to improve instability, low bioavailability and synergistic effects with no toxicity. Thus, more in-vivo research is needed to explore the gastrointestinal fate of different pentacyclic triterpenoids as bioactive delivery systems so that their safety and efficacy can be established. The bioavailability assessment of bioactive pentacyclic triterpenoids in various food products still needs to be conducted. In addition, a strict regulatory system managing their applications in the food field would be addressed. (2) There is an increasing focus on exploring precise treatments for the public. Designing "smart" nanoformulations, that can release bioactive compounds from the vehicles with stimulus, is one of the future directions [95]. For example, the chitosan at the surface of nanovehicles facilitated the rapid release of the bioactive compounds to a low-pH environment, a microenvironmental characteristic of the tumors, thus enhancing the therapy efficiency [96]. Pentacyclic triterpenoids-derived NPs endow tissues with targeting ability, allowing for targeted delivery with no additional chemical modifications. The cancer-targeting was reported to be achieved by combining folate, lactoferrin or lactobionic acid residues with the nanocarriers, increasing the cellular uptake [97,98]. Deng et al. reported that betulinic acid could interact with the CB1 receptor, which facilitated efficient brain penetration [23]. Reversible non-covalent interactions could highly respond to stimuli such as pH, light and heat. In addition, the targeting of nanovehicles to the proper tissues or organs could decrease the side effects of treatment. Thus, pentacyclic triterpenoids can be designed as vehicle platforms with stimuli-responsiveness or active targeting to enlarge the anticancer, antimicrobial, antiviral and other biological activities. (3) A large amount of research has reported that self-assembled/co-assembled pentacyclic triterpenoids could enhance their efficiency in antitumor treatment. Surprisingly, cytotoxicity was not observed when incubated with normal cell lines, showing superior selectivity for cancer therapy. Future scientists could not only concentrate on their anticancer effects but also pay more attention to their other biological activities. Owing to the self-assembly and co-assembly properties, it would be possible to develop combined prevention or treatments for metabolic syndrome such as hyperglycemia and hyperlipidemia. A few bioactive pentacyclic triterpenoids could be released in a changing environment, which would also help decrease the dosages, reduce the multi-drug resistance and achieve synergistic therapeutic effects. Further in-vivo and interdisciplinary studies are needed to assess the pros and cons of such bioactive delivery systems over extended periods. All efforts would contribute to the enhancement of natural pentacyclic triterpenoids as a bioactive delivery system for synergetic biological applications, therefore improving the life quality of human beings. (4) Nowadays, self-assembled pentacyclic triterpenoids are almost prepared via conventional or modified emulsion-solvent evaporation method, which limits more pentacyclic triterpenoids with assembled properties and bioactivities to be discovered, considering their wide structural diversities. Larger scales, lack of delivery practicability, low drug loading efficiency and poor aqueous solubility also largely limited their further applications. Thus, developing multiple efficient emerging techniques or modifications to existing methods that could improve scalability, drug loading efficiency, and aqueous solubility, and exploring the physicochemical properties and biological activities of more pentacyclic triterpenoids are urgently needed in the near future, and can significantly help expand the delivery development.

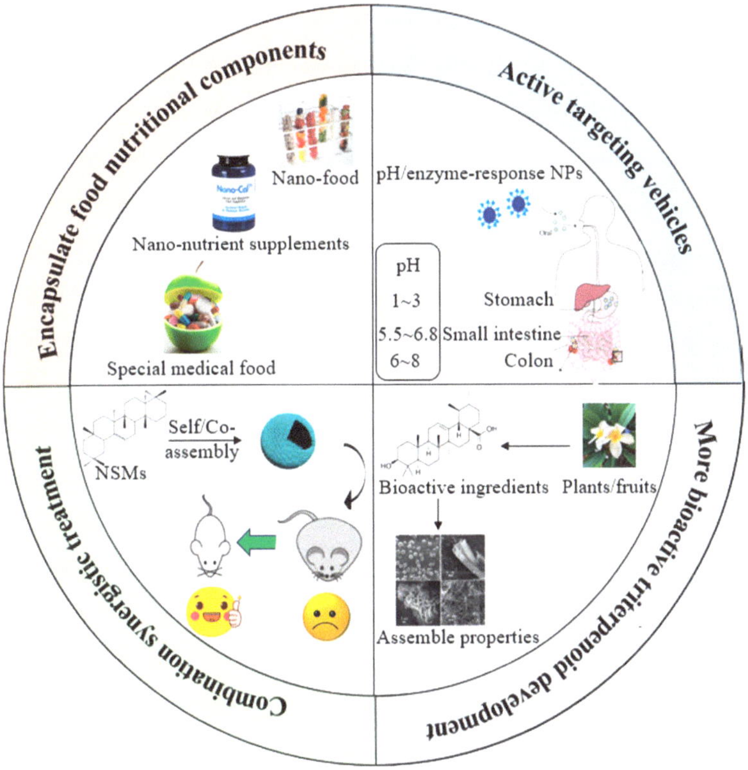

Figure 3. Overview of several research perspectives of pentacyclic triterpenoids in food science [21,76].

Author Contributions: Conceptualization, W.T. and Q.G.; methodology, writing—original draft preparation, Z.Z.; writing—review and editing, W.T. and J.C.; visualization, W.T. and Q.G.; supervision, J.C. and Q.G.; project administration, W.T. and Q.G.; funding acquisition, W.T. and Q.G. All authors have read and agreed to the published version of the manuscript.

Funding: This work was supported by the Open Project Program of the State Key Laboratory of Food Nutrition and Safety (No. SKLFNS-KF-202301) and the Ministry of Science and Technology of China, the fund of Cultivation Project of Double First-Class Disciplines of Food Science and Engineering, Beijing Technology & Business University (No. BTBUYXTD202202).

Institutional Review Board Statement: Not applicable.

Informed Consent Statement: Not applicable.

Data Availability Statement: No new data were created or analyzed in this study. Data sharing is not applicable to this article.

Conflicts of Interest: The authors declare that there are no conflicts of interest.

References

1. Oh, Y.S. Bioactive compounds and their neuroprotective effects in diabetic complications. *Nutrients* **2016**, *8*, 472. [CrossRef] [PubMed]
2. Datta, S.; Bhattacharya, S. Multifarious facets of sugar-derived molecular gels: Molecular features, mechanisms of self-assembly and emerging applications. *Chem. Soc. Rev.* **2015**, *44*, 5596–5637. [CrossRef]
3. Wani, T.A.; Shah, A.G.; Wani, S.M.; Wani, I.A.; Masoodi, F.A.; Nissar, N.; Shagoo, M.A. Suitability of different food grade materials for the encapsulation of some functional foods well reported for their advantages and susceptibility. *Crit. Rev. Food Sci. Nutr.* **2016**, *56*, 2431–2454. [CrossRef]

4. Huang, L.; Huang, X.H.; Yang, X.; Hu, J.Q.; Zhu, Y.Z.; Yan, P.Y.; Xie, Y. Novel nano-drug delivery system for natural products and their application. *Pharmacol. Res.* **2024**, *201*, 107100. [CrossRef] [PubMed]
5. Zhong, H.; Zhu, Y.; Yu, J.; Xu, X.; Gleeson, J.P.; Ryan, S.M.; Brayden, D.J. Oral delivery strategies for nutraceuticals: Delivery vehicles and absorption enhancers. *Trends Food Sci. Technol.* **2016**, *53*, 90–101.
6. Furr, H.C.; Clark, R.M. Intestinal absorption and tissue distribution of carotenoids. *J. Nutr. Biochem.* **1997**, *8*, 364–377. [CrossRef]
7. Teng, W.; Zhao, L.; Yang, S.; Zhang, C.; Liu, M.; Luo, J.; Jin, J.; Zhang, M.; Bao, C.; Li, D.; et al. The hepatic-targeted, resveratrol loaded nanoparticles for relief of high fat diet-induced nonalcoholic fatty liver disease. *J. Control. Release* **2019**, *307*, 139–149. [CrossRef] [PubMed]
8. Borel, P.; Dangles, O.; Kopec, R.E. Fat-soluble vitamin and phytochemical metabolites: Production, gastrointestinal absorption, and health effects. *Prog. Lipid Res.* **2023**, *90*, 101220. [CrossRef]
9. Gonnet, M.; Lethuaut, L.; Boury, F. New trends in encapsulation of liposoluble vitamins. *J. Control. Release* **2010**, *146*, 276–290. [CrossRef]
10. Katouzian, I.; Jafari, S.M. Nano-encapsulation as a promising approach for targeted delivery and controlled release of vitamins. *Trends Food Sci. Technol.* **2016**, *53*, 34–48. [CrossRef]
11. Guo, Y.; Qiao, D.; Zhao, S.; Zhang, B.; Xie, F. Starch-based materials encapsulating food ingredients: Recent advances in fabrication methods and applications. *Carbohyd. Polym.* **2021**, *270*, 118358. [CrossRef]
12. Chen, S.; McClements, D.J.; Jian, L.; Han, Y.; Dai, L.; Mao, L.; Gao, Y. Core–shell biopolymer nanoparticles for co-delivery of curcumin and piperine: Sequential electrostatic deposition of hyaluronic acid and chitosan shells on the zein core. *ACS Appl. Mater.* **2019**, *41*, 38103–38115. [CrossRef]
13. Huang, H.; Belwal, T.; Liu, S.; Duan, Z.; Luo, Z. Novel multi-phase nano-emulsion preparation for co-loading hydrophilic arbutin and hydrophobic coumaric acid using hydrocolloids. *Food Hydrocoll.* **2019**, *93*, 92–101. [CrossRef]
14. Wei, Z.; Gao, Y. Physicochemical properties of β-carotene bilayer emulsions coated by milk proteins and chitosan-EGCG conjugates. *Food Hydrocoll.* **2016**, *52*, 590–599. [CrossRef]
15. Kim, J.Y.; Huber, K.C. Preparation and characterization of corn starch β-carotene composites. *Carbohyd. Polym.* **2016**, *136*, 394–401. [CrossRef]
16. Zhu, F. Encapsulation and delivery of food ingredients using starch based systems. *Food Chem.* **2017**, *229*, 542–552. [CrossRef]
17. Cui, F.; Han, S.; Wang, J.; McClements, D.J.; Liu, X.; Liu, F. Co-delivery of curcumin and epigallocatechin gallate in W/O/W emulsions stabilized by protein fibril-cellulose complexes. *Colloids Surf. B Biointerfaces* **2023**, *222*, 113072. [CrossRef]
18. Fu, S.; Yang, X. Recent advances in natural small molecules as drug delivery systems. *J. Mater. Chem. B* **2023**, *11*, 4584–4599. [CrossRef] [PubMed]
19. Cheng, J.; Wang, S.; Zhao, H.; Liu, Y.; Yang, X. Exploring the self-assembly mechanism and effective synergistic antitumor chemophototherapy of a biodegradable and glutathione responsive ursolic acid prodrug mediated photosensitive nanodrug. *Biomater. Sci.* **2021**, *9*, 3762–3775. [CrossRef]
20. Lin, B.; Peng, X.; Cheng, J.; Wang, J. Natural gambogic acid-tuned self-assembly of nanodrugs towards synergistic chemophototherapy against breast cancer. *J. Mater. Chem. B* **2024**, *12*, 5940–5949. [CrossRef]
21. Yang, X.; Ma, C.; Chen, Z.; Liu, J.; Liu, Y.; Xie, R.; Zhao, H.; Deng, G.; Chen, A.T.; Gong, N.B.; et al. Single small molecule-assembled nanoparticles mediate efficient oral drug delivery. *Nano Res.* **2019**, *12*, 2468–2476. [CrossRef]
22. Wang, J.; Zhao, H.; Qiao, W.; Cheng, J.; Han, Y.; Yang, X. Nanomedicine-cum-carrier by co-assembly of natural small products for synergistic enhanced antitumor with tissues protective actions. *ACS Appl. Mater.* **2020**, *12*, 42537–42550. [CrossRef] [PubMed]
23. Deng, G.; Ma, C.; Zhao, H.; Zhang, S.; Liu, J.; Liu, F.; Chen, Z.; Chen, A.T.; Yang, X.; Avery, J.; et al. Anti-edema and antioxidant combination therapy for ischemic stroke via glyburide-loaded betulinic acid nanoparticles. *Theranostics* **2019**, *9*, 6991–7002. [CrossRef] [PubMed]
24. Ghante, M.H.; Jamkhande, P.G. Role of pentacyclic triterpenoids in chemoprevention and anticancer treatment: An overview on targets and underling mechanisms. *J. Pharmacopunct.* **2019**, *22*, 55–67. [CrossRef] [PubMed]
25. Li, Y.; Wang, J.; Li, L.; Song, W.; Li, M.; Hua, X.; Wang, Y.; Yuan, J.; Xue, Z. Natural products of pentacyclic triterpenoids: From discovery to heterologous biosynthesis. *Nat. Prod. Rep.* **2023**, *40*, 1303–1353. [CrossRef] [PubMed]
26. Zheng, X.Q.; Song, L.X.; Han, Z.Z.; Yang, Y.B.; Zhang, Y.; Gu, L.H.; Yang, L.; Chou, G.X.; Wang, Z.T. Pentacyclic triterpenoids from spikes of *Prunella vulgaris* L. with thyroid tumour cell cytostatic bioactivities. *Nat. Prod. Res.* **2023**, *37*, 1518–1526. [CrossRef] [PubMed]
27. Zhang, S.; Peng, B.; Chen, Z.; Yu, J.; Deng, G.; Bao, Y.; Ma, C.; Du, F.; Sheu, W.C.; Kimberly, W.T.; et al. Brain-targeting, acid-responsive antioxidant nanoparticles for stroke treatment and drug delivery. *Bioact. Mater.* **2022**, *16*, 57–65. [CrossRef] [PubMed]
28. Wang, C.; Wu, P.; Tian, S.; Xue, J.; Xu, L.; Li, H.; Wei, X. Bioactive pentacyclic triterpenoids from the leaves of Cleistocalyx operculatus. *J. Nat. Prod.* **2016**, *79*, 2912–2923. [CrossRef] [PubMed]
29. Xu, R.; Fazio, G.C.; Matsuda, S.P.T. On the origins of triterpenoid skeletal diversity. *Phytochemistry* **2004**, *65*, 261–291. [CrossRef]
30. Wagle, A.; Seong, S.H.; Jung, H.A.; Choi, J.S. Identifying an isoflavone from the root of Pueraria lobata as a potent tyrosinase inhibitor. *Food Chem.* **2019**, *15*, 383–389. [CrossRef]
31. Yang, Y.H.; Dai, S.Y.; Deng, F.H.; Peng, L.H.; Li, C.; Pei, Y.H. Recent advances in medicinal chemistry of oleanolic acid derivatives. *Phytochemistry* **2022**, *203*, 113397. [CrossRef]

32. Guo, S.; Duan, J.; Tang, Y.; Qian, Y.; Zhao, J.; Qian, D.; Su, S.; Shang, E. Simultaneous qualitative and quantitative analysis of triterpenic acids, saponins and flavonoids in the leaves of two Ziziphus species by HPLC-PDA-MS/ELSD. *J. Pharm. Biomed. Anal.* **2011**, *56*, 264–270. [CrossRef]
33. Laszczyk, M.N. Pentacyclic triterpenes of the lupane, oleanane and ursane group as tools in cancer therapy. *Planta Med.* **2009**, *75*, 1549–1560. [CrossRef]
34. Salvador, J.A.R.; Leal, A.S.; Valdeira, A.S.; Gonçalves, B.M.F.; Alho, D.P.S.; Figueiredo, S.A.C.; Silvestre, S.M.; Mendes, V.I.S. Oleanane-, ursane-, and quinone methide friedelane-type triterpenoid derivatives: Recent advances in cancer treatment. *Eur. J. Med. Chem.* **2017**, *142*, 95–130. [CrossRef] [PubMed]
35. Yousef, B.A.; Hassan, H.M.; Zhang, L.Y.; Jiang, Z.Z. Anticancer potential and molecular targets of pristimerin: A mini-review. *Curr. Cancer Drug Targets* **2017**, *17*, 100–108. [CrossRef] [PubMed]
36. Ahmed, M.; Qin, P.; Ji, M.; An, R.; Guo, H.; Shafi, J. Spinasterol, 22,23-Dihydrospinasterol and Fernenol from *Citrullus colocynthis* L. with Aphicidal Activity against *Cabbage aphid Brevicoryne brassicae* L. *Molecules* **2020**, *25*, 2184. [CrossRef] [PubMed]
37. Li, X.; Huang, X.; Feng, Y.; Wang, Y.; Guan, J.; Deng, B.; Chen, Q.; Wang, Y.; Chen, Y.; Wang, J.; et al. Cylindrin from Imperata cylindrica inhibits M2 macrophage formation and attenuates renal fibrosis by downregulating the LXR-α/PI3K/AKT pathway. *Eur. J. Pharmacol.* **2023**, *950*, 175771. [CrossRef] [PubMed]
38. Li, X.Y.; Lv, J.M.; Cao, Z.Q.; Wang, G.Q.; Lin, F.L.; Chen, G.D.; Qin, S.Y.; Hu, D.; Gao, H.; Yao, X.S. Biosynthetic characterization of the antifungal fernane-type triterpenoid polytolypin for generation of new analogues via combinatorial biosynthesis. *Org. Biomol. Chem.* **2023**, *21*, 851–857. [CrossRef] [PubMed]
39. Wang, W.; Liu, P.; Zhang, Y.; Li, J.; Tao, H.; Gu, Q.; Zhu, W. 2-Hydroxydiplopterol, a new cytotoxic pentacyclic triterpenoid from the halotolerant fungus Aspergillus variecolor B-17. *Arch. Pharm. Res.* **2009**, *32*, 1211–1214. [CrossRef]
40. Carrero, J.J.; Ortiz, A.G.; Avesani, C.M.; Bakker, S.J.L.; Bellizzi, V.; Chauveau, P.; Clase, C.M.; Cupisti, A.; Cuevas, A.E.; Molina, P.; et al. Plant-based diets to manage the risks and complications of chronic kidney disease. *Nat. Rev. Nephrol.* **2020**, *16*, 525–542. [CrossRef]
41. Ghosh, A.; Panda, C.K. Role of pentacyclic triterpenoid acids in the treatment of bladder cancer. *Mini-Rev. Med. Chem.* **2022**, *22*, 1331–1340. [CrossRef]
42. Mandal, A.; Ghosh, S.; Bothra, A.K.; Nanda, A.K.; Ghosh, P. Synthesis of friedelan triterpenoid analogs with DNA topoisomerase IIα inhibitory activity and their molecular docking studies. *Eur. J. Med. Chem.* **2012**, *54*, 137–143. [CrossRef]
43. Li, W.; Song, Y.; Zhang, P.; Zhu, H.; Chen, L.; Xiao, Y.; Xing, Y. Oleanolic acid inhibits cell survival and proliferation of prostate cancer cells in vitro and in vivo through the PI3K/Akt pathway. *Tumor. Biol.* **2016**, *37*, 7599–7613. [CrossRef]
44. Shyu, M.H.; Kao, T.C.; Yen, G.C. Oleanolic acid and ursolic acid induce apoptosis in HuH7 human hepatocellular carcinoma cells through a mitochondrial-dependent pathway and downregulation of XIAP. *J. Agric. Food. Chem.* **2010**, *58*, 6110–6118. [CrossRef]
45. Mlala, S.; Oyedeji, A.O.; Gondwe, M.; Oyedeji, O.O. Ursolic acid and its derivatives as bioactive agents. *Molecules* **2019**, *24*, 2751. [CrossRef]
46. Zuco, V.; Supino, R.; Righetti, S.C.; Cleris, L.; Marchesi, E.; Gambacorti-Passerini, C.; Formelli, F. Selective cytotoxicity of betulinic acid on tumor cell lines, but not on normal cells. *Cancer Lett.* **2002**, *175*, 17–25. [CrossRef]
47. Lin, C.; Wen, X.; Sun, H. Oleanolic acid derivatives for pharmaceutical use: A patent review. *Expert Opin. Ther. Pat.* **2016**, *26*, 643–655. [CrossRef]
48. Wu, S.Y.; Wang, W.J.; Dou, J.H.; Gong, L.K. Research progress on the protective effects of licorice-derived 18β-glycyrrhetinic acid against liver injury. *Acta Pharmacol.* **2021**, *42*, 18–26. [CrossRef]
49. Xiao, Y.; Xu, J.; Mao, C.; Jin, M.; Wu, Q.; Zou, J.; Gu, Q.; Zhang, Y.; Zhang, Y. 18β-glycyrrhetinic acid ameliorates acute Propionibacterium acnes-induced liver injury through inhibition of macrophage inflammatory protein-1alpha. *J. Biol. Chem.* **2010**, *285*, 1128–1137. [CrossRef]
50. Pollier, J.; Goossens, A. Oleanolic acid. *Phytochemistry* **2012**, *77*, 10–15. [CrossRef]
51. Aragón, S.M.; Heras, B.; Reus, M.I.S.; Benedi, J. Pharmacological modification of endogenous antioxidant enzymes by ursolic acid on tetrachloride-induced liver damage in rats and primary cultures of rat hepatocytes. *Exp. Toxicol. Pathol.* **2001**, *53*, 199–206. [CrossRef]
52. Saravanan, R.; Viswanathan, P.; Pugalendi, K.V. Protective effect of ursolic acid on ethanol-mediated experimental liver damage in rats. *Life Sci.* **2006**, *78*, 713–718. [CrossRef]
53. Manna, P.; Sil, P.C. Impaired redox signaling and mitochondrial uncoupling contributes vascular inflammation and cardiac dysfunction in type 1 diabetes: Protective role of arjunolic acid. *Biochimie* **2012**, *94*, 786–797. [CrossRef]
54. Zheng, J.; He, J.; Ji, B.; Li, Y.; Zhang, X. Antihyperglycemic effects of platycodon grandiflorum (Jacq.) A. DC. extract on streptozotocin-induced diabetic mice. *Plant Foods Hum. Nutr.* **2007**, *62*, 7–11. [CrossRef]
55. Lee, H.; Kang, R.; Kim, Y.S.; Chung, S.I.; Yoon, Y. Platycodin D inhibits adipogenesis of 3T3-L1 cells by modulating kruppel-like factor 2 and peroxisome proliferator-activated receptor γ. *Phytother Res.* **2010**, *24*, S161–S167. [CrossRef]
56. Li, S.; Liao, X.; Meng, F.; Wang, Y.; Sun, Z.; Guo, F.; Li, X.; Meng, M.; Li, Y.; Sun, C. Therapeutic role of ursolic acid on ameliorating hepatic steatosis and improving metabolic disorders in high-fat diet-induced nonalcoholic fatty liver disease rats. *PLoS ONE* **2014**, *9*, e86724.
57. Jung, S.H.; Ha, Y.J.; Shim, E.K.; Choi, S.Y.; Jin, J.L.; Yun-Choi, H.S.; Lee, J.R. Insulin-mimetic and insulin-sensitizing activities of a pentacyclic riterpenoid insulin receptor activator. *Biochem. J.* **2007**, *403*, 243–250. [CrossRef]

58. Seo, D.Y.; Lee, S.R.; Heo, J.W.; No, M.H.; Rhee, B.D.; Ko, K.S.; Kwak, H.B.; Han, J. Ursolic acid in health and disease. *Korean J. Physiol. Pharmacol.* **2018**, *22*, 235–248. [CrossRef]
59. Ajala-Lawal, R.A.; Aliyu, N.O.; Ajiboye, T.O. Betulinic acid improves insulin sensitivity, hyperglycemia, inflammation and oxidative stress in metabolic syndrome rats via PI3K/Akt pathways. *Arch. Physiol. Biochem.* **2020**, *126*, 107–115. [CrossRef]
60. Melo, C.L.; Queiroz, M.G.R.; Filho, A.C.V.A.; Rodrigues, A.M.; Sousa, D.F.; Almeida, J.G.L.; Pessoa, O.D.L.; Silveira, E.R.; Menezes, D.B.; Melo, T.S.; et al. Betulinic acid, a natural pentacyclic triterpenoid, prevents abdominal fat accumulation in mice fed a high-fat diet. *J. Agric. Food Chem.* **2009**, *57*, 8776–8781. [CrossRef]
61. Wu, H.F.; Natschke, S.L.M.; Xu, X.D.; Yang, M.H.; Cheng, Y.Y.; Yu, S.S.; Lee, K.H. Recent advances in natural anti-HIV triterpenoids and analogs. *Med. Res. Rev.* **2020**, *40*, 2339–2385. [CrossRef]
62. Isah, M.B.; Ibrahim, M.A.; Mohammed, A.; Aliyu, A.B.; Masola, B.; Coetzer, T.H.T. A systematic review of pentacyclic triterpenes and their derivatives as chemotherapeutic agents against tropical parasitic diseases. *J. Parasitol.* **2016**, *143*, 1219–1231. [CrossRef]
63. Zhang, Z.; Jiang, M.; Xie, X.; Yang, H.; Wang, X.; Xiao, L.; Wang, N. Oleanolic acid ameliorates high glucose-induced endothelial dysfunction via PPAR delta activation. *Sci. Rep.* **2017**, *7*, 40237. [CrossRef]
64. Jo, E.; Choi, M.H.; Kim, H.S.; Park, S.N.; Lim, Y.K.; Kang, C.K.; Kook, J.K. Antimicrobial effects of oleanolic acid, ursolic acid, and Sophoraflavanone G against enterococcus faecalis and propionibacterium acnes. *Int. J. Oral Biol.* **2014**, *39*, 75–79. [CrossRef]
65. Liu, R.; Zhang, L.; Xiao, S.; Chen, H.; Han, Y.; Niu, B. Ursolic acid, the main component of blueberry cuticular wax, inhibits Botrytis cinerea growth by damaging cell membrane integrity. *Food Chem.* **2023**, *415*, 135753. [CrossRef]
66. Micota, B.; Sadowska, B.; Podsędek, A.; Redzynia, M.; Różalska, B. *Leonurus cardiaca* L. herb--a derived extract and an ursolic acid as the factors affecting the adhesion capacity of Staphylococcus aureus in the context of infective endocarditis. *Acta Biochim. Pol.* **2014**, *61*, 385–388. [CrossRef]
67. Zhi, K.; Wang, J.; Zhao, H.; Yang, X. Self-assembled small molecule natural product gel for drug delivery: A breakthrough in new application of small molecule natural products. *Acta Pharm. Sin. B* **2020**, *10*, 913–927. [CrossRef]
68. Yadav, S.; Sharma, A.K.; Kumar, P. Nanoscale self-assembly for therapeutic delivery. *Front. Bioeng. Biotechnol.* **2020**, *8*, 127. [CrossRef]
69. Das, D.; Kar, T.; Das, P.K. Gel-nanocomposites: Materials with promising applications. *Soft Matter* **2012**, *8*, 2348–2365. [CrossRef]
70. Bag, B.G.; Paul, K. Vesicular and fibrillar gels by self-assembly of nanosized oleanolic acid. *Asian J. Org. Chem.* **2012**, *1*, 150–154. [CrossRef]
71. Bag, B.G.; Hasan, S.N.; Ghorai, S.; Panja, S.K. First self-assembly of dihydroxy triterpenoid maslinic acid yielding vesicles. *ACS Omega* **2019**, *4*, 7684–7690. [CrossRef]
72. Bag, B.G.; Dash, S.S. First self-assembly study of betulinic acid, a renewable nano-sized, 6-6-6-6-5 pentacyclic monohydroxy triterpenic acid. *Nanoscale* **2011**, *3*, 4564–4566. [CrossRef] [PubMed]
73. Zhi, K.; Zhao, H.; Yang, X.; Zhang, H.; Wang, J.; Wang, J.; Regenstein, J.M. Natural product gelators and a general method for obtaining them from organisms. *Nanoscale* **2018**, *10*, 3639–3643. [CrossRef] [PubMed]
74. Bag, B.G.; Majumdar, R. Self-assembly of a renewable nano-sized triterpenoid 18β-glycyrrhetinic acid. *RSC Adv.* **2012**, *2*, 8623–8626. [CrossRef]
75. Lu, J.; Wu, X.; Liu, L.; Chen, H.; Liang, Y. First organogelation study of ursolic acid, a natural ursane triterpenoid. *Chem. Lett.* **2016**, *45*, 860–862. [CrossRef]
76. Bag, B.G.; Das, S.S.; Hasana, S.N.; Barai, A.C. Nanoarchitectures by hierarchical self-assembly of ursolic acid: Entrapment and release of fluorophores including anticancer drug doxorubicin. *RSC Adv.* **2017**, *7*, 18136–18143. [CrossRef]
77. Hu, X.; Hou, B.; Xu, Z.; Saeed, M.; Sun, F.; Gao, Z.; Lai, Y.; Zhu, T.; Zhang, F.; Zhang, W.; et al. Supramolecular prodrug nanovectors for active tumor targeting and combination immunotherapy of colorectal cancer. *Adv. Sci.* **2020**, *7*, 1903332. [CrossRef] [PubMed]
78. Hoppe, P.P.; Krennrich, G. Bioavailability and potency of natural-source and all-racemic alpha-tocopherol in the human: A dispute. *Eur. J. Nutr.* **2000**, *39*, 183–193. [CrossRef] [PubMed]
79. Zhang, J.; Cui, Y.X.; Feng, X.N.; Cheng, M.; Tang, A.N.; Kong, D.M. pH-controlled intracellular in situ reversible assembly of photothermal agent for smart chemo-photothermal synergetic therapy and ATP imaging. *ACS Appl. Mater.* **2019**, *11*, 39624–39632. [CrossRef]
80. Liu, S.; Zhang, J.; Fu, R.; Feng, H.; Chu, Y.; Huang, D.; Liu, H.; Li, C.; Ma, C.; El-Aty, A.M.A. Improved stability, epithelial permeability and cellular antioxidant activity of β-carotene via encapsulation by self-assembled α-lactalbumin micelles. Improved stability and aqueous solubility of β-carotene via encapsulation in self-assembled bioactive oleanolic acid nanoparticles. *Food Chem.* **2022**, *373 Pt B*, 131498.
81. Wang, J.; Qiao, W.; Zhao, H.; Cheng, J.; Han, Y.; Yang, X. A highly atom-economical bioactive nanocarrier for synergistically enhanced antitumor with reduced liver injury. *New J. Chem.* **2020**, *44*, 16741–16751. [CrossRef]
82. Wang, J.; Zhao, H.; Zhi, K.; Yang, X. Exploration of the natural active small-molecule drug-loading process and highly efficient synergistic antitumor efficacy. *ACS Appl. Mater.* **2020**, *12*, 6827–6839. [CrossRef] [PubMed]
83. Simard, J.M.; Chen, M.; Tarasov, K.V.; Bhatta, S.; Ivanova, S.; Melnitchenko, L.; Tsymbalyuk, N.; West, G.A.; Gerzanich, V. Newly expressed SUR1-regulated NC(Ca-ATP) channel mediates cerebral edema after ischemic stroke. *Nat. Med.* **2016**, *12*, 433–440. [CrossRef] [PubMed]
84. Wang, J.; Qiao, W.; Zhao, H.; Yang, X. Paclitaxel and betulonic acid synergistically enhance antitumor efficacy by forming co-assembled nanoparticles. *Biochem. Pharmacol.* **2020**, *182*, 114232. [CrossRef] [PubMed]

85. Cheng, J.; Zhao, H.; Wang, J.; Han, Y. Bioactive natural small molecule-tuned coassembly of photosensitive drugs for highly efficient synergistic and enhanced Type I photochemotherapy. *ACS Appl. Mater.* **2020**, *12*, 43488–43500. [CrossRef]
86. Bag, B.; Majumdar, R. Vesicular self-assembly of a natural triterpenoid arjunolic acid in aqueous medium: Study of entrapment properties and in situ generation of gel-gold nanoparticle hybrid material. *RSC Adv.* **2014**, *4*, 53327–53334. [CrossRef]
87. Zheng, J.; Fan, R.; Wu, H.; Yao, H.; Yan, Y.; Liu, J.; Ran, L.; Sun, Z.; Yi, L.; Dang, L.; et al. Directed self-assembly of herbal small molecules into sustained release hydrogels for treating neural inflammation. *Nat. Commun.* **2019**, *10*, 1604. [CrossRef] [PubMed]
88. Bag, B.G.; Garai, C.; Ghorai, S. Vesicular self-assembly of a natural ursane-type dihydroxy-triterpenoid corosolic acid. *RSC Adv.* **2019**, *9*, 15190–15195. [CrossRef] [PubMed]
89. Wang, J.; Qiao, W.; Li, X.; Zhao, H.; Zhang, H.; Dong, A.; Yang, X. A directed co-assembly of herbal small molecules into carrier-free nanodrugs for enhanced synergistic antitumor efficacy. *J. Mater. Chem. B* **2021**, *9*, 1040–1048. [CrossRef] [PubMed]
90. Zhang, B.; Jiang, J.; Wu, P.; Zou, J.; Le, J.; Lin, J.; Li, C.; Luo, B.; Zhang, Y.; Huang, R.; et al. A smart dual-drug nanosystem based on co-assembly of plant and food-derived natural products for synergistic HCC immunotherapy. *Acta Pharm. Sin. B* **2021**, *11*, 246–257. [CrossRef]
91. Kaps, A.; Gwiazdoń, P.; Chodurek, E. Nanoformulations for delivery of pentacyclic triterpenoids in anticancer therapies. *Molecules* **2021**, *26*, 1764–1782. [CrossRef]
92. Garanti, T.; Stasik, A.; Burrow, A.J.; Alhnan, M.A.; Wan, K.W. Anti-glioma activity and the mechanism of cellular uptake of asiatic acid-loaded solid lipid nanoparticles. *Int. J. Pharmaceut.* **2016**, *500*, 305–315. [CrossRef] [PubMed]
93. Das, S.; Ghosh, S.; De, A.K.; Bera, T. Oral delivery of ursolic acid-loaded nanostructured lipid carrier coated with chitosan oligosaccharides: Development, characterization, in vitro and in vivo assessment for the therapy of leishmaniasis. *Int. J. Biol. Macromol.* **2017**, *102*, 996–1008. [CrossRef] [PubMed]
94. Dai, L.; Cao, X.; Liu, K.; Li, C.; Zhang, G.; Deng, L.; Si, C.; He, J.; Lei, J. 2015. Self-assembled targeted folate-conjugated eight-arm-polyethylene glycol-betulinic acid nanoparticles for co-delivery of anticancer drugs. *J. Mater. Chem. B* **2015**, *3*, 3754–3766. [CrossRef] [PubMed]
95. Saharkhiz, S.; Zarepour, A.; Zarrabi, A. A new theranostic pH-responsive niosome formulation for doxorubicin delivery and bio-imaging against breast cancer. *Int. J. Pharm.* **2023**, *637*, 122845. [CrossRef] [PubMed]
96. Zhu, Y.; Ling, J.; Xu, X.; Ouyang, X.K.; Wang, N. Redox and pH dual sensitive carboxymethyl chitosan functionalized polydopamine nanoparticles loaded with doxorubicin for tumor chemo-photothermal therapy. *Int. J. Biol. Macromol.* **2023**, *240*, 124488. [CrossRef] [PubMed]
97. Guo, M.; Lu, X.; Wang, Y.; Brodelius, P.E. Comparison of the interaction between lactoferrin and isomeric drugs. *Spectrochim. Acta A Mol. Biomol. Spectrosc.* **2017**, *173*, 593–607. [CrossRef]
98. Wu, Z.C.; Liu, X.Y.; Liu, J.Y.; Piao, J.S.; Piao, M.G. Preparation of betulinic acid galactosylated chitosan nanoparticles and their effect on liver fibrosis. *Int. J. Nanomed.* **2022**, *17*, 4195–4210. [CrossRef]

foods

MDPI

Article

Lentinula edodes Sing Polysaccharide: Extraction, Characterization, Bioactivities, and Emulsifying Applications

Yan Dai [1], Lei Wang [1,*], Xingyi Chen [1], Angxin Song [1], Laping He [1,2], Lingyuan Wang [1] and Diandian Huang [1]

[1] College of Liquor and Food Engineering, Guizhou University, Guiyang 550025, China; xychen3@gzu.edu.cn (X.C.); axsong@gzu.edu.cn (A.S.); helaping@163.com (L.H.)

[2] Key Laboratory of Agricultural and Animal Products Storage & Processing of Guizhou Province, Guizhou University, Guiyang 550025, China

* Correspondence: lwang25@gzu.edu.cn or wanglei6166@126.com; Tel./Fax: +86-0851-88236702

Abstract: In the present work, the optimization of extraction, emulsifying properties, and biological activities of polysaccharides from *Lentinula edodes* Sing (LES) were studied. The results showed LES polysaccharides extracted by hot water or ultrasonication are a group of β-glucan. Among all the samples, the one extracted by hot water showed the best emulsifying capacity. In addition, the results demonstrated that LES polysaccharide had strong scavenging activities in vitro on DPPH and ABTS radicals, which reached the highest level for the one extracted by 90 min ultrasonication ($p < 0.05$). Overall, *Lentinula edodes* Sing polysaccharides (LESPs) may have potential applications as emulsifying agents in food industries.

Keywords: *Lentinula edodes* Sing; polysaccharide; emulsifying properties; antioxidant properties; hypoglycemic activity

Citation: Dai, Y.; Wang, L.; Chen, X.; Song, A.; He, L.; Wang, L.; Huang, D. *Lentinula edodes* Sing Polysaccharide: Extraction, Characterization, Bioactivities, and Emulsifying Applications. *Foods* **2023**, *12*, 3289. https://doi.org/10.3390/foods12173289

Academic Editors: Qing Guo and Shuai Chen

Received: 30 July 2023
Revised: 21 August 2023
Accepted: 21 August 2023
Published: 1 September 2023

1. Introduction

Lentinula, a genus of the family Marasmiaceae, which is best known for the edible and medicinal *Lentinula edodes* originating from East Asia, is widely cultivated in Asia, Europe, Australia, and North and South America [1,2]. *Lentinula edodes* is now the second most popular edible fungi in the world owing to its desirable taste and multiple functional activities [3]. One of the major ingredients which contribute to *Lentinula edodes*' bioactive function is the polysaccharide fraction, which has shown antioxidant, antibacterial, immunomodulatory, antitumor, and antiinflammation effects, etc. [4]. For example, M et al. extracted and isolated linear (1→6)-β-d-glucan G-1 from *Lentinula edodes*. The IC50 of 2.2-diphenyl;-1 picrylhydrazine (DPPH) is 183.8 μg/mL and its antioxidant activity is higher than other fungal polysaccharides [5]. Moreover, Murphy et al. found through an in vitro lung injury model that Lentinan effectively attenuated the production of proinflammatory cytokines (IL-2, IL-6, IL-8, and TNF-α), while significantly increasing the secretion of anti-inflammatory cytokine IL-10 [6]. HPLPS, another polysaccharide isolated and purified from the fruiting bodies of *Lentinula edodes*, and reduced the proliferation of HepG2 and HeLa cells in a dose-dependent manner [7]. However, as a subfamily member of *Lentinula*, *Lentinula edodes* Sing (LES, referred to as "Masang fungus" in Chinese), is a kind of wild mushroom unique to Southwest China that grows on the dead branches of the *Coriaria nepalensis* Wall in spring and autumn [8,9]. LES has only been domesticated and planted in recent years, so there are few related studies, especially of its polysaccharides, which have not been studied [10,11]. The extraction of edible fungi polysaccharide using hot water is a conventional and simple method but it possesses some disadvantages like long extraction time, high energy requirement, and less extraction yield [12,13]. Ultrasound-assisted extraction (UAE) is currently considered an efficient and environmentally friendly method for extracting polysaccharides from various plant sources [14,15]. Ultrasonic waves can disrupt the edible fungi cell walls by the cavitation effect, resulting in the mass transfer between the solvent and the material to extract [16]. However, polysaccharides in natural products are extracted with hot water and UAE, which have different molecular weights, chemical compositions, and molecular conformations, such as the presence and location of functional groups and linkage patterns [17,18]. There are parameters that significantly influence polysaccharide extraction efficiency and biological activity through different ultrasonic conditions, such as ultrasonic power [19]. The result is a difference in the physiological activities of polysaccharides derived from different extraction methods [20]. Currently, UAE has widely used the extraction process of polysaccharides from various natural products (such as mushrooms) to improve the yield of target compounds [16].

Alzorqi et al. reported on the polysaccharides obtained by UAE from the *Ganoderma lucidum* that have a higher molecular weight and a better branching degree, as well as better in vitro antioxidant activity compared to traditional hot water extraction [21]. The lentinan normally has a high molecular weight, ranging from 1.16×10^3 to 6.84×10^3 kDa, which includes glucose, galactose, mannose, and arabinose with different molar ratios [1]. Additionally, the polysaccharides that contained β-glycoside bonds or with low molecular weight showed higher antioxidant bioactivity than other fungal polysaccharides [5,22]. Until now, there is little information on the physiochemical characteristics and the functional activities of the LES polysaccharide. The polysaccharides were extracted through hot water extraction and UAE, after which four polysaccharides were obtained, namely H90, U10, U20, and U60. Through the use of chemical and structural characteristics, antioxidant activity, hypoglycemic activity, and antilipid digestion studies, it is expected to provide a theoretical basis for the LES polysaccharide.

2. Materials and Methods

2.1. Materials and Reagents

The dried *Lentinula edodes* Sing was obtained from Guangzhou Extraction Biotechnology Co., Ltd., Guangzhou, China. A 30-g sample was pulverized with a miniature plant grinder (FSJ-A05N6, Guangdong Bear Electric Co., Ltd., Guangzhou, China) at low speed for one minute and the next pulverization was continued after the temperature was cooled to room temperature, and so repeated several times. The powder is passed through a 100-mesh sieve (aperture about 150 μm), and the part larger than the pore size in the sample is recrushed. Finally, the sample powder with a particle size of less than 150 μm was vacuum-packed and stored at room temperature until analysis and extraction. The sample powder was then vacuum packed and stored at room temperature until analysis and extraction.

Phenol, KBr (spectroscopic-grade), D-galacturonic acid, D-glucose, ABTS [2,2′-Azino-bis (3-ethylbenzothiazoline-6-sulfonic acid) 2,2-diphenyl-1-picrylhydrazyl (DPPH), and monosaccharide standards were purchased from Sigma-Aldrich (Shanghai) Trading Co., Ltd. (Shanghai, China). The acarbose, α-amylase (from the porcine pancreas), α-Glucosidase (from yeast), and 4-nitrophenyl-α-D-glucopyr-anoside (pNPG) were obtained from Shanghai Yuanye Bio-Technology Co., Ltd. (Shanghai, China). Other chemical reagents used in this study were of analytical grade, most of which were purchased from Sinopharm Chemical Reagent Co., Ltd. (Shanghai, China).

2.2. Ultrasound-Assisted Extraction

In this study, the ultrasound-assisted extraction of polysaccharides from LES was based on the method of Mao et al. with appropriate modifications [23]. Briefly, the pretreated sample powder was mixed with distilled water at the liquid–solid ratio of 30:1 and then extracted in a SCIENTZ-IID processor with a maximum output power of 600 W (Ningbo Scientz Biotechnology Co., Ltd., Ningbo, China). The ultrasound treatment of LES was performed at a fixed processor power of 80% amplitude (480 W) for a selected effective ultrasonic time of 10, 20, and 60 min. During the ultrasound treatment, one hundred milliliters of the LES solution was contained in a 250 mL glass beaker, which was immersed in ice water to keep the solution temperature below 50 °C.

2.3. Hot Water Extraction

As the comparison group, hot water extraction of the polysaccharides from LES followed the method previously reported by Akram et al., with slight modifications [24]. Briefly, ten grams of pretreated sample were mixed with 300 mL of water and then extraction with hot water was carried out directly for 1.5 h in a water bath at 90 °C (repeated two times) without being pretreated with the ultrasound processor.

The insoluble material was separated from the solution by centrifugation for 15 min at 8000 rpm. The supernatant was concentrated to one quarter of its original volume and precipitated overnight at 4 °C with four volumes of ethanol. After another centrifugation (8000 rpm for 15 min), the precipitate was dissolved by the addition of a tenfold volume of water and lyophilized (Model FD-1C-50, Shanghai Shun Zhi Instrument Manufacturing Co., Ltd., Shanghai, China) to obtain polysaccharides. The samples were named H90, U10, U20, and U60 according to the ultrasonic time from small to large. The extraction yield (%) of polysaccharides was calculated according to Equation (1):

$$\text{Yield (\%)} = \frac{\text{weight of the dried crude polysaccharide (g)}}{\text{weight of powder (g)}} \times 100\% \quad (1)$$

2.4. Chemical Composition Analysis

The contents of neutral sugar, uronic acid, total phenolic, and protein were determined according to methods including the phenol–sulfuric acid method [25], 3-hydroxy diphenyl [26], Folin–Ciocalteu [27], and bicinchoninic acid [28].

The monosaccharide composition of LESPs was determined by ion chromatography (ICS 5000$^+$, Thermo Fisher Scientific, Waltham, MA, USA) equipped with Dionex™ CarboPac™ PA20 (150 × 3.0 mm, 10 μm) liquid chromatography column and the injection volume was 5 μL. The appropriate amount of sample was hydrolyzed with 1 mL trifluoroacetic acid (TFA, 2 M) at 121 °C for 2 h and blown dry with nitrogen. Add 99.99% methanol to the wash, then blow dry, and repeat the methanol wash 2 to 3 times. Add sterile water to dissolve and transfer to the chromatographic bottle for testing. For mobile phase A (H_2O), mobile phase B (0.1 M NaOH), and mobile phase C (0.1 M NaOH, 0.2 M NaAC), the column temperature was 30 °C and the monosaccharide components were analyzed and detected by an electrochemical detector.

2.5. Determination of Color Differences

A colorimeter (NH350, Zhejiang Top Cloud-Agri Technology Co., Ltd., Hangzhou, China) was used to measure the color of different *Lentinula edodes* Sing polysaccharide samples [29,30]. The instrument was calibrated using black and white tiles. The L* (lightness), a* (redness), and b* (yellowness) values were measured three times. The calculation method of the whiteness index (WI) and total color difference (ΔE*) of each sample is as follows:

$$\text{WI} = 100 - \sqrt{(100 - L^*)^2 + a^{*2} + b^{*2}} \quad (2)$$

$$\Delta E^* = \sqrt{(L^* - L_0^*)^2 + (a^* - a_0^*)^2 + (b^* - b_0^*)^2} \quad (3)$$

where L_0^*, a_0^*, and b_0^* were the chromaticity values of the blank sample.

2.6. Fourier Transform Infrared (FTIR) Spectroscopy

The FTIR spectra of LES polysaccharides were analyzed by FT-IR spectroscopy (PerkinElmer Frontier, Waltham, MA, USA). The polysaccharide sample was mixed with the dried spectroscopic-grade KBr powder (baked at 400 °C for 30 min, the ratio of the KBr/sample is 100/1) and then pressed into a tablet for FT-IR spectroscopy measurement. The scanning range is 4000 to 400 cm^{-1}, the resolution is 4 cm^{-1}, and the number of scans is 32.

2.7. Scanning Electron Microscope (SEM)

Dried samples prepared by scanning thin gold films were scanned using a scanning electron microscope (TM3030, HITACHI, Tokyo, Japan) with an accelerating voltage of 15.0 kV; then, the polysaccharide samples were photographed and observed with 800× and 3000× magnifications.

2.8. Emulsifying Properties

The determination of emulsifying properties followed the previous method carried out by Xu et al., with appropriate modification [31]. Briefly, ten milliliters of samples (10 mg/mL, soaked overnight) were mixed with 10 mL of sunflower oil in a 50 mL centrifuge tube, using a high-speed shear homogenizer at 10,000 rpm, for 2 min. The mixture was placed at room temperature for 30 min. The emulsion volume was measured and the microscopic morphology was observed. The emulsion capacity was calculated according to Equation (4):

$$\text{The emulsion capacity } (\%) = \frac{\text{Height of emulsion layer}}{\text{total height of fluid}} \times 100 \quad (4)$$

2.9. Congo Red Test and Iodine-Potassium Iodide Analysis

The helical structure of polysaccharides was determined by the Congo red test, which was based on the method of Mutaillifu et al. with appropriate modifications [32]. Briefly, one milliliter of the Congo red solution (100 µmol/L) was mixed with 1.0 mL of polysaccharide sample solution (1.0 mg/mL); then, water and NaOH solution (1.0 mol/L) were added to the total volume of 4 mL by adjusting the NaOH final concentration of 0, 0.05, 0.10, 0.15, 0.20, 0.25, 0.30, 0.35, and 0.40 mol/L, respectively. As a control, Congo red solution was mixed with the same concentration of NaOH solution. After incubating at room temperature for 5 min, the solutions were scanned at the wavelength of 400–600 nm using a microplate reader (SuperMax-3100, Shanghai Flash Spectrum Biological Technology Co., Ltd., Shanghai, China). Iodine-potassium iodide analysis was performed by reacting polysaccharides with 0.2% potassium iodide solution (containing 0.02% iodine) and measuring the sweeping spectrum at 300–700 nm [33].

2.10. Antioxidant Activity

The antioxidant activity of the sample was measured using a DPPH, $ABTS^+$, and TPTZ assay. The tested samples were prepared at different concentrations (0.05, 0.1, 0.2, 0.4, 0.8, and 1.6 mg/mL). VC was used as the positive control.

2.10.1. Scavenging Capacity of DPPH Free Radicals

The determination of DPPH free radical scavenging activity followed the previous method carried out by Zhou et al., with appropriate modification [34]. Briefly, three hundred microliters of DPPH· methanol solution (0.1 mM) were mixed with 100 µL of samples and incubated at room temperature for 30 min in the dark. Then, the absorbance was measured at 517 nm using a microplate reader (SuperMax-3100, Shanghai Flash Spectrum Biological Technology Co., Ltd., Shanghai, China). The DPPH· scavenging activity was calculated according to Equation (2):

$$\text{The DPPH· scavenging rate } (\%) = \left(1 - \frac{A_2 - A_1}{A_0}\right) \times 100 \quad (5)$$

where A_2 was the absorbance of the mixtures, with A_1 without DPPH·, and A_0 without sample. The absence was substituted by an equivalent solvent.

2.10.2. Scavenging Capacity of ABTS Free Radicals

The determination of ABTS free radical scavenging activity followed the previous method carried out by Wang et al., with appropriate modification [35]. Briefly, the $ABTS^+$

solution (7.0 mM) was mixed with an equal volume of $K_2S_2O_8$ solution (2.45 mM). The $ABTS^+$ working solution was configured following incubation at 4 °C for 16 h in darkness and was diluted with PBS (pH = 7.4) to the absorbance value at 734 nm as 0.70 ± 0.05. Fifty microliters of sample solution were mixed with 150 μL of $ABTS^+$ working fluids and incubated at room temperature for 5 min in the dark and the absorbance was measured at 734 nm. The $ABTS^+$ scavenging activity was calculated according to Equation (3):

$$\text{The ABTS}^{+}\text{ scavenging rate }(\%) = \left(1 - \frac{A_2 - A_1}{A_0}\right) \times 100 \quad (6)$$

where A_2 was the absorbance of mixtures, with A_1 without $ABTS^+$ and A_0 without sample. The absence was substituted by an equivalent solvent.

2.10.3. Ferric-Reducing Antioxidant Power (FRAP)

The FRAP of polysaccharides was determined by a T-AOC Assay Kit (Macklin Biochemical Co., Ltd., Shanghai, China) [21]. Briefly, the FRAP reagent was freshly prepared by mixing acetate buffer, TPTZ solution, and ferric chloride at the volume ratio of 7:1:1. Then, thirty microliters of sample solution and 90 μL of distilled water were added to 900 μL of the FRAP reagent. The mixture was kept at 37 °C for 10 min in the dark and the absorbance was measured at 593 nm. An $FeSO_4$ solution (0.003125–0.1 μM) was used for the standard curve (y = 11.119x − 0.0074, R^2 = 0.9988) and the FRAP value was calculated according to the Equation (4):

$$\text{FRAP value of polysaccharides}(\mu\text{mol/mg}) = 34 \times \frac{x}{\text{Cpr}} \quad (7)$$

where x is calculated from the standard curve, and Cpr is the sample concentration.

2.11. *Assay for Hypoglycemic Activity*

The hypoglycemic activity of the LESPs was characterized by measuring α-amylase, α-glucosidase inhibitory activity, and the glucose adsorption capacity. Acarbose was used as the positive control.

2.11.1. Inhibition of the α-Amylase Activity

The inhibitory activity of LESPs on α-amylase was determined according to Niu et al., with appropriate modification [36]. Briefly, five hundred microliters of sample solutions of different concentrations (0.1, 0.2, 0.4, 0.8, 1.6, and 3.2 mg/mL) were mixed with 500 μL of α-amylase solution (1 U/mL) and incubated at 25 °C for 10 min. Then, 500 μL of starch solution (1%, *w/v*) were added. The enzyme reaction was terminated by adding 1 mL of DNS reagent after incubation for 10 min at 25 °C, followed by incubation in a boiling water bath for 5 min. The reaction mixture was diluted by adding 10 mL of distilled water. The absorbance of the mixture was measured at 540 nm using a multimode microplate reader. The inhibition rate was calculated using the following formula:

$$\alpha\text{-Amylase activity inhibition }(\%) = (1 - \frac{A_2 - A_1}{A_0}) \times 100 \quad (8)$$

where A_2 was the absorbance of mixtures, with A_1 without α-Amylase, and A_0 without sample. The absence was substituted by equivalent PBS.

2.11.2. Inhibition of the α-Glucoside Activity

The inhibitory activity of LESPs on α-amylase was determined according to Niu et al., with appropriate modification [37]. Briefly, fifty microliters of sample solutions of different concentrations (0.1, 0.2, 0.4, 0.8, 1.6, and 3.2 mg/mL) were mixed with 100 μL of 5 mM α-glucosidase solution (0.5 U/mL) and incubated at 37 °C for10 min. Then, 100 μL of 4-nitrophenyl-α-D-glucopyranoside (pNPG) were added. The enzyme reaction was terminated by adding 1 mL of 1 M sodium carbonate solution after incubation for 20 min at 37 °C. Measure the absorbance of the mixture at 405 nm using a multimode microplate reader. The inhibition rate was calculated using the following formula:

$$\alpha\text{-glucosidase activity inhibition } (\%) = (1 - \frac{A_2 - A_1}{A_0}) \times 100 \quad (9)$$

where A_2 was the absorbance of mixtures, with A_1 without α-glucosidase, and A_0 without sample. The absence was substituted by equivalent PBS.

2.11.3. Glucose Adsorption Capacity (GAC)

Two-hundred fifty mg of polysaccharide were dissolved in 25 mL of glucose solution (100 mmol/L) and placed in a water bath at 37 °C for 6 h [38]. Then, it was centrifuged at 3000 rpm for 15 min and the glucose content in the supernatant was measured by the DNS method.

In the same way, distilled water was used instead of glucose solution as the sample blank. A glucose solution without a sample was used as the control experiment. Guar gum was used as a positive control. The GAC was calculated using the following formula:

$$GAC(mg/g) = \frac{G_1 - (G_2 - G_3)}{W} \times V \quad (10)$$

where G1, G2, and G3 are the glucose concentration in the supernatant of the control experiment, the sample experiment, and the blank group, respectively (mg/mL); W is the weight of the sample (g); V is the volume of the mixture (25 mL).

2.12. *Effects of Polysaccharides on the Lipolysis In Vitro Digestion Model*

2.12.1. Polysaccharide Stock Solutions Preparation

Two-hundred fifty mg of each polysaccharide sample (H90, U10, U20, and U60) were dispersed into 50 mL of distilled water to prepare a stock solution (0.5%, *w/v*). The solution was then stirred overnight at room temperature at 300 rpm to ensure complete dissolution of polysaccharide.

2.12.2. Stock Emulsion Preparation

Two hundred g of 2.5% (*w/v*) tween 80 emulsifier solution were used to emulsify 50 g of corn oil with a high-speed homogenizer at 16,000 rpm for 10 min (homogenization for 2 min at intervals of 3 min and homogenization for 5 times in total) to prepare a crude oil emulsion containing 20% (*w/w*) corn oil.

2.12.3. Polysaccharide–Emulsion Mixture Preparation

The polysaccharide solution was mixed with a crude oil emulsion (9:1, *w/w*) and stirred at 500 rpm for 2 h at room temperature to obtain a polysaccharide–lipid emulsion. The lipid emulsion prepared with distilled water instead of polysaccharide solution was used as a control.

2.12.4. Static In Vitro Digestion Model

In vitro digestion using a simulated electrolyte solution and enzymes according to methods reported in the literature [39,40]. Simulated saliva fluid (*SSF*) was simulated by dissolving 0.375 g NaCl, 0.075 g $CaCl_2$, and 0.75 g KCl in 500 mL deionized water and adjusting the pH to 6.80 with a 0.1 M NaOH solution. Simulated gastric fluid (*SGF*), comprising 1.6 g of sodium chloride, 0.075 g of $CaCl_2$, 0.55 g of potassium chloride, and 0.3 g of $NaHCO_3$, was dissolved in 500 mL of deionized water and, then, the pH was adjusted to 2.5 with 1 M HCl solution. Simulated intestine fluid (*SIF*) was prepared by dissolving 2.7 g of sodium chloride, 0.175 g of $CaCl_2$, and 0.3265 g of KCl in 500 mL of deionized water and, then, the pH value was adjusted to 7.0 with 0.1 M NaOH solution.

The digestion model consists of three stages of simulated oral digestion, gastric digestion, and intestinal digestion [41]. First, mix the polysaccharide–lipid emulsion with SSF (1:1, *w/w*), adjust the pH to 6.8 using 0.1 M sodium hydroxide or hydrochloric acid solution, add α-amylase (50 U/mL), and incubate with constant agitation at 100 rpm for 10 min at 37 °C. Mix the saliva phase obtained after incubation with SGF (1:1, *w/w*), adjust the pH to 2.5 using 0.1 M hydrochloric acid solution, add pepsin (400 U/mL), and incubate at 37 °C with continuous stirring at 100 rpm for 2 h. The gastric phase obtained after incubation was then mixed with SIF (1:1, *w/w*), bile salts (10 mM) were added, the pH was adjusted to 7.0 using a 0.1 M NaOH solution, and incubated at 37 °C with constant agitation at 100 rpm for 2 h. Meanwhile, sodium hydroxide solution (0.05 M) was added in a dropwise manner to maintain a constant pH of 7.0, and the volume of sodium hydroxide solution added changed with time; then, the concentration of free fatty acid (FFA) produced by lipolysis was calculated. The release of FFA (%, *w/w*) was calculated using the following formula [41]:

$$\mathrm{FFA}(\%) = \left(\frac{\mathrm{V_{NaOH}} \times \mathrm{C_{NaOH}} \times \mathrm{MW_{Lipid}}}{2 \times \mathrm{W_{Lipid}}}\right) \times 100 \quad (11)$$

where V_{NaOH} is the volume used to neutralize FFA with sodium hydroxide (0.05 M) during the titration reaction; C_{NaOH} is the concentration of sodium hydroxide solution (0.05 M); MW_{Lipid} is the average molecular weight of corn oil (872 g/mol), and W_{Lipid} is the initial weight of corn oil in the enteric phase.

2.13. Statistical Analysis

All experiments were performed in triplicate and the corresponding data were expressed as means ± standard deviation (SD). One-way ANOVA was performed by *t*-test and Tukey's test using SPSS (Version 16; SPSS Inc., Chicago, IL, USA), where the probability value of $p < 0.05$ was considered to be statistically significant.

3. Results

3.1. Physicochemical Characterizations Analysis

The content of reducing sugar, polyphenol, protein, uronic acid, and yield of H90, U10, U20, and U60 are given in Table 1.

Table 1. The yield of *Lentinula edodes Sing* polysaccharides and its chemical composition between four extraction methods.

Sample	Reducing Sugar (%)	Polyphenol (%)	Protein (%)	Uronic Acid (%)	Yield (%)
H90	46.08 ± 0.22^{a}	1.08 ± 0.01^{b}	5.32 ± 0.18	4.31 ± 0.09	6.24 ± 0.10^{d}
U10	45.08 ± 0.28^{ab}	1.11 ± 0.04^{b}	5.35 ± 0.25	4.71 ± 0.08	7.83 ± 0.17^{c}
U20	44.26 ± 0.15^{b}	1.17 ± 0.02^{b}	5.76 ± 0.17	4.78 ± 0.10	8.48 ± 0.15^{b}
U60	45.13 ± 0.19^{ab}	1.38 ± 0.05^{a}	6.22 ± 0.61	5.09 ± 0.26	9.43 ± 0.19^{a}

Note: Data are expressed as the means ± standard deviation of at least three independent replicates. Means followed by different small letters for the same column are significantly different ($p < 0.05$).

3.2. Monosaccharide Composition Analysis

Table 2 shows the monosaccharide composition of the polysaccharides obtained by four extraction conditions. According to the table, it can be seen that PSP1 contains a rich monosaccharide composition. Among them, glucose, galactose, and mannose are the main monosaccharides, which is consistent with the lentinan extracted by Zhao et al. [2].

Table 2. The monosaccharide composition (molar ratio, %) of *Lentinula edodes Sing* polysaccharides.

Sample	Fucose	Rhamnose	Arabinose	Galactose	Glucose	Xylose	Mannose	Glucuronic Acid
H90	2.11%	0.28%	0.19%	8.58%	80.06%	1.04%	6.92%	0.84%
U10	2.26%	0.14%	0.10%	8.94%	80.38%	0.51%	6.84%	0.82%
U20	2.33%	0.14%	0.09%	9.07%	80.15%	0.51%	6.87%	0.85%
U60	2.32%	0.19%	0.10%	9.14%	79.79%	0.44%	6.51%	1.50%

3.3. Determination of Appearance Difference

Polysaccharides could be incorporated into food systems due to their many excellent properties to produce functional foods. However, the dark color of polysaccharides could be unfavorable to the quality of food, and dark-colored polysaccharides should be avoided. Table 3 shows the color differences of the polysaccharide freeze-dried powder obtained at different ultrasonic treatment times during the polysaccharide extraction process. The ultrasonic treatment times significantly influenced the color ($p < 0.05$). The highest whiteness index is the ultrasonic treatment for 20 min (82.96 ± 0.37), and the darkest color is the ultrasonic treatment for 60 min (79.91 ± 0.25). The difference in the color of extracted polysaccharides might be related to the length of the sonication time of the extraction process. The ultrasonic treatment can contribute to the whiteness index. With the increase of ultrasonic time, the temperature of the center of the system gradually increases, which promotes the progress of chemical reactions such as Maillard reactions. Eventually, the polysaccharide darkens in color. Furthermore, the total color difference (ΔE^*) between the standard color plate and polysaccharide samples was calculated; it was observed that the smaller the whiteness index, the larger the value of ΔE^*.

Table 3. Effect of ultrasonic treatment on the color values of *Lentinula edodes Sing* polysaccharides.

Sample	L^*	a^*	b^*	WI	ΔE^*
H90	83.78 ± 0.14 [b]	3.58 ± 0.17 [c]	5.50 ± 0.34 [c]	82.50 ± 0.21 [b]	13.08 ± 0.21 [b]
U10	86.02 ± 0.18 [a]	4.18 ± 0.06 [b]	8.94 ± 0.12 [b]	82.88 ± 0.17 [ab]	12.83 ± 0.17 [b]
U20	85.88 ± 0.24 [a]	4.06 ± 0.14 [b]	8.62 ± 0.30 [b]	82.96 ± 0.37 [a]	12.72 ± 0.38 [b]
U60	83.32 ± 0.26 [c]	4.84 ± 0.04 [a]	10.11 ± 0.12 [a]	79.91 ± 0.25 [c]	15.77 ± 0.25 [a]

Data are expressed as the means ± standard deviation of at least three independent replicates. Means followed by different small letters for the same column are significantly different ($p < 0.05$).

3.4. Analysis of Fourier Transform Infrared (FTIR) Spectroscopy, Congo Red Analysis, and Iodine-Potassium Iodide

As shown in Figure 1A, the FT-IR spectra of H90, U10, U20, and U60 exhibit similar characteristic peaks. All these peaks are characteristics of polysaccharides. Peaks at 3380 and 2931 cm^{-1} are connected to the absorption of the O–H and C–H stretching vibration, respectively [20]. The absorption peak at 1632 cm^{-1} was attributed to the C=O bond, while the peak at 1407 cm^{-1} was attributed to the C–H stretching vibration of the aldehyde group [3,37]. The results were consistent with the results of the monosaccharide composition analysis, which confirmed that the four polysaccharides contained glucuronic acid. In this region of 1028–1246 cm^{-1}, weak peaks can be attributed to the stretching vibration of the C–O–C or C–O–H group and indicated that pyran rings could be present in these polysaccharides [42]. The peaks at 847 and 933 cm^{-1} are attributed to α-d-glucans, while the weak absorbance band 889 cm^{-1} symbolizes β-d-glucans [43]. At different ultrasonic times, the FTIR spectra of the four polysaccharides were not significantly different, which may be because the degraded sugar units still existed in the samples.

When the solution contains low NaOH concentrations, the triple-helix conformation of the polysaccharide to which Congo red binds causes a redshift of maximum absorption wavelength (λmax) [44]. The shifts in λmax of the four polysaccharide–Congo red complexes with different concentrations (0–0.40 mol/L) of NaOH were shown in Figure 1B. Compared with the blank solution of Congo red, λmax of the complexes showed an obvious redshift with the change of NaOH concentration, indicating that the polysaccharides have triple-helix conformation (Figure 1A).

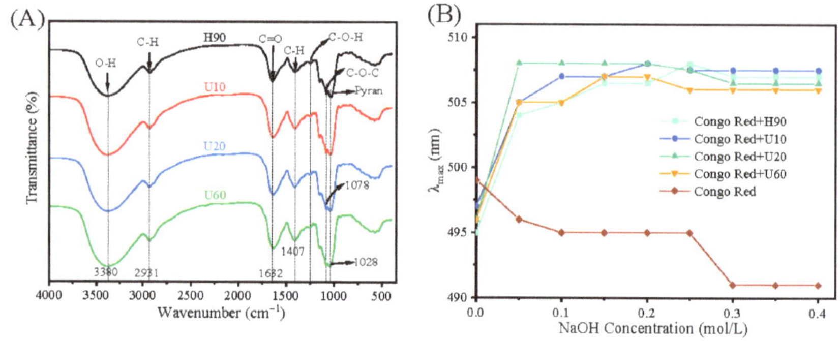

Figure 1. FTIR spectra (**A**) and triple-helical conformation (**B**) analysis of polysaccharides.

Figure 2 shows the scanning spectrum of polysaccharide–iodine–potassium iodide complexes. In each of the four polysaccharides, the maximum absorption peak is mainly concentrated near 350 nm, whereas there is no peak at 565 nm, indicating a long side chain and more branches [33].

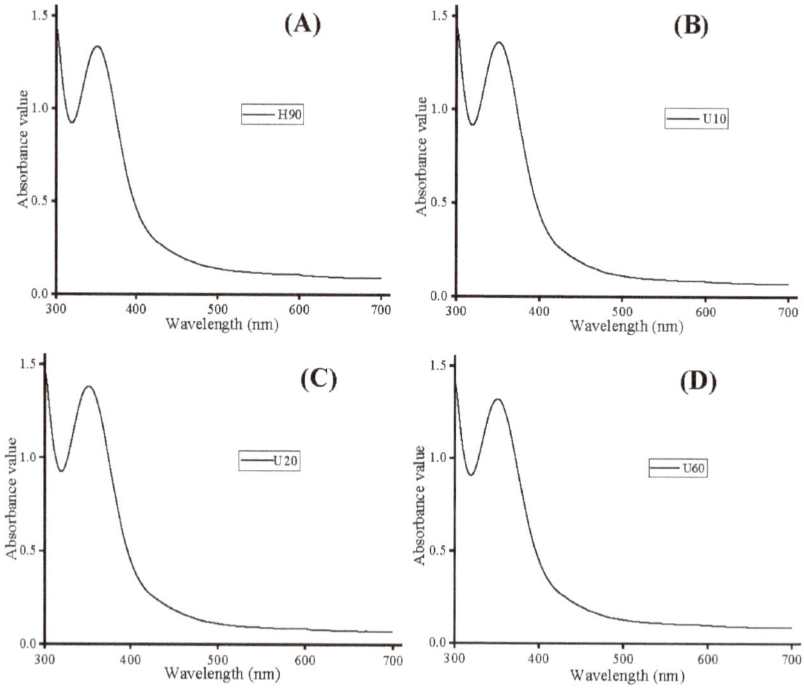

Figure 2. Iodine-potassium iodide analysis of polysaccharides. (**A–D**) correspond to samples H90, U10, U20 and U60, respectively.

3.5. Emulsifying Properties

The digital photos and optical micrographs (Figures 3 and 4) of the prepared emulsion can visually show the emulsifying properties of LESPs. After 1 h of preparation, the H90 extracted by hot water showed slight phase separation, while the polysaccharide emulsified solutions extracted by ultrasonic assistance showed different degrees of phase separation and demulsification. After 7 days, compared with the ultrasonically extracted polysaccharide emulsion, the hot water-extracted polysaccharide solution still maintained a higher emulsion layer height, showing better emulsion stability. Compared with pectin, lentinan has weaker emulsifying activity and stability. It may be because lentinan has a three-dimensional helical structure, has more hydrophilic groups, and lacks surface activity at the oil–water interface [45]. In addition, lentinan is a neutral polysaccharide with less surface charge, which reduces the charge repulsion between droplets, which is not conducive to the formation and stability of emulsions [46].

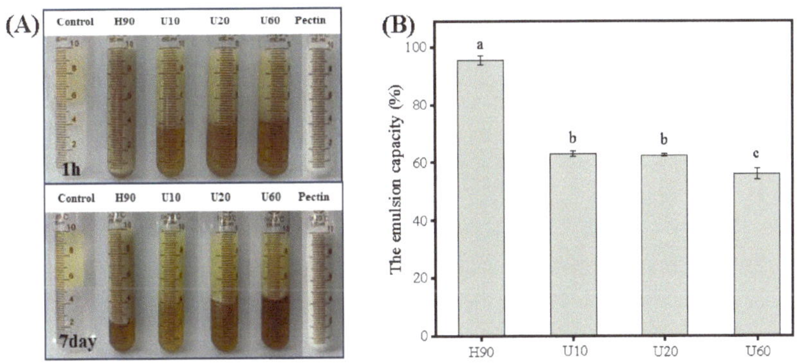

Figure 3. The digital photos (**A**) and emulsion capacity (**B**) of the emulsion. Different letters above columns show significant differences in emulsifying capacity among different samples ($p < 0.05$). Error bars represent standard deviation ($n = 3$).

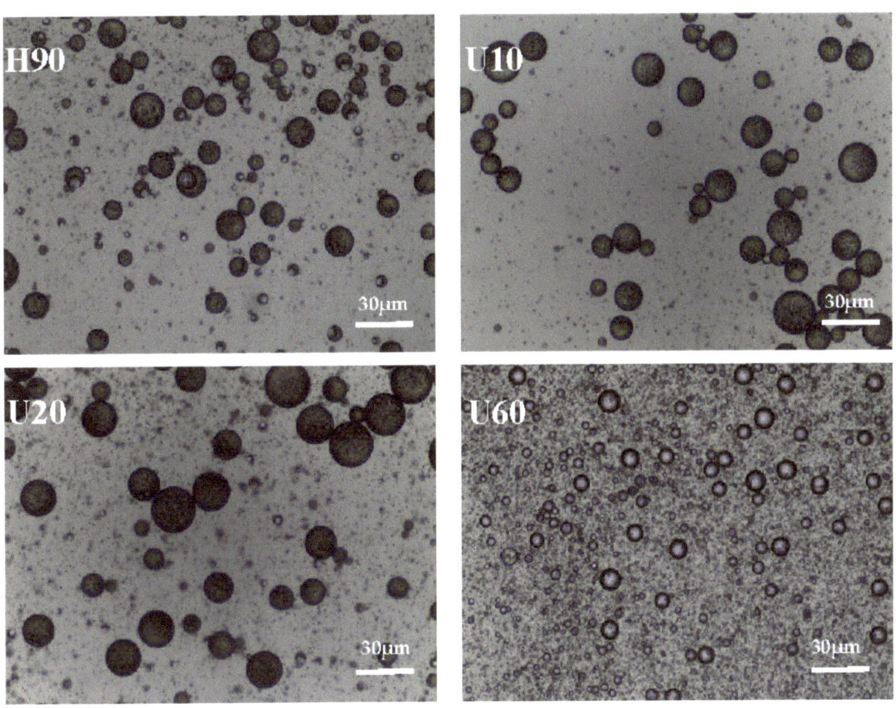

Figure 4. The optical micrographs of emulsions stabilized by LESP.

3.6. Scanning Electron Microscope (SEM)

Scanning electron microscopy was used to determine the microscopic characteristics of the samples obtained under different extraction conditions. SEM micrographs of LESP samples at different magnifications are presented in Figure 5a–l. The H90 (Figure 5a–c) surface is rough and irregular and its surface is bestrewed dense pores and networks when compared with U10, U20, and U60, indicating that H90 may have a smaller intermolecular force to promote the formation of loose structure [3]. In addition, some bead-like structures were observed from the surfaces of U10, U20, and U60, indicating that U10, U20, and U60 may be used for encapsulation purposes in drug delivery system design [47].

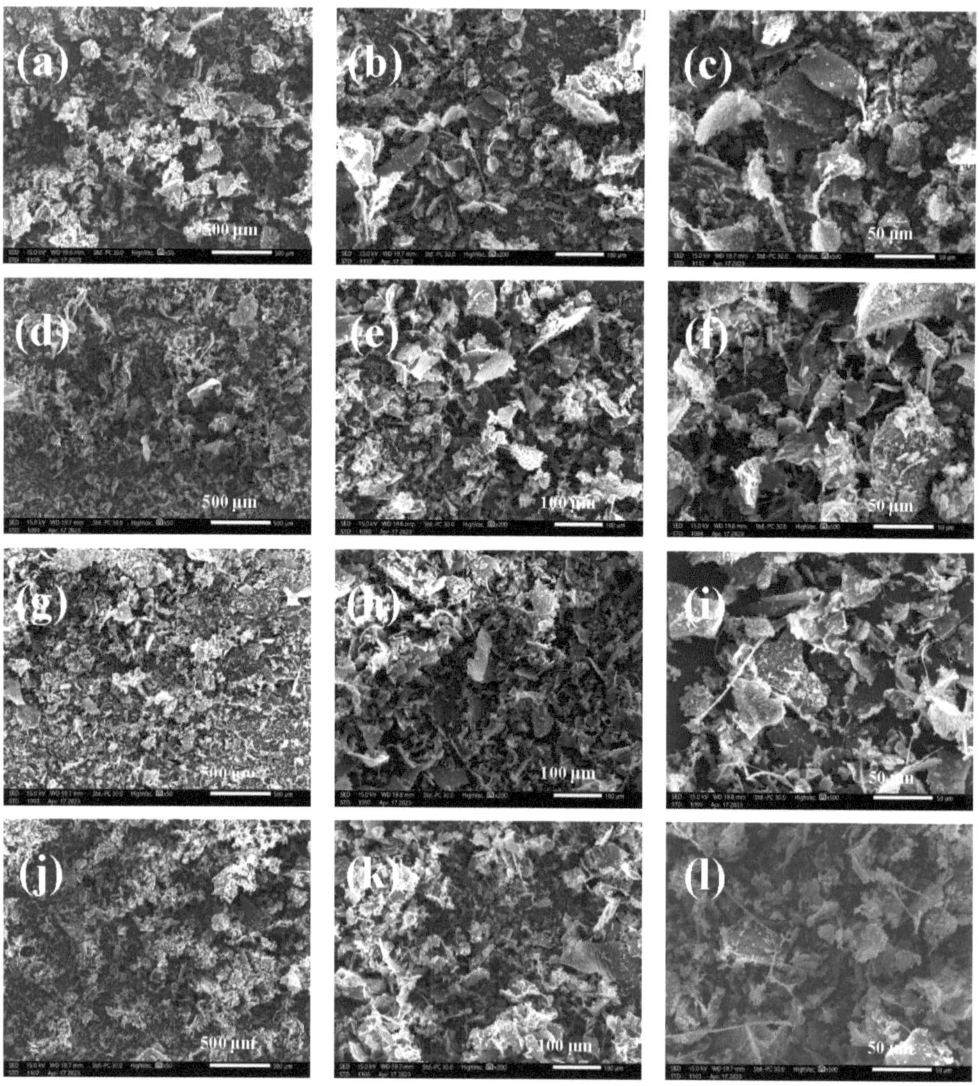

Figure 5. (**a–c**) SEM with different magnified images of LESP without ultrasonication treatments and (**d–l**) SEM with LESP obtained under different ultrasonication conditions.

3.7. Antioxidant Activity In Vitro

The activities of scavenging DPPH free radicals and ABTS free radicals are two common indicators to reflect the antioxidant capacity [35]. The scavenging activity of LESPs on DPPH free radicals and ABTS free radicals are shown in Figure 5a,b, respectively. All LESPs exhibited concentration-dependent scavenging activity on DPPH free radicals and ABTS free radicals.

When the DPPH free radicals were reduced, the methanol solution of DPPH's purple color faded to light yellow and the absorbance at 517 nm decreased. The degree of absorbance variation reflects the antioxidant capacity of a substance [18]. As shown in Figure 6A, all LESPs exhibited concentration-dependent scavenging activity on the DPPH radical. But it is not particularly noticeable, and it is very low compared to Vc.

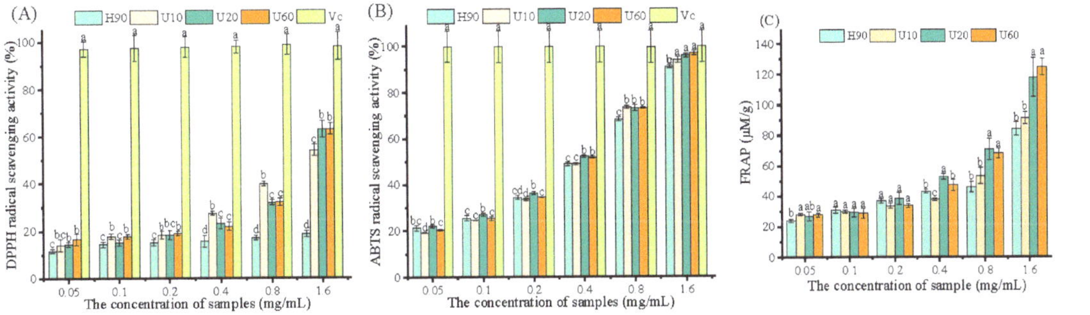

Figure 6. DPPH (**A**), ABTS (**B**), and FRAP (**C**) radical scavenging rate. Different letters above columns show significant differences in antioxidant capacity among different samples ($p < 0.05$). Error bars represent standard deviation ($n = 3$).

As can be seen from Figure 6B, with the increase of the polysaccharide concentration, the ABTS free radical scavenging rate increased significantly. When the concentration is 1.60 mg/mL, the ABTS free radical scavenging ability is close to that of Vc at the same concentration. Furthermore, the scavenging activity of LESPs on ABTS free radicals showed a descending order of U20 (IC50 = 0.29 mg/mL) > U60 (IC50 = 0.30 mg/mL) > U10 (IC50 = 0.32 mg/mL) > H90 (IC50 = 0.33 mg/mL).

The antioxidant capacity of the iron reduction is expressed in the ferrous sulfate equivalent. As shown in Figure 6C, the iron reduction abilities of LESPs obtained by the four methods are concentration dependent. When the concentration is 1.60 mg/mL, the antioxidant capacity of the iron reduction is 80.27, 91.27, 117.90, and 124.85 μmol/g respectively. It can be seen from the results that, with the increase of ultrasonic time, the iron reduction and antioxidation capacity also gradually increased.

3.8. Inhibitory Effects of LES Polysaccharides on α-Amylase and α-Glucosidase Activities

α-amylase is widespread among living organisms and α-amylase can hydrolyze starch into oligosaccharides such as maltose. The byproducts of α-amylase hydrolysis are ultimately broken down by other enzymes (such as α-glucosidase) into molecules of glucose, which are rapidly absorbed through the intestinal wall [48]. Inhibition of α-glucosidase, α-amylase, and other carbohydrate-hydrolyzing enzyme activities could reduce the hydrolysis of postprandial carbohydrates, thereby reducing blood glucose levels [49,50]. As shown in Figure 7, the inhibitory effects of LESPs and acarbose on α-amylase and α-glucosidase were concentration-dependent. As documented, the inhibitory effects of natural polysaccharides on α-glucosidase and α-amylase were positively correlated with the contents of some monosaccharides, such as arabinose, xylose, mannose, and uronic acid [51–53]. In addition, the hydroxyl and ketone groups of polysaccharides can promote changes in the polarity and molecular conformation of the enzyme, resulting in a partial loss of enzyme activity [36,54]. It can be seen from the Figure that LESPs exerted a better in vitro effect on hypoglycemia. Furthermore, the inhibitory effect of polysaccharide on α-amylase showed a descending order of acarbose (IC_{50} = 0.16 mg/mL) > H90 (IC_{50} = 0.32 mg/mL) > U60 (IC_{50} = 0.36 mg/mL) > U20 (IC_{50} = 0.40 mg/mL) > U10 (IC_{50} = 0.50 mg/mL) (Figure 7A). Furthermore, the inhibitory effect of polysaccharide on α-glucosidase showed a descending order of acarbose (IC_{50} = 0.27 mg/mL) > H90 (IC_{50} = 0.34 mg/mL) > U10 (IC_{50} = 0.38 mg/mL) > U60 (IC_{50} = 0.39 mg/mL) > U20 (IC_{50} = 0.48 mg/mL) (Figure 7B). As indicated by IC50, the inhibitory activity of H90 on α-amylase and α-glucosidase was stronger than U10, U20, and U60. This may be mainly related to the higher content of arabinose, xylose, and mannose in H90. In addition, according to the results of the positive control and the IC50, it was shown that the inhibitory activity of LESPs on α-glucosidase was stronger than α-amylase. It can be seen that polysaccharides have different mechanisms of inhibitory effects on α-glucosidase and α-amylase [36], which may be related to the composition and content of monosaccharides such as arabinose [49,51].

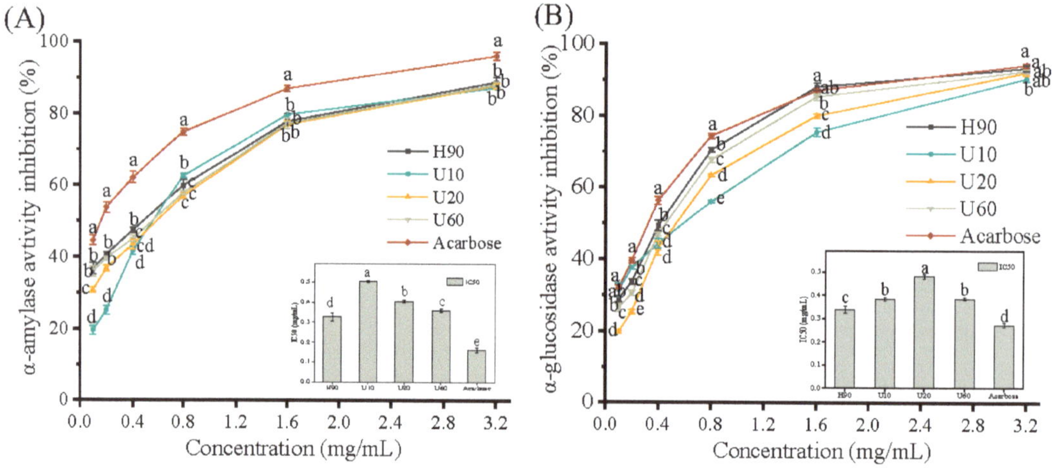

Figure 7. The inhibition effects of LESPs on α-amylase ability (**A**) and α-glucosidase ability (**B**). Different letters above columns show significant differences in enzyme inhibitory activities among different samples ($p < 0.05$). Error bars represent standard deviation ($n = 3$).

3.9. *Effects of LES Polysaccharides on Glucose Adsorption Capacity*

Samples absorb glucose and reduce the digestion and absorption of glucose by the human intestine. It is an effective way to reduce postprandial blood sugar and is widely used to evaluate the hypoglycemic effect of polysaccharides [36]. The glucose adsorption capacity of four polysaccharide samples at the same glucose concentration (100 mM) was shown in Figure 8A, with the guar gum as a positive control. It can be seen from the Figure that all samples can effectively adsorb and bind glucose. Among them, H90 has the strongest glucose adsorption capacity and there is no significant difference with the positive control (guar gum). The GAC of the polysaccharides may be related to the physical structure, such as the internal porosity and surface area of polysaccharides [55]. With the increase of ultrasonic time, the adsorption capacity of polysaccharides to glucose decreased significantly. Combined with the SEM analysis, this may be because the ultrasonic treatment reduces the internal porosity of polysaccharides, thereby reducing the total surface area and the adsorption capacity to glucose. Therefore, we concluded that LESPs could adsorb glucose and reduce its utilization for intestinal absorption, thereby suppressing postprandial hyperglycemia.

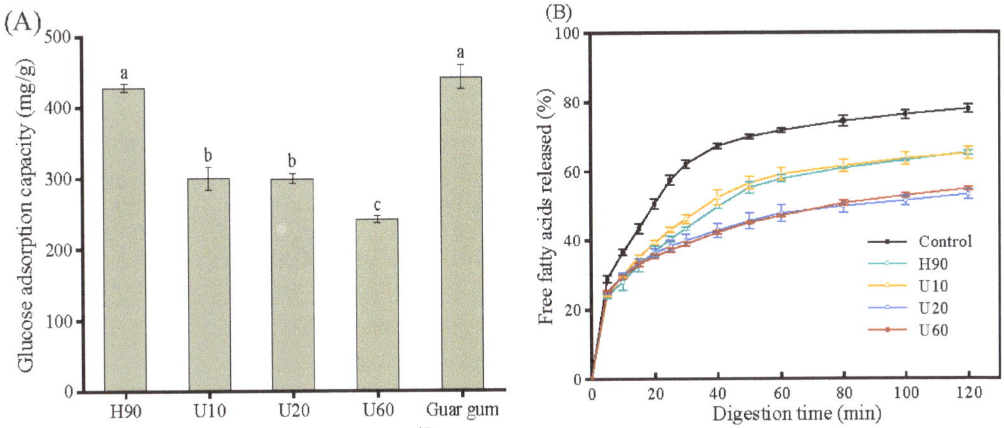

Figure 8. Influence of polysaccharides obtained under different extraction conditions on glucose adsorption capacity (**A**) and in vitro digestion of lipid droplets (**B**). Different letters above columns show significant differences in glucose adsorption capacity among different samples ($p < 0.05$). Error bars represent standard deviation ($n = 3$).

3.10. Effects of LES Polysaccharides on Lipolysis

Reducing the speed rate and extent of lipid digestion within the small intestine is an effective strategy for the treatment of postprandial spikes in blood lipid levels that occur after the intake of high-fat foods [41]. We, therefore, examined the effect of LESP on lipid digestion. As revealed by the iodine-potassium iodide experiment, our polysaccharide samples do not contain starch. In addition, since the lipase activity in the gastric lumen is significantly lower than that in the duodenal tract, assays in the oral and gastric phases can be omitted [41,56]. The relationship between the release degree of free fatty acids in the reaction system and the digestion time was detected by the titration method during the incubation in the intestinal stage, to judge the influence of polysaccharides on the rate and degree of lipid digestion. The lipolysis of emulsions in the presence and absence of LESPs is shown in Figure 8B. Overall, FFA release increased rapidly at the onset of lipolysis and, then, gradually increased over time. At 120 min, the final degree of lipid digestion was as follows: 77.87 ± 1.27%, 65.21 ± 0.63%, 64.94 ± 1.79%, 53.17 ± 1.43%, and 54.75 ± 0.56% for the control, H90, U10, U20, and U60 samples, respectively. Lipid digestibility and extent were significantly reduced in the presence of lentinan compared to the control group ($p < 0.05$).

4. Conclusions

A novel source of polysaccharide from *Lentinula edodes* Sing (LES) was obtained using hot water extraction and ultrasonication for different intervals. According to the results of Congo red, iodine-potassium iodide, and infrared spectrum, all four polysaccharides have rich branched chain structures and three-dimensional helical structures with similar FT-IR spectra. Moreover, it revealed that the extraction condition could significantly ($p < 0.05$) affect the extraction yield and emulsifying properties, as well as other physiochemical and biological properties of LES polysaccharide. To be specific, the polysaccharides obtained by hot water extraction showed a stronger emulsifying ability and stability, as well as stronger hypoglycemic activity in vitro. On the other hand, ultrasonication exhibited a positive impact on the antioxidant capacity of LES polysaccharides when the interval was extended from 10 to 60 min. In addition, ultrasonication increased the resistance of polysaccharides to lipid digestion. The present study showed that *Lentinula edodes* Sing can be a great potential source of polysaccharides, especially the one extracted by hot water for hydrocolloids.

Author Contributions: Conceptualization, L.W. (Lei Wang); methodology, L.W. (Lei Wang), Y.D., L.W. (Lingyuan Wang) and D.H.; software, Y.D.; validation, A.S. and X.C.; formal analysis, L.W. (Lei Wang); investigation, Y.D.; resources, L.W. (Lei Wang) and L.H.; data curation, Y.D. and X.C.; writing—original draft preparation, Y.D.; writing—review and editing, L.W. (Lei Wang); visualization, A.S. and X.C.; supervision, L.W. (Lei Wang) and L.H.; project administration, Y.D.; funding acquisition, L.W., L.H., A.S. and X.C. All authors have read and agreed to the published version of the manuscript.

Funding: This research was funded by the Natural Science Foundation of China (32101900), Qiankehe project (QKHJC ZK[2022] General 107, ZK[2023] General 092 and ZK[2022] General 043), the High-Level Innovative Talents Training Project of Guizhou Province (QKHPTRC-GCC[2022]026-1), Guizhou University (Guidarenjihe [2020] 007 and Guidateganghezi [2021] 41).

Data Availability Statement: The data used to support the findings of this study can be made available by the corresponding author upon request.

Conflicts of Interest: The authors declare no conflict of interest.

References

1. Sheng, K.; Wang, C.; Chen, B.; Kang, M.; Wang, M.; Liu, K.; Wang, M. Recent advances in polysaccharides from *Lentinus edodes* (Berk.): Isolation, structures and bioactivities. *Food Chem.* **2021**, *358*, 129883. [CrossRef]
2. Zhao, Y.M.; Yang, J.M.; Liu, Y.H.; Zhao, M.; Wang, J. Ultrasound assisted extraction of polysaccharides from *Lentinus edodes* and its anti-hepatitis B activity in vitro. *Int. J. Biol. Macromol.* **2018**, *107 Pt. B*, 2217–2223. [CrossRef]
3. Wang, K.; Guo, J.; Cheng, J.; Zhao, X.; Ma, B.; Yang, X.; Shao, H. Ultrasound-assisted extraction of polysaccharide from spent *Lentinus edodes* substrate: Process optimization, precipitation, structural characterization and antioxidant activity. *Int. J. Biol. Macromol.* **2021**, *191*, 1038–1045. [CrossRef]
4. Mwangi, R.W.; Macharia, J.M.; Wagara, I.N.; Bence, R.L. The antioxidant potential of different edible and medicinal mushrooms. *Biomed. Pharmacother.* **2022**, *147*, 112621. [CrossRef]
5. Morales, D.; Rutckeviski, R.; Villalva, M.; Abreu, H.; Soler-Rivas, C.; Santoyo, S.; Iacomini, M.; Smiderle, F.R. Isolation and comparison of alpha- and beta-D-glucans from shiitake mushrooms (*Lentinula edodes*) with different biological activities. *Carbohydr. Polym.* **2020**, *229*, 115521. [CrossRef]
6. Murphy, E.J.; Masterson, C.; Rezoagli, E.; O'Toole, D.; Major, I.; Stack, G.D.; Lynch, M.; Laffey, J.G.; Rowan, N.J. β-Glucan extracts from the same edible shiitake mushroom *Lentinus edodes* produce differential in-vitro immunomodulatory and pulmonary cytoprotective effects—Implications for coronavirus disease (COVID-19) immunotherapies. *Sci. Total Environ.* **2020**, *732*, 139330. [CrossRef] [PubMed]
7. Li, W.; Wang, J.; Chen, Z.; Gao, X.; Chen, Y.; Xue, Z.; Guo, Q.; Ma, Q.; Chen, H. Physicochemical properties of polysaccharides from *Lentinus edodes* under high pressure cooking treatment and its enhanced anticancer effects. *Int. J. Biol. Macromol.* **2018**, *115*, 994–1001. [CrossRef] [PubMed]
8. Xue, X.; Peng, L.; Xiaofeng, L.; Zhun, X.; Chunying, D.; Jianwen, W. Identification and Biological Characteristics of Wild *Lentinus edodes*. *North Hortic.* **2021**, *4*, 118–123.
9. Dong, Y.; Liao, X.; Jin, T.; Xie, Y. Nutrient Analysis and Safety Evaluation of *Lentinus edodes* (Beak.) *Sing. Edible Fungi China* **2021**, *40*, 91–96+102.
10. Hanwu, L.; Gaochao, P.; Dongsheng, Q.; Fanglun, Z.; Yan, S. Analysis of Aroma Components of a Wild Sangria Mushroom in Guizhou Province. *Edible Fungi China* **2019**, *38*, 107–108.
11. Xie, Y.; Li, P.; Lu, L.; Liao, X.; Zhang, L.; Yang, L.; Kang, C. A Kind of Cultivation Method of Masang Mushroom Returning to Nature. CN106665124A [P/OL], 17 May 2017.
12. Tang, W.; Liu, C.; Liu, J.; Hu, L.; Huang, Y.; Yuan, L.; Liu, F.; Pan, S.; Chen, S.; Bian, S.; et al. Purification of polysaccharide from *Lentinus edodes* water extract by membrane separation and its chemical composition and structure characterization. *Food Hydrocoll.* **2020**, *105*, 105851. [CrossRef]
13. Zhang, Y.; Wang, D.; Chen, Y.; Liu, T.; Zhang, S.; Fan, H.; Liu, H.; Li, Y. Healthy function and high valued utilization of edible fungi. *Food Sci. Hum. Wellness* **2021**, *10*, 408–420. [CrossRef]
14. Wu, J.; Chen, R.; Tan, L.; Bai, H.; Tian, L.; Lu, J.; Gao, M.; Bai, C.; Sun, H.; Chi, Y. Ultrasonic disruption effects on the extraction efficiency, characterization, and bioactivities of polysaccharides from *Panax notoginseng* flower. *Carbohydr. Polym.* **2022**, *291*, 119535. [CrossRef] [PubMed]
15. Sharmila, G.; Muthukumaran, C.; Suriya, E.; Muppidathi Keerthana, R.; Kamatchi, M.; Kumar, N.M.; Anbarasan, T.; Jeyanthi, J. Ultrasound aided extraction of yellow pigment from *Tecoma castanifolia* floral petals: Optimization by response surface method and evaluation of the antioxidant activity. *Ind. Crops Prod.* **2019**, *130*, 467–477. [CrossRef]
16. Leong, Y.K.; Yang, F.-C.; Chang, J.-S. Extraction of polysaccharides from edible mushrooms: Emerging technologies and recent advances. *Carbohydr. Polym.* **2021**, *251*, 117006. [CrossRef]
17. Tabarsa, M.; Anvari, M.; Joyner, H.S.; Behnam, S.; Tabarsa, A. Rheological behavior and antioxidant activity of a highly acidic gum from *Althaea officinalis* flower. *Food Hydrocoll.* **2017**, *69*, 432–439. [CrossRef]

18. Guo, X.; Shang, X.; Zhou, X.; Zhao, B.; Zhang, J. Ultrasound-assisted extraction of polysaccharides from *Rhododendron aganniphum*: Antioxidant activity and rheological properties. *Ultrason. Sonochem.* **2017**, *38*, 246–255. [CrossRef]
19. AlYammahi, J.; Hai, A.; Krishnamoorthy, R.; Arumugham, T.; Hasan, S.W.; Banat, F. Ultrasound-assisted extraction of highly nutritious date sugar from date palm (*Phoenix dactylifera*) fruit powder: Parametric optimization and kinetic modeling. *Ultrason. Sonochem.* **2022**, *88*, 106107. [CrossRef]
20. Chen, S.; Qin, L.; Xie, L.; Yu, Q.; Chen, Y.; Chen, T.; Lu, H.; Xie, J. Physicochemical characterization, rheological and antioxidant properties of three alkali-extracted polysaccharides from mung bean skin. *Food Hydrocoll.* **2022**, *132*, 107867. [CrossRef]
21. Alzorqi, I.; Sudheer, S.; Lu, T.-J.; Manickam, S. Ultrasonically extracted beta-D-glucan from artificially cultivated mushroom, characteristic properties and antioxidant activity. *Ultrason. Sonochem.* **2017**, *35*, 531–540. [CrossRef]
22. Lin, Y.; Zeng, H.; Wang, K.; Lin, H.; Li, P.; Huang, Y.; Zhou, S.; Zhang, W.; Chen, C.; Fan, H. Microwave-assisted aqueous two-phase extraction of diverse polysaccharides from *Lentinus edodes*: Process optimization, structure characterization and antioxidant activity. *Int. J. Biol. Macromol.* **2019**, *136*, 305–315. [CrossRef] [PubMed]
23. Mao, Y.H.; Song, A.X.; Li, L.Q.; Yang, Y.; Yao, Z.P.; Wu, J.Y. A high-molecular weight exopolysaccharide from the Cs-HK1 fungus: Ultrasonic degradation, characterization and in vitro fecal fermentation. *Carbohydr. Polym.* **2020**, *246*, 116636. [CrossRef] [PubMed]
24. Akram, K.; Shahbaz, H.M.; Kim, G.R.; Farooq, U.; Kwon, J.H. Improved Extraction and Quality Characterization of Water-Soluble Polysaccharide from Gamma-Irradiated *Lentinus edodes*. *J. Food Sci.* **2017**, *82*, 296–303. [CrossRef] [PubMed]
25. Dubois, M.; Gilles, K.; Hamilton, J.K.; Rebers, P.A.; Smith, F. A colorimetric method for the determination of sugars. *Nature* **1951**, *168*, 167. [CrossRef]
26. Blumenkrantz, N.; Asboe-Hansen, G. New method for quantitative determination of uronic acids. *Anal. Biochem.* **1973**, *54*, 484–489. [CrossRef]
27. Bach, F.; Ferreira Zielinski, A.A.; Helm, C.V.; Maciel, G.M.; Pedro, A.C.; Stafussa, A.P.; Avila, S.; Isidoro Haminiuk, C.W. Bio compounds of edible mushrooms: In vitro antioxidant and antimicrobial activities. *LWT Food Sci. Technol.* **2019**, *107*, 214–220. [CrossRef]
28. Smith, P.K.; Krohn, R.I.; Hermanson, G.T.; Mallia, A.K.; Gartner, F.H.; Provenzano, M.D.; Fujimoto, E.K.; Goeke, N.M.; Olson, B.J.; Klenk, D.C. Measurement of protein using bicinchoninic acid. *Anal. Biochem.* **1985**, *150*, 76–85. [CrossRef]
29. Tao, L.; Wang, J.; Wei, Y.; Xiong, B.; Zhang, J.; Lan, W.; Yu, L. Efficient decolorization of oligosaccharides in ginseng (*Panax ginseng*) residue using ultrasound-assisted macroporous resin. *Food Chem.* **2023**, *419*, 136098. [CrossRef]
30. Alboofetileh, M.; Rezaei, M.; Tabarsa, M.; You, S.; Mariatti, F.; Cravotto, G. Subcritical water extraction as an efficient technique to isolate biologically-active fucoidans from *Nizamuddinia zanardinii*. *Int. J. Biol. Macromol.* **2019**, *128*, 244–253. [CrossRef]
31. Xu, K.; Martinez, M.M.; Yang, B.; Guo, M. Fine structure, physicochemical and antioxidant properties of LM-pectins from okra pods dried under different techniques. *Carbohydr. Polym.* **2020**, *241*, 116272. [CrossRef]
32. Mutaillifu, P.; Bobakulov, K.; Abuduwaili, A.; Huojiaaihemaiti, H.; Nuerxiati, R.; Aisa, H.A.; Yili, A. Structural characterization and antioxidant activities of a water soluble polysaccharide isolated from *Glycyrrhiza glabra*. *Int. J. Biol. Macromol.* **2020**, *144*, 751–759. [CrossRef] [PubMed]
33. Wang, D.; Wang, D.; Yan, T.; Jiang, W.; Han, X.; Yan, J.; Guo, Y. Nanostructures assembly and the property of polysaccharide extracted from *Tremella Fuciformis* fruiting body. *Int. J. Biol. Macromol.* **2019**, *137*, 751–760. [CrossRef] [PubMed]
34. Zhou, S.; Huang, G.; Chen, G. Extraction, structural analysis, derivatization and antioxidant activity of polysaccharide from Chinese yam. *Food Chem.* **2021**, *361*, 130089. [CrossRef]
35. Wang, S.; Li, G.; Zhang, X.; Wang, Y.; Qiang, Y.; Wang, B.; Zou, J.; Niu, J.; Wang, Z. Structural characterization and antioxidant activity of *Polygonatum sibiricum* polysaccharides. *Carbohydr. Polym.* **2022**, *291*, 119524. [CrossRef]
36. Niu, G.; You, G.; Zhou, X.; Fan, H.; Liu, X. Physicochemical properties and in vitro hypoglycemic activities of hsian-tsao polysaccharide fractions by gradient ethanol precipitation method. *Int. J. Biol. Macromol.* **2023**, *231*, 123274. [CrossRef]
37. Tang, Y.; He, X.; Liu, G.; Wei, Z.; Sheng, J.; Sun, J.; Li, C.; Xin, M.; Li, L.; Yi, P. Effects of different extraction methods on the structural, antioxidant and hypoglycemic properties of red pitaya stem polysaccharide. *Food Chem.* **2023**, *405 Pt. A*, 134804. [CrossRef]
38. Lin, D.; Long, X.; Xiao, L.; Wu, Z.; Chen, H.; Zhang, Q.; Wu, D.; Qin, W.; Xing, B. Study on the functional properties and structural characteristics of soybean soluble polysaccharides by mixed bacteria fermentation and microwave treatment. *Int. J. Biol. Macromol.* **2020**, *157*, 561–568. [CrossRef] [PubMed]
39. Ma, G.; Xu, Q.; Du, H.; Muinde Kimatu, B.; Su, A.; Yang, W.; Hu, Q.; Xiao, H. Characterization of polysaccharide from *Pleurotus eryngii* during simulated gastrointestinal digestion and fermentation. *Food Chem.* **2022**, *370*, 131303. [CrossRef]
40. Brodkorb, A.; Egger, L.; Alminger, M.; Alvito, P.; Assuncao, R.; Ballance, S.; Bohn, T.; Bourlieu-Lacanal, C.; Boutrou, R.; Carriere, F.; et al. INFOGEST static in vitro simulation of gastrointestinal food digestion. *Nat. Protoc.* **2019**, *14*, 991–1014. [CrossRef]
41. Espinal-Ruiz, M.; Parada-Alfonso, F.; Restrepo-Sanchez, L.P.; Narvaez-Cuenca, C.E.; McClements, D.J. Impact of dietary fibers [methyl cellulose, chitosan, and pectin] on digestion of lipids under simulated gastrointestinal conditions. *Food Funct.* **2014**, *5*, 3083–3095. [CrossRef]
42. Yin, C.; Fan, X.; Fan, Z.; Shi, D.; Gao, H. Optimization of enzymes-microwave-ultrasound assisted extraction of *Lentinus edodes* polysaccharides and determination of its antioxidant activity. *Int. J. Biol. Macromol.* **2018**, *111*, 446–454. [CrossRef]

43. Baeva, E.; Bleha, R.; Lavrova, E.; Sushytskyi, L.; Čopíková, J.; Jablonsky, I.; Klouček, P.; Synytsya, A. Polysaccharides from Basidiocarps of Cultivating Mushroom *Pleurotus ostreatus*: Isolation and Structural Characterization. *Molecules* **2019**, *24*, 2740. [CrossRef]
44. Guo, X.; Kang, J.; Xu, Z.; Guo, Q.; Zhang, L.; Ning, H.; Cui, S.W. Triple-helix polysaccharides: Formation mechanisms and analytical methods. *Carbohydr. Polym.* **2021**, *262*, 117962. [CrossRef] [PubMed]
45. Tang, Q.; Huang, G. Improving method, properties and application of polysaccharide as emulsifier. *Food Chem.* **2021**, *376*, 131937. [CrossRef] [PubMed]
46. Xu, X.; Luo, L.; Liu, C.; McClements, D.J. Utilization of anionic polysaccharides to improve the stability of rice glutelin emulsions: Impact of polysaccharide type, pH, salt, and temperature. *Food Hydrocoll.* **2017**, *64*, 112–122. [CrossRef]
47. Zhang, X.; Bi, C.; Shi, H.; Li, X. Structural studies of a mannoglucan from *Cremastra appendiculata* (Orchidaceae) by chemical and enzymatic methods. *Carbohydr. Polym.* **2021**, *272*, 118524. [CrossRef] [PubMed]
48. Goff, H.D.; Repin, N.; Fabek, H.; El Khoury, D.; Gidley, M.J. Dietary fibre for glycaemia control: Towards a mechanistic understanding. *Bioact. Carbohydr. Diet. Fibre* **2018**, *14*, 39–53. [CrossRef]
49. Li, F.; Wei, Y.; Liang, L.; Huang, L.; Yu, G.; Li, Q. A novel low-molecular-mass pumpkin polysaccharide: Structural characterization, antioxidant activity, and hypoglycemic potential. *Carbohydr. Polym.* **2021**, *251*, 117090. [CrossRef]
50. Chen, J.; Li, L.; Zhang, X.; Zhang, Y.; Zheng, Q.; Lan, M.; Li, B. Structural characteristics and antioxidant and hypoglycemic activities of a heteropolysaccharide from *Anemarrhena asphodeloides* Bunge. *Int. J. Biol. Macromol.* **2023**, *236*, 123843. [CrossRef]
51. Zhu, J.; Zhou, H.; Zhang, J.; Li, F.; Wei, K.; Wei, X.; Wang, Y. Valorization of Polysaccharides Obtained from Dark Tea: Preparation, Physicochemical, Antioxidant, and Hypoglycemic Properties. *Foods* **2021**, *10*, 2276. [CrossRef]
52. Zheng, Q.; Jia, R.B.; Ou, Z.R.; Li, Z.R.; Zhao, M.; Luo, D.; Lin, L. Comparative study on the structural characterization and alpha-glucosidase inhibitory activity of polysaccharide fractions extracted from *Sargassum fusiforme* at different pH conditions. *Int. J. Biol. Macromol.* **2022**, *194*, 602–610. [CrossRef] [PubMed]
53. Yan, J.K.; Yu, Y.B.; Wang, C.; Cai, W.D.; Wu, L.X.; Yang, Y.; Zhang, H.N. Production, physicochemical characteristics, and in vitro biological activities of polysaccharides obtained from fresh bitter gourd (*Momordica charantia* L.) via room temperature extraction techniques. *Food Chem.* **2021**, *337*, 127798. [CrossRef] [PubMed]
54. Zhang, J.Q.; Li, C.; Huang, Q.; You, L.J.; Chen, C.; Fu, X.; Liu, R.H. Comparative study on the physicochemical properties and bioactivities of polysaccharide fractions extracted from *Fructus Mori* at different temperatures. *Food Funct.* **2019**, *10*, 410–421. [CrossRef] [PubMed]
55. Pan, Y.; Wang, M.; Wang, Z.; Huang, X.; Hu, X.; Wang, Q.; Nie, S. A novel soluble powder containing high *Dendrobium huoshanense* polysaccharide and its in vitro hypoglycemic activities evaluation. *Bioact. Carbohydr. Diet. Fibre* **2023**, *30*, 100362. [CrossRef]
56. Chen, C.; Huang, Q.; You, L.J.; Fu, X. Chemical property and impacts of different polysaccharide fractions from *Fructus* Mori. on lipolysis with digestion model in vitro. *Carbohydr. Polym.* **2017**, *178*, 360–367.

foods

MDPI

Article

Effect of Different Temperatures on the Storage Stability of Flaxseed Milk

Chen Meng [1,2], Yashu Chen [2], Xintian Wang [2], Hongjian Chen [3,*] and Qianchun Deng [2,*]

1 School of Food and Biological Engineering, Hubei University of Technology, Wuhan 430068, China; m18407119868@163.com
2 Key Laboratory of Oilseeds Processing, Ministry of Agriculture, Oil Crops Research Institute, Chinese Academy of Agricultural Sciences, Wuhan 430062, China; yashuchen@sina.com (Y.C.); wxin_t@163.com (X.W.)
3 College of Health Science and Engineering, Hubei University, Wuhan 430062, China
* Correspondence: chenhongjian@hubu.edu.cn (H.C.); dengqianchun@caas.cn (Q.D.)

Abstract: In this study, the physical and oxidative stability of flaxseed milk without food additives at different temperatures (25 °C and 37 °C) was assessed. Over in 206 days in storage, the particle size, Turbiscan stability index (TSI), centrifugal sedimentation rate, and primary and secondary oxidation products of flaxseed milk increased, viscosity decreased, and the absolute value of the potential first decreased and then increased. These phenomena indicated a gradual decrease in the physical stability of flaxseed milk, accompanied by drastic oxidative changes. The antioxidant capacity of flaxseed milk was related to the location of the physical distribution of flaxseed lignin, which was more effective in the aqueous phase compared to the non-aqueous phase. Interestingly, after 171 days in storage at 37 °C, the particle size of flaxseed milk was approximately doubled (6.98 μm → 15.27 μm) and the absolute value of the potential reached its lowest point (−13.49 mV), when the content of primary oxidation products reached its maximum (8.29 mmol/kg oil). The results showed that temperature had a significant effect on the stability of flaxseed milk and that stability decreased with increasing temperature and shortened shelf life. This work provides a theoretical basis for elucidating the stabilization–destabilization mechanism of flaxseed milk.

Keywords: flaxseed; plant-based milk substitutes; oxidative stability; physical stability

Citation: Meng, C.; Chen, Y.; Wang, X.; Chen, H.; Deng, Q. Effect of Different Temperatures on the Storage Stability of Flaxseed Milk. *Foods* **2023**, *12*, 3571. https://doi.org/10.3390/foods12193571

Academic Editor: Youn Young Shim

Received: 4 September 2023
Revised: 20 September 2023
Accepted: 23 September 2023
Published: 26 September 2023

1. Introduction

With the continuous upgrading of food production technology and the increasing health and environmental awareness of consumers, plant-based foods are thriving [1]. Plant-based milk substitutes are new products developed in response to this trend, as there is growing interest due to its low saturated fat, low cholesterol content, and high dietary fiber [2].

Plant-based milk substitutes have been reported to play an important role in improving or managing the immune system [3–5]. They may also reduce the risk of low bone mass and have high levels of antioxidants while improving physiological function and possessing free radical scavenging properties, etc. [6]. These alternatives are also referred to as non-dairy alternatives, which mainly include soya, almond, peanut, walnut, and coconut milk. Various other sources of crops are also used to produce plant-based milk, but in relatively small quantities, such as flaxseed.

Flaxseed (*Linum usitatissimum* L.), also known as caraway seed, is one of the major oil crops in the word and is considered to be a major functional food [7]. Research has confirmed the significant role of flaxseed in the prevention and treatment of cardiovascular disease, regulation of human estrogen levels, and enhancement of intestinal health [4,8]. The plant-based milk substitute made from flaxseed not only has excellent nutritional and health characteristics by integrating the macro- and micronutrients of flaxseed but can also be green and additive-free with low food safety risks. Plant-based foods are designed to meet the needs of the general public for a new era of healthy products. As a new milk alternative, they address the disadvantages of animal protein drinks that are high in saturated fatty acids and cholesterol; however, some protein and lipid deficiencies and environmental factors make plant-based milk substitutes much less physically stable than milk [9]. Jeske et al. found that most plant-based milk substitutes have separation rates 14 times quicker than milk [10]. The systems of plant-based milk substitutes are complex in that they contain emulsions and other particles, including residual plant cells, soluble and insoluble proteins, dietary fiber, lipids, and carbohydrates [9]. These components may interact with each other within the system in suspension. During storage, plant-based milk substitutes are prone to stratification, flocculation, and sedimentation, which can affect the quality and sensory aspects of the product. Plant-based milk substitutes as a colloid may exist in the form of an emulsion or suspension. The stability of plant-based milk substitutes are influenced by various factors. These factors include the characteristics of the colloidal particles present, such as their size, density, surface charge, interfacial chemistry, and surface hydrophobicity. The nature of the surrounding aqueous phase, including its pH, ionic strength, composition, density, and viscosity, also plays a role. Additionally, the stability of the product can be influenced by environmental conditions during its service life, such as temperature changes, storage duration, and mechanical forces. Previous attempts have been made to explain the interactions within colloidal systems using the double-layer theory and DLVO theory, but hydration and hydrophobic and steric interactions may also exist in plant-based milk substitutes [11,12]. Electrostatic interactions may also arise from interactions between some components, and ζ-potential values can be indicative of the electrostatic behavior of particles in colloids [13]. Flaxseed milk is enriched with n-3 polyunsaturated fatty acids, which are prone to oxidative mass transfer triggered by the oil core, thus changing interfacial composition and charge distribution. This further alters the balance between electrostatic repulsion and van der Waals gravitational forces on the surface of the oil droplets, which affects the stability of flaxseed milk. Flaxseed milk with no food additives has a strong relationship to stability from its physical state, chemical oxidation and decomposition.

In the current work, whole flaxseed was processed and used to prepare a flaxseed milk without food additives. The study explored the changes in the physicochemical stability of flaxseed milk at different temperatures and the interfacial distribution of the main antioxidant substances in flaxseed milk. This work may serve as a theoretical foundation for clarifying the stabilization–destabilization mechanism and provide theoretical support for expanding the market applications of flaxseed milk. It is also expected to be a reference for elaborating on and improving the storage stability of plant-based milk substitutes.

2. Materials and Methods

2.1. Materials and Chemicals

Flaxseed was purchased from Gansu Academy of Agricultural Sciences (Lanzhou, China). 2-thiobarbituric acid (TBA), 1,1,3,3-tetraethoxypropane, sodium azide, and cumene hydroperoxide were purchased from Sigma-Aldrich (Sigma-Aldrich, Saint Louis, MS, USA). All other reagents were acquired from Sinopharm Chemical Reagent Co., Ltd. (Sinopharm Chemical Reagent Co., Ltd., Shanghai, China).

2.2. Flaxseed Milk Preparation

A 9.5 cm-diameter Petri dish filled with flaxseed (19 g) was placed in a microwave oven (CEM Mars-6, North Carolina, USA) at 700 W for 5 min [14]. The microwave-treated flaxseeds were degummed in an aqueous solution at 40 °C (1:10, *w*/*v*) for 2 h. The flaxseeds were vacuum-precooled at 4 °C with deionized water (1:8, *w*/*v*; pH = 7.0) and ground (Horizontal-60, Shenzhen, China). The resulting emulsion was hydrolyzed with 1% cellulase and 2% glucoamylase for 1 h at 50 °C. The emulsion was filtered through a 200 mesh filter cloth and collected. High-pressure homogenization was performed after inactivation of the enzyme at 95 °C for 15 s (GYB40-10S, Shanghai, China). Finally, after

UHT sterilization at 137 °C for 15 s, the flaxseed milk was stored for further analysis (ST21-4338-1, Shanghai, China).

2.3. Storage Experiment

Two quantities of flaxseed milk were stored in the dark at 25 °C and 37 °C, respectively, for 206 days. At each sample point, three bottles of flaxseed milk were tested. The flaxseed milk stored at a temperature of 25 °C was sampled every two weeks, resulting in a total of six samples. Additionally, the flaxseed milk stored at 37 °C was sampled once a week, yielding a total of ten samples.

2.4. Measurement of Particle Size and ζ-Potential

The mean particle size and size distribution of the flaxseed milk were measured via static light scattering (Mastersizer 2000, Malvern Instruments, Westborough, MA). The refractive indexes of flaxseed oil and water phases were 1.490 and 1.330, respectively, which were used in the calculation of particle size. The ζ-potential of emulsions were determined via particle electrophoresis instrument (ZetaSizer-Pro, Malvern Instruments, Worcestershire, UK). The flaxseed milk samples were diluted with buffer solution (v/v, 1:250).

2.5. Determination of Viscosity

The static experimental samples were cooled to 25 °C, and the viscosity of the samples was measured with a DV-II digital viscometer (DV-II, Brookfield, MA, USA). After the corresponding rotor and speed were selected, the amount of liquid was taken by the diffuse viscometer rotor.

2.6. Evaluation of Gravity Separation Stability

A Turbiscan Lab Expert stability analyzer (Formulaction Inc., Toulouse, France) was used to monitor the change to the overall instability of the emulsion over time, to clarify the gravity separation stability of the emulsion, and to provide a basis for further evaluation of physical stability during storage. Turbiscan software determined the Turbiscan stability index (TSI).

After the emulsion was processed for the corresponding time, it was shaken well, and 10 g of the sample was accurately weighed and centrifuged at 1900× g (J-25, Beckman Coulter, CA, USA) for 15 min. The precipitate weight at the bottom of the centrifuge tube was recorded to calculate the centrifugal precipitation rate:

$$\text{Centrifugal precipitation rate } (\%) = \frac{Precipitate\ weight\ (\text{g})}{Centrifugal\ sample\ weight\ (\text{g})} \times 100$$

2.7. Observation of Microstructure

The microstructure of emulsions was observed via confocal laser microscopy (CLSM) (Nikon D-Eclipse C1 80i, Nikon, Melville, New York). A 20 µL quantity of Nile red-ethanol solution (1 mg/mL) was added to 200 µL of emulsion and buffer (v/v, 1:1) mixed solution [15]. The excitation and emission wavelengths of 543 nm and 605 nm were used to record the oil phase images (NISElements, Nikon, Melville, NY, USA).

A conductive carbon gel was applied to the sample stage, and the emulsion was dripped onto the conductive carbon gel. The sample stage with the sample adhered to it was snap-frozen in liquid nitrogen slush and transferred to the sample preparation chamber for sublimation and gold plating under vacuum using a cryogenic frozen preparation transfer system (PP3000T, Quorum, East Sussex, UK). Samples were sublimated at −90 °C for 10 min, followed by sputtering gold-plating at 10 mA for 60 s, then were sent to the scanning electron microscopy (SEM) sample chamber for observation (Quanta 450, FEI, OR, USA) (cold stage temperature −140 °C, accelerating voltage 5 kV).

2.8. Determination of Lipid Hydroperoxide

The primary oxidation product in the emulsion was mainly lipid hydroperoxide. The determination method is described by Kiralan [16]. A 0.3 mL quantity of flaxseed milk was mixed with 1.5 mL isooctane/isopropanol (v/v, 3:1), and the emulsion was destroyed by vortex. Then, the mixed solution was centrifuged at 2968× g for 10 min to obtain the supernatant. The upper solution (200 μL) was mixed with 2.8 mL methanol / butanol (v/v, 2:1), 15 μL 3.94 M ammonium thiocyanate and 15 μL Fe^{2+} were added. The Fe^{2+} solution was prepared from an equal amount of 0.132 M $BaCl_2$ and 0.144 M $FeSO_4$. The 200 uL supernatant was replaced with isooctane/isopropanol (v/v, 3:1) as blank control. The above mixed solution was vortexed and incubated at room temperature for 20 min. After the reaction was completed, it was measured at 510 nm with a spectrophotometer (DU 800, Beckman Coulter, CA, USA). The standard curve was made with cumene hydroperoxide as standard to determine the concentration of hydrogen peroxide. All values are expressed as mmol cumene hydroperoxide equivalent per kilogram of oil.

2.9. Determination of Lipid TBARS

The determination method of secondary reaction products of lipid oxidation was described by Liang [17]. A 1 mL quantity of flaxseed milk was mixed with 2 mL of TBA reagent, destroyed by vortex. Then, it was placed in boiling water above 90 °C for 15 min and cooled to room temperature in a cold water bath. The supernatant was centrifuged at 2968× g for 15 min, which was measured for absorbance at 532 nm. The TBARS concentration was calculated from the standard curve of 1,1,3,3-tetraethoxypropane (TEP) to be mmol TEP equivalent per kilogram of oil. The TBARS solution was prepared according to the method previously described by Cheng et al. [15].

2.10. Distribution of Flaxseed Lignans in Flaxseed Milk

The physical location of flaxseed lignins in the emulsion was determined according to the method described by Panya et al. [18] with some modifications. A 1 mL quantity of emulsion was centrifuged at 5180× g for 1 h at 4 °C using a high-speed centrifuge. The flaxseed milk was divided into three layers, and the aqueous phase was carefully collected using a needle and syringe. Afterward, 1 mL of methanol was added to the remaining portion and vortexed vigorously. The supernatant was obtained after centrifugation and carefully collected using a needle and syringe, as above. The concentration of flaxseed lignans in the aqueous and non-aqueous phases was determined via HPLC (Waters, MA, USA) using an ACQUITY UPLC® BEH Shield RP 18 column (2.1 × 100 mm, 1.7 μm). The solvents were 100% methanol and 0.5% acetic acid solution, and the injection volume was 2 μL. The non-linear gradient elution was performed at an elution flow rate of 0.1 mL/min. The detector wavelength was set to observe the sample peak at 280 nm. Flaxseed lignan macromolecules (purity > 95%) were used as the standard to configure the concentration gradient using 70% methanol water, and the sample was injected according to the above method. The standard curve was plotted for quantification with $R^2 > 0.999$.

2.11. Statistical Analysis

All test results were taken as the average of three parallel measurements, which were presented as mean ± standard deviations. The statistical significance of the results was determined via analysis of variance (ANOVA) and the Tukey test, using SPSS 21.0 software (SPSS, Inc., Chicago, IL, USA). A threshold of $p < 0.05$ was employed to determine statistical significance.

3. Results and Discussion

Over 150 days in storage, the flaxseed milk underwent relatively little change in the indicators, so the data after 150 days was selected for graphical analysis.

3.1. Particle Size and ζ-Potential

The variations in particle size distribution and $D_{4,3}$ of flaxseed milk during storage are presented in Figure 1A,B. Fresh flaxseed milk (day 0) showed a moderately narrow single-peak distribution with a $D_{4,3}$ of 6.98 μm. The particle size distribution of flaxseed milk changed from initially narrow to a broad single peak, with a small peak at 0.1 μm to 1 μm, after 164 days in storage at 25 °C. However, after 178 days, there was another small peak at 10 μm to 100 μm, which indicated particle flocculation or aggregation. At 178 days in storage at 37 °C, the flaxseed milk showed a large peak at 100 μm to 1000 μm. As reported by other researchers, the 0.1–10 μm peak might represent protein particles and oil bodies [19]. Oil bodies consist of surface proteins surrounding oil droplets, which stabilize flaxseed milk [20]. The intermediate particle size peaks of 10–100 μm might have included oil droplets and protein aggregates [19,21]. Maximum particle size peaks between 100 and 1000 μm might have consisted of aggregates of oil droplets or particles. Some plant-based milk substitutes, such as almond, cashew, oat, quinoa and rice beverages, present themselves as a polydisperse system [10]. Their suspension system contains a diverse range of interactions due to the presence of both large and small particles, including residual decomposed plant cells, soluble and insoluble proteins, lipids, and carbohydrates [22]. With increasing storage time at 37 °C, various interactions might have occurred between the various components of flaxseed milk, such as proteins, polysaccharides (flaxseed gum), and polyphenols (flaxseed lignans). At 199 days, the flaxseed milk showed a single peak distribution, presumably breaking the stable three-dimensional gel network structure formed by flaxseed milk and causing the originally stable large particles complex to break up into smaller particles.

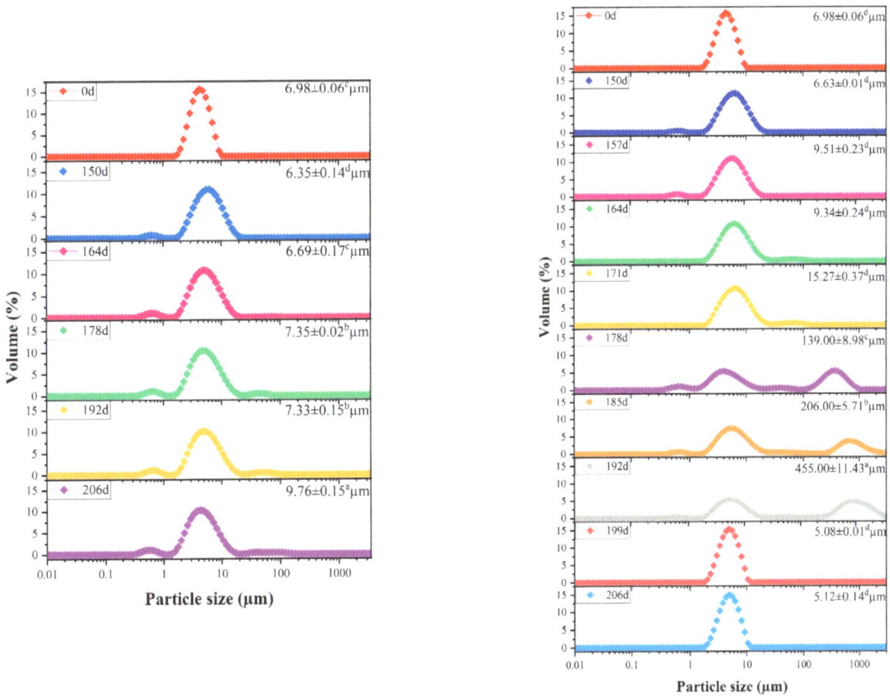

(A) (B)

Figure 1. *Cont.*

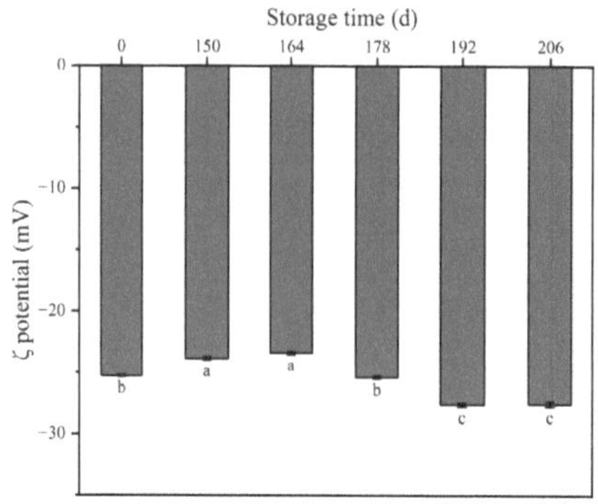

(C)

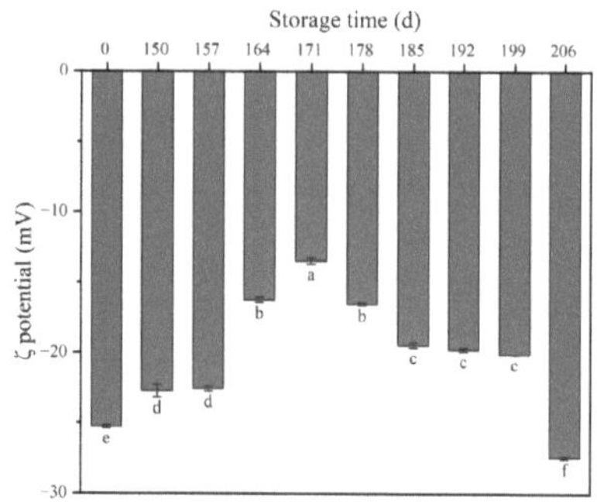

(D)

Figure 1. Particle size distribution of flaxseed milk at 25 °C (**A**) and 37 °C (**B**) at 206 days in storage. ζ-potential of flaxseed milk at 25 °C (**C**) 37 °C (**D**) at 206 days in storage. Different letters following the mean in the same column indicate significant differences ($p < 0.05$).

The ζ-potential of the droplet affects the physical stability and the oxidative stability of flaxseed milk [23]. The variation in potential of flaxseed milk during storage is shown in Figure 1C,D. The flaxseed milk that was stored at 25 °C for 206 days showed an increase in potential to 2.21 mV. This suggests that flaxseed milk was exposed to some lipid droplet and interfacial oxidation during storage, during which oxidation of lipids or anionic reactions formed via hydrolysis might have led to a change in ζ-potential [24]. The absolute values of the potentials of the flaxseed milk stored at 37 °C increased after decreasing to the lowest point after 171 days in storage. This might be due to the aggregation of large particles weakening the electrostatic interaction between the particles. Generation of more small particles increased the specific surface area of the particles, which resulted in an increase of negative charge, causing a jump in potential values in the later stages. This was consistent with the particle size distribution results.

3.2. *Visual Appearance and Microstructure*

The overall appearance of flaxseed plant-based milk at different temperatures during storage was shown in Figure 2A. Originally, the flaxseed milk had a homogeneous, cream-like appearance, indicating a uniform distribution of oil droplets. The color of the flaxseed milk was bright and homogeneous, mainly presenting as creamy and slightly yellow. There was little overall change to the color of flaxseed milk stored at 25 °C for 206 days. The flaxseed milk appeared to aggregate after 157 days in storage at 37 °C, darkened in color by 178 days, and oiling-off occurred at 185 days. A slight amount of oil phase precipitated from the upper layer of the flaxseed milk at 171 days in storage. The flaxseed milk gradually decreased in liquid level between days 178 and 220 in storage. Throughout storage, the interfaces between the flocculated particles and the continuous phase were disrupted, and the dispersed phase fused. The disappearance of the small droplet interface and the formation of large droplets led to aggregation and ultimately to the complete separation of the oil and aqueous phases. It has been shown that protein and oil droplets in almond milk flocculate and form oil–protein clusters. The aggregation of oil–protein clusters facilitated the phase separation process [25].

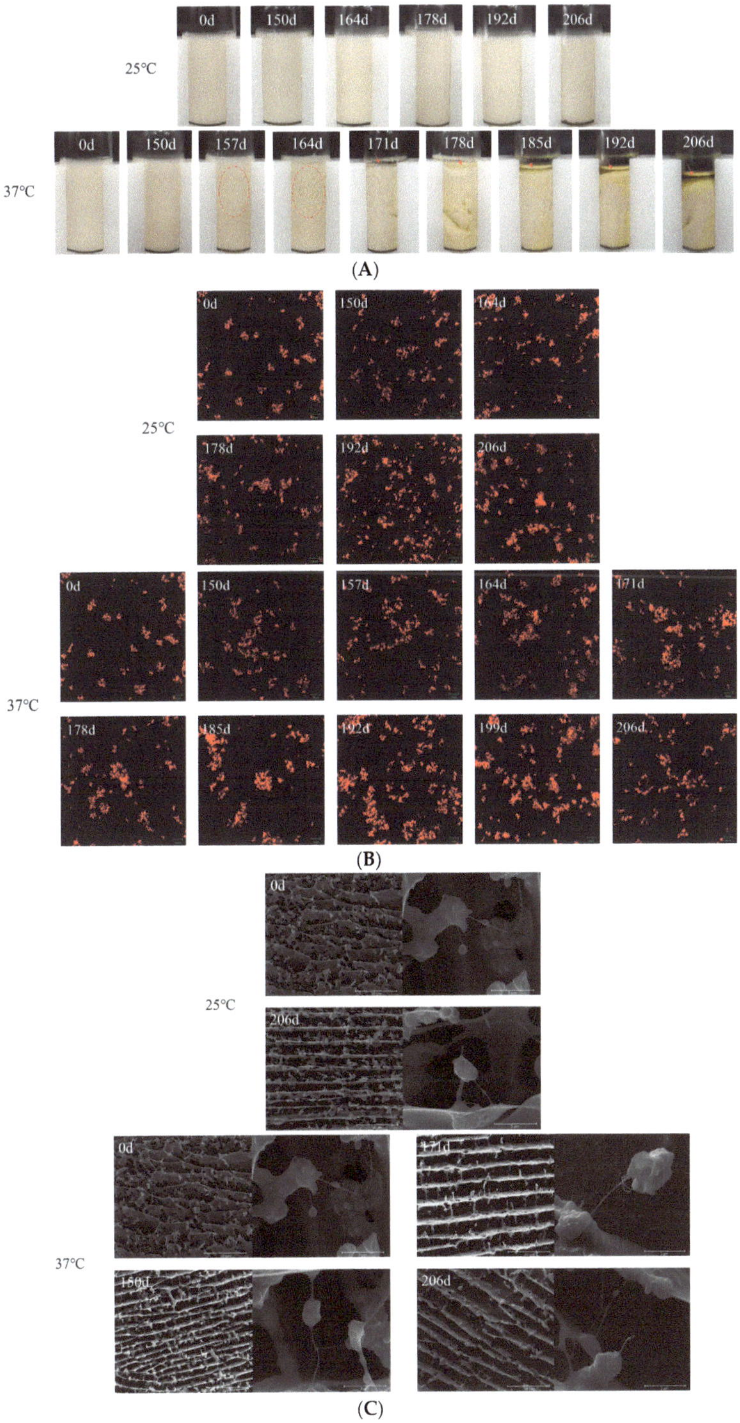

Figure 2. The visual appearance of flaxseed milk during storage at different temperatures for 0–206 days (**A**) and CLSM plots and (**B**) cryo-SEM plots (**C**). The scale bar in (**B**) is 10 μm.

The microstructure of lipid droplets of flaxseed milk was illustrated via CLSM (Figure 2B). The fresh flaxseed milk (day 0) had lipid droplets (red) uniformly distributed throughout the aqueous phase (black). The lipid droplet particles had moved closer together by 150 days in storage at 25 °C. After 206 days in storage, the particle size of the lipid droplets increased, which could have been due to flocculation and agglomeration. In the flaxseed milk stored at 37 °C for 150 days, it was clearly observed that the small lipid droplet particles had aggregated together, but there was no significant change in particle size (Figure 1) at this time. This was presumably due to different measurement methods. Measurement of particle size required multiple dilutions, and if oil separation had occurred, the measured size of particles retained in the emulsion was small [26]. Then, large clusters of lipid droplets were observed. After 199 days, the previously tightly bound lipid droplets were observed to disperse, which was probably due to phase separation.

The cryo-SEM results of flaxseed milk (Figure 2C) revealed the presence of a well-arranged three-dimensional network structure. The flaxseed gum and flaxseed protein entangled in the oil droplets were linked to the flaxseed membrane layer structure, resulting in a stable network structure. After 206 days at 25 °C and 150 days at 37 °C, the flaxseed gum filaments between the membrane layer structures showed significant fractures, and the shape of the oil body changed from smooth and full to uneven. However, the aggregation of flaxseed protein or oil bodies on the lamellar structure might have been due to the precipitation of oil from the oil bodies. After 206 days in storage at 37 °C, more small pores appeared in the lamellar structure of the flaxseed milk, and the adhesion between the lamellar structure and the oil body was reduced, which could also serve as an explanation for the fact that the originally stable large particle complex had broken down into smaller particles. The overall morphology of the lamellar structure would not change significantly during storage. In whole peanut milk, increased exposure of peanut proteins to hydrophobic groups improved water-holding and oil-mixing capacity. Peanut proteins were adsorbed on the surface of oil droplets to form complexes which formed a dense three-dimensional network structure and improved physical stability [27].

3.3. Viscosity

Viscosity was one of the most important factors affecting the stability of flaxseed plant-based milk. The viscosity of the flaxseed milk during storage is shown in Figure 3. The initial viscosity of flaxseed milk was 43.25 mPa·s. Studies have reported that the viscosities of commercial coconut milk and almond milk are 47.80 and 26.23 mPa·s, respectively [10]. Unlike these commercial plant-based milk substitutes, the high viscosity of the flaxseed milk in this study was caused by flaxseed gum and protein. The high viscosity of the flaxseed milk led to a high surface tension, which effectively impeded the movement of suspended particles. It largely prevented collisions and agglomeration in the flaxseed milk and slowed down the settling of flaxseed milk, enhancing system stability [9]. The viscosity of the flaxseed milk stored at 25 °C for 206 days remained essentially unchanged. The viscosity of the flaxseed milk stored at 37 °C tended to decrease after 157 days. Subsequently, the viscosity of the flaxseed milk kept decreasing slowly. This might have been due to the disruption of the three-dimensional reticulation in the flaxseed milk over the increasing storage time. This also indicates that storage temperature has a strong influence on the viscosity of flaxseed milk, but without a significant effect at room temperature.

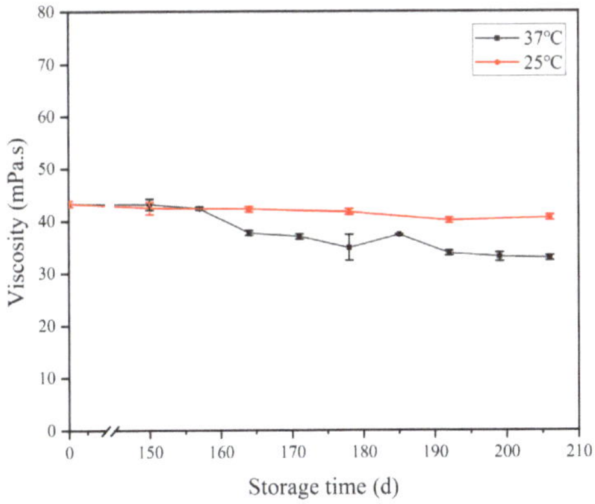

Figure 3. The viscosity of flaxseed milk stored at different temperatures.

3.4. Gravity Separation Stability

The stability of flaxseed milk was further characterized via monitoring of the centrifugal sedimentation rate and TSI during storage (Figure 4). Generally, centrifugal precipitation and TSI increased progressively during the storage period, which indicated a corresponding decrease in flaxseed milk stability. The centrifugal sedimentation rate and TSI values of the flaxseed milk stored at 25 °C increased slightly with increasing storage time. The centrifugal sedimentation rate and TSI growth rate of the flaxseed milk began to increase at 164 days in storage at 37 °C. At this point, the average particle size of flaxseed milk was 9.34 μm, and the presence of flocculation led to an increase in particle size. The flaxseed milk was an O/W emulsion, which is a thermodynamically unstable system. The large positive free energy at the interface of its oil droplets drove the droplets to coalesce. According to Stokes' law, the larger the radius of the droplet, the greater the difference in density between the two phases and the faster the delamination or settling rate [28]. As storage time increased, the suspended particles in the flaxseed milk proceeded in Brownian motion, and aggregation occurred as particles collided with each other. The large particles were precipitated via gravity, which caused an increase in the centrifugal sedimentation rate of the flaxseed milk and an increase in the TSI coefficient.

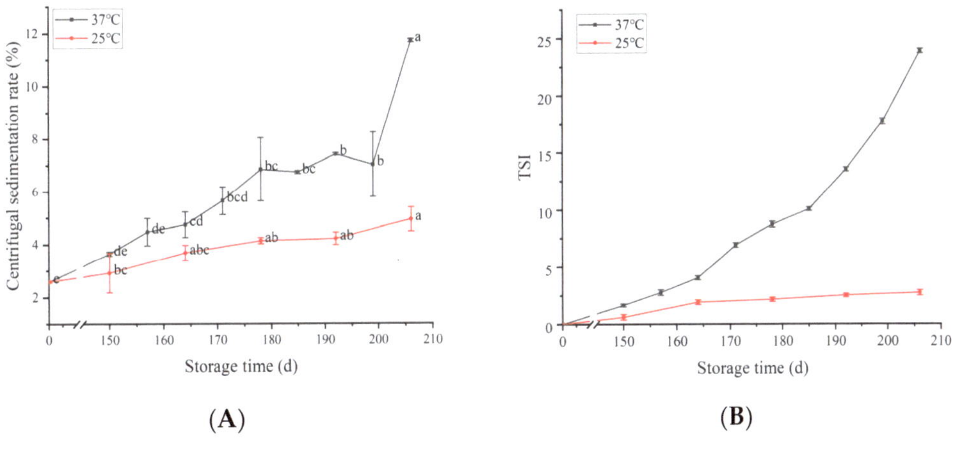

Figure 4. Centrifugal sedimentation rate (**A**) and TSI (**B**) of flaxseed milk stored at different temperatures. Different letters following the mean in the same column indicate significant differences ($p < 0.05$).

3.5. Oxidative Stability

The concentration of primary (hydroperoxide) and secondary (TBARS) lipid oxidation products was monitored during storage to assess the oxidative stability of flaxseed milk. As shown in Figure 5A, the hydroperoxide concentration was highest at 8.29 mmol/kg oil during storage at 25 °C for 192 days. Figure 5B shows that the concentration of hydroperoxides of flaxseed milk stored at 37 °C initially increased with time, peaking at 171 days, and then decreased. The increase in TBARS concentration was rapid for 150 days and then decreased gradually over the next 60 days. The relative effect of different temperatures on the rate of lipid oxidation was analyzed via comparison of the linear slopes of the curves. The rate of lipid oxidation in the flaxseed milk stored at 37 °C was significantly higher compared to that of the milk stored at 25 °C. Hydroperoxide concentrations were maximal for the 25 °C milk at 192 days, with similar levels in the 37 °C milk at 171 days, which was probably their oxidative inflection point at this time. The absolute value of the potential was highest at the lowest hydroperoxide content for flaxseed milk stored at 37 °C for 171 days. It might be that the aggregation of large particles weakened the electrostatic interaction between the particles, thus leading to the change in potential, which also indicated a close relationship between physical and oxidative stability. Many studies have also demonstrated that remaining lipid oxidation products are generally captured at the interface and adversely affect the physical stability of the emulsion [24]. During storage, the level of hydroperoxides produced increased for 171 days and then decreased over a longer storage period. This might be due to the fact that the rate of decomposition of the primary reaction products was faster than their formation over long storage times. At the same time, TBARS levels steadily increased to a maximum and then decreased over a longer storage time. This indicated that further degradation led to the formation of volatile products, which would cause products to sour, which is unacceptable to consumers.

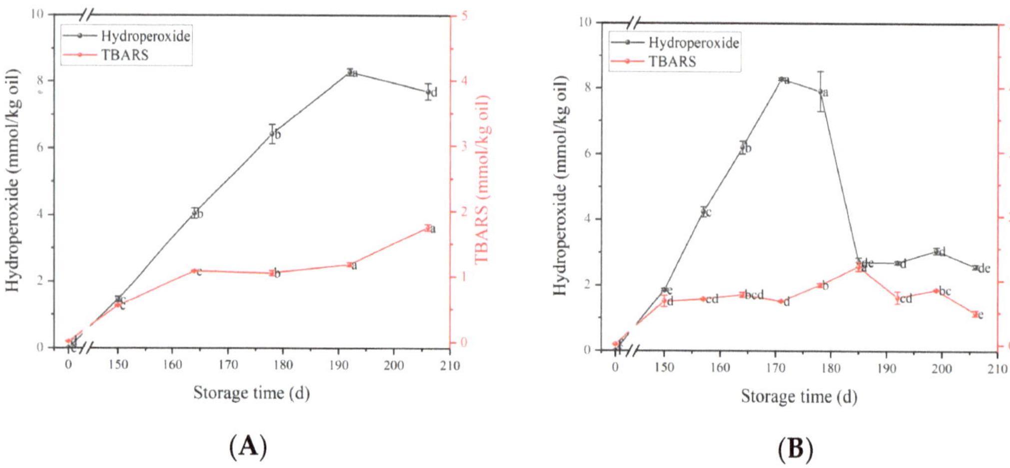

Figure 5. The primary and secondary products of flaxseed milk stored at 25 °C (**A**) and 37 °C (**B**). Different letters following the mean in the same column indicate significant differences ($p < 0.05$).

3.6. The Influence of Polyphenol Partitioning on Oxidative Stability of Flaxseed Plant-Based Milk

Flaxseed lignans are phenolic compounds, which mainly present in polymer complexes in flaxseed. Studies have demonstrated the antioxidant effect of flaxseed lignans in flaxseed oil emulsion [15,29]. The distribution of the flaxseed lignan macromolecules (FLM) in flaxseed milk is shown in Figure 6. Initially, the flaxseed lignans were distributed evenly in the oil–water phase of the flaxseed milk. Overall, the flaxseed lignin content all decreased with increasing storage time. However, it was obvious that the rate of reduction in lignin content in the aqueous phase of the flaxseed milk stored at both temperatures was greater than that in the non-aqueous phase. This might be due to partial depletion or migration of FLM in the aqueous phase to the oil phase or to particle surfaces. Previous studies have calculated the topological polar surface area (TPSA) of different flaxseed lignins, which suggests potential penetration into the interior of lipid droplets due to the lower TPSA values of flaxseed lignans [30]. That was why the flaxseed lignin content in the non-aqueous phase remained essentially constant between 150 and 164 days in storage. In addition, the number and position of phenolic hydroxyl groups and hydrophobic groups, such as phenylmethyl, played an important role in the stability and antioxidant activity of phenolic compounds in O/W emulsions [31,32]. Interestingly, the flaxseed lignin content in the non-aqueous phase also started to decrease in the flaxseed milk stored at 37 °C at 171 days. Corresponding to what has been mentioned above, oiling-off in the upper layer of flaxseed milk could be clearly observed and the content of hydroperoxide values peaked at this time. This might be explained by the leakage of lignin from the non-aqueous phase due to the breaking of flaxseed milk, which was heavily depleted for antioxidant purposes. This also showed that the relationship between chemical oxidation and physical destabilisation of flaxseed milk was inextricably linked.

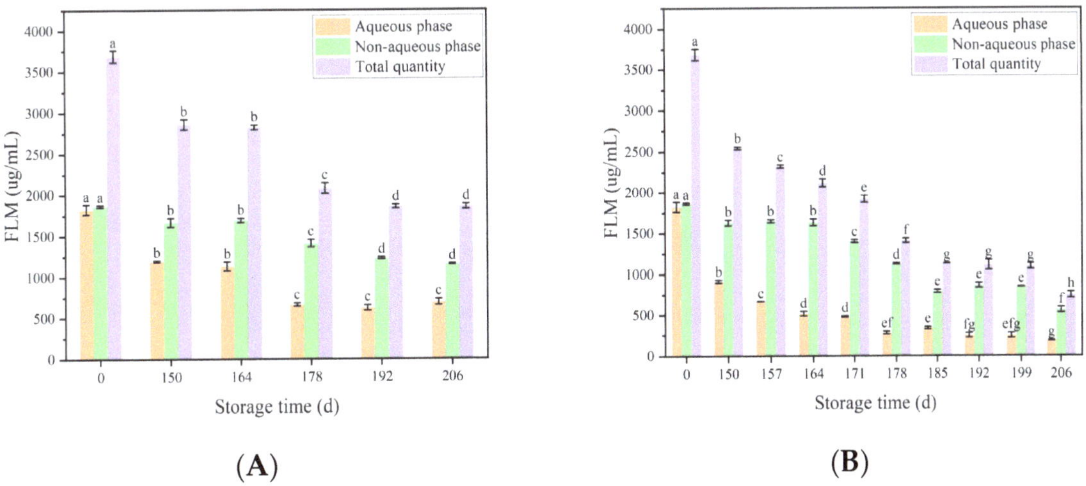

Figure 6. Flaxseed lignans migration in aqueous and non-aqueous phases of flaxseed milk at 25 °C (**A**) 37 °C (**B**). Different letters following the mean in the same column indicate significant differences ($p < 0.05$).

3.7. Statistical Correlation Analysis

The results of the statistical correlation analysis of the physical and chemical stability of the flaxseed milk are shown in Figure 7. The particle size of the flaxseed milk was positively correlated with the centrifugal sedimentation rate, TSI coefficient, POV, and TBARS content and negatively correlated with the flaxseed lignin content. These results suggest that the chemical oxidation of flaxseed milk is closely related to physical stability. The larger the particle size, the more unstable the whole system, which was also consistent with Stokes' law. Furthermore, the concentrations of POV and TBARS were negatively correlated with the flaxseed lignin content of each phase and the total lignin content. This also suggests that flaxseed lignans have antioxidant properties for flaxseed milk. There was some variation in the effect and correlation of physical stability and chemical oxidation rates when stored at different temperatures. Physical destabilization and chemical oxidation reactions in flaxseed milk stored at 37 °C were more intense than those stored at 25 °C.

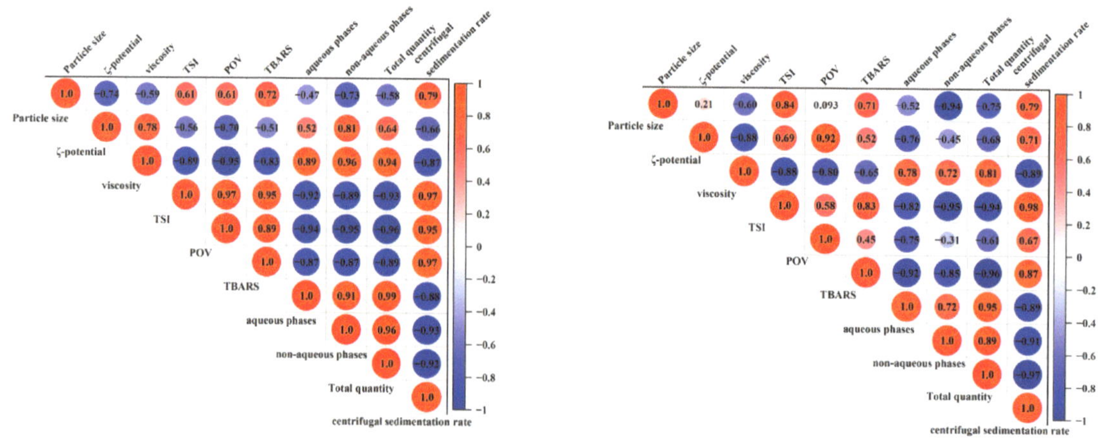

(A) (B)

Figure 7. Results of statistical correlation analysis of the physical and chemical stability of flaxseed milk in storage at 25 °C (**A**) 37 °C (**B**).

4. Conclusions

We studied the storage stability of flaxseed milk at different temperatures. It showed that chemical oxidation was closely linked to physical destabilization. The values of particle size, TSI, and centrifugal sedimentation rate of flaxseed milk gradually increased with increasing storage time, which indicated a decrease in physical stability. At the same time, the change in the potential of the flaxseed milk coincided with a change in the hydroperoxide content, showing that the change in physical stability was accompanied by a drastic oxidative change. The flaxseed lignans in the flaxseed milk were resistant to oxidation, and the degree of influence depended on the physical location and content of the flaxseed lignans. The degree of oxidation was also proven to affect the microstructure of the flaxseed milk, by causing stabilization–destabilization the flaxseed milk interface to collapse. The study was important for exploring the stabilization–destabilization mechanism of flaxseed milk. It would also hopefully be useful in elucidating the changes in storage stability of other plant-based milk substitutes.

Author Contributions: Conceptualization, methodology, software, validation, formal analysis, investigation, data curation, visualization, writing—original draft preparation, C.M.; writing—review and editing, Y.C.; writing—review and editing, X.W.; writing—review and editing, H.C.; supervision, H.C.; resources, project administration, funding acquisition: H.C. and Q.D. All authors have read and agreed to the published version of the manuscript.

Funding: This research was funded by National Natural Science Foundation of China (32001740), the Key R&D Program of Hubei Province of China (2021BBA224), and the Earmarked Fund for China Agriculture Research System (CARS-14).

Data Availability Statement: The data used to support the findings of this study can be made available by the corresponding author upon request.

Conflicts of Interest: The authors declare no conflict of interest.

References

1. Munekata, P.E.S.; Domínguez, R.; Budaraju, S.; Roselló-Soto, E.; Barba, F.J.; Mallikarjunan, K.; Roohinejad, S.; Lorenzo, J.M. Effect of Innovative Food Processing Technologies on the Physicochemical and Nutritional Properties and Quality of Non-Dairy Plant-Based Beverages. *Foods* **2020**, *9*, 288. [CrossRef] [PubMed]
2. Sethi, S.; Tyagi, S.K.; Anurag, R.K. Plant-based milk alternatives an emerging segment of functional beverages: A review. *J. Food Sci. Technol.* **2016**, *53*, 3408–3423. [CrossRef]

3. Iorio, M.C.; Bevilacqua, A.; Corbo, M.R.; Campaniello, D.; Sinigaglia, M.; Altieri, C. A case study on the use of ultrasound for the inhibition of Escherichia coli O157:H7 and Listeria monocytogenes in almond milk. *Ultrason. Sonochem.* **2019**, *52*, 477–483. [CrossRef] [PubMed]
4. Parikh, M.; Maddaford, T.G.; Austria, J.A.; Aliani, M.; Netticadan, T.; Pierce, G.N. Dietary Flaxseed as a Strategy for Improving Human Health. *Nutrients* **2019**, *11*, 1171. [CrossRef] [PubMed]
5. Lee, J.; Townsend, J.A.; Thompson, T.; Garitty, T.; De, A.; Yu, Q.; Peters, B.M.; Wen, Z.T. Analysis of the Cariogenic Potential of Various Almond Milk Beverages using a Streptococcus mutans Biofilm Model in vitro. *Caries Res.* **2018**, *52*, 51–57. [CrossRef] [PubMed]
6. Vagadia, B.H.; Vanga, S.K.; Singh, A.; Gariepy, Y.; Raghavan, V. Comparison of Conventional and Microwave Treatment on Soymilk for Inactivation of Trypsin Inhibitors and In Vitro Protein Digestibility. *Foods* **2018**, *7*, 6. [CrossRef]
7. Singh, K.K.; Mridula, D.; Rehal, J.; Barnwal, P. Flaxseed: A Potential Source of Food, Feed and Fiber. *Crit. Rev. Food Sci. Nutr.* **2011**, *51*, 210–222. [CrossRef]
8. Kajla, P.; Sharma, A.; Sood, D.R. Flaxseed-a potential functional food source. *J. Food Sci. Technol.* **2015**, *52*, 1857–1871. [CrossRef]
9. Patra, T.; Rinnan, Å.; Olsen, K. The physical stability of plant-based drinks and the analysis methods thereof. *Food Hydrocoll.* **2021**, *118*, 106770. [CrossRef]
10. Jeske, S.; Zannini, E.; Arendt, E.K. Evaluation of Physicochemical and Glycaemic Properties of Commercial Plant-Based Milk Substitutes. *Plant Foods Hum. Nutr.* **2017**, *72*, 26–33. [CrossRef]
11. Derjaguin, B.; Landau, L. Theory of the stability of strongly charged lyophobic sols and of the adhesion of strongly charged particles in solutions of electrolytes. *Prog. Surf. Sci.* **1993**, *43*, 30–59. [CrossRef]
12. Butler, J.A.V. Theory of the Stability of Lyophobic Colloids. *Nature* **1948**, *162*, 315–316. [CrossRef]
13. Cano-Sarmiento, C.; Téllez-Medina, D.I.; Viveros-Contreras, R.; Cornejo-Mazón, M.; Figueroa-Hernández, C.Y.; García-Armenta, E.; Alamilla-Beltrán, L.; García, H.S.; Gutiérrez-López, G.F. Zeta Potential of Food Matrices. *Food Eng. Rev.* **2018**, *10*, 113–138. [CrossRef]
14. Yu, X.; Nie, C.; Zhao, P.; Zhang, H.; Qin, X.; Deng, Q.; Huang, F.; Zhu, Y.; Geng, F. Influences of microwave exposure to flaxseed on the physicochemical stability of oil bodies: Implication of interface remodeling. *Food Chem.* **2022**, *368*, 130802. [CrossRef] [PubMed]
15. Cheng, C.; Yu, X.; McClements, D.J.; Huang, Q.; Tang, H.; Yu, K.; Xiang, X.; Chen, P.; Wang, X.; Deng, Q. Effect of flaxseed polyphenols on physical stability and oxidative stability of flaxseed oil-in-water nanoemulsions. *Food Chem.* **2019**, *301*, 125207. [CrossRef]
16. Kiralan, S.S.; Doğu-Baykut, E.; Kittipongpittaya, K.; McClements, D.J.; Decker, E.A. Increased Antioxidant Efficacy of Tocopherols by Surfactant Solubilization in Oil-in-Water Emulsions. *J. Agric. Food Chem.* **2014**, *62*, 10561–10566. [CrossRef]
17. Liang, L.; Chen, F.; Wang, X.; Jin, Q.; Decker, E.A.; McClements, D.J. Physical and Oxidative Stability of Flaxseed Oil-in-Water Emulsions Fabricated from Sunflower Lecithins: Impact of Blending Lecithins with Different Phospholipid Profiles. *J. Agric. Food Chem.* **2017**, *65*, 4755–4765. [CrossRef]
18. Panya, A.; Laguerre, M.; Bayrasy, C.; Lecomte, J.; Villeneuve, P.; McClements, D.J.; Decker, E.A. An Investigation of the Versatile Antioxidant Mechanisms of Action of Rosmarinate Alkyl Esters in Oil-in-Water Emulsions. *J. Agric. Food Chem.* **2012**, *60*, 2692–2700. [CrossRef]
19. Sheth, T.; Seshadri, S.; Prileszky, T.; Helgeson, M.E. Multiple nanoemulsions. *Nat. Rev. Mater.* **2020**, *5*, 214–228. [CrossRef]
20. Hao, J.; Wang, Q.; Li, X.; Duo, X. Extraction of structurally intact and well-stab ilized rice bran oil bodies as natural pre-emulsified O/W emulsions and investigation of their rheological properties and components interaction. *Food Res. Int.* **2023**, *164*, 112457. [CrossRef]
21. Zaaboul, F.; Raza, H.; Cao, C.; Yuanfa, L. The impact of roasting, high pressure homogenization and sterilization on peanut milk and its oil bodies. *Food Chem.* **2019**, *280*, 270–277. [CrossRef] [PubMed]
22. Mukherjee, D.; Chang, S.K.C.; Zhang, Y.; Mukherjee, S. Effects of Ultra-High Pressure Homogenization and Hydrocolloids on Physicochemical and Storage Properties of Soymilk. *J. Food Sci.* **2017**, *82*, 2313–2320. [CrossRef] [PubMed]
23. Yi, J.; Qiu, M.; Liu, N.; Tian, L.; Zhu, X.; Decker, E.A.; McClements, D.J. Inhibition of Lipid and Protein Oxidation in Whey-Protein-Stabilized Emulsions Using a Natural Antioxidant: Black Rice Anthocyanins. *J. Agric. Food Chem.* **2020**, *68*, 10149–10156. [CrossRef] [PubMed]
24. Berton-Carabin, C.C.; Ropers, M.-H.; Genot, C. Lipid Oxidation in Oil-in-Water Emulsions: Involvement of the Interfacial Layer. *Compr. Rev. Food Sci. Food Saf.* **2014**, *13*, 945–977. [CrossRef]
25. Wang, X.; Ye, A.; Singh, H. Structural and physicochemical changes in almond milk during in vitro gastric digestion: Impact on the delivery of protein and lipids. *Food Funct.* **2020**, *11*, 4314–4326. [CrossRef]
26. Wang, X.; Yu, K.; Cheng, C.; Peng, D.; Yu, X.; Chen, H.; Chen, Y.; Julian McClements, D.; Deng, Q. Effect of sesamol on the physical and chemical stability of plant-based flaxseed oil-in-water emulsions stabilized by proteins or phospholipids. *Food Funct.* **2021**, *12*, 2090–2101. [CrossRef] [PubMed]
27. Dai, T.; Shuai, X.; Chen, J.; Li, C.; Wang, J.; Liu, W.; Liu, C.; Wang, R. Whole peanut milk prepared by an industry-scale microfluidization system: Physical stability, microstructure, and flavor properties. *LWT* **2022**, *171*, 114140. [CrossRef]
28. McClements, D.J. Development of Next-Generation Nutritionally Fortified Plant-Based Milk Substitutes: Structural Design Principles. *Foods* **2020**, *9*, 421. [CrossRef]

29. Socrier, L.; Quéro, A.; Verdu, M.; Song, Y.; Molinié, R.; Mathiron, D.; Pilard, S.; Mesnard, F.; Morandat, S. Flax phenolic compounds as inhibitors of lipid oxidation: Elucidation of their mechanisms of action. *Food Chem.* **2019**, *274*, 651–658. [CrossRef]
30. Cheng, C.; Yu, X.; Huang, F.; Peng, D.; Chen, H.; Chen, Y.; Huang, Q.; Deng, Q. Effect of different structural flaxseed lignans on the stability of flaxseed oil-in-water emulsion: An interfacial perspective. *Food Chem.* **2021**, *357*, 129522. [CrossRef]
31. Sorensen, A.D.M.; Villeneuve, P.; Jacobsen, C. Alkyl caffeates as antioxidants in O/W emulsions: Impact of emulsifier type and endogenous tocopherols. *Eur. J. Lipid Sci. Tech.* **2017**, *119*, 2–14. [CrossRef]
32. Shahidi, F.; Zhong, Y. Revisiting the Polar Paradox Theory: A Critical Overview. *J. Agric. Food Chem.* **2011**, *59*, 3499–3504. [CrossRef] [PubMed]

foods

Article

The Emulsifying Properties, In Vitro Digestion Characteristics and Storage Stability of High-Pressure-Homogenization-Modified Dual-Protein-Based Emulsions

Meishan Wu [1,2,†], Xiaoye He [1,2,†], Duo Feng [1,2], Hu Li [1,2], Di Han [1,2], Qingye Li [1,2], Boya Zhao [1,2], Na Li [1,2], Tianxin Liu [1,2] and Jing Wang [1,2,*]

1 Institute of Food and Nutrition Development, Ministry of Agriculture and Rural Affairs, Beijing 100081, China
2 The Key Lab of Food Resources Monitoring and Nutrition Evaluation, Ministry of Agriculture and Rural Affairs, Beijing 100081, China
* Correspondence: wangjing07@caas.cn
† These authors contributed equally to this work.

Abstract: The droplet size, zeta potential, interface protein adsorption rate, physical stability and microrheological properties of high-pressure-homogenization (HPH)-modified, dual-protein-based whey–soy (whey protein isolate—soy protein isolate) emulsions containing different oil phase concentrations (5%, 10% and 15%; *w*/*w*) were compared in this paper. The in vitro digestion characteristics and storage stability of the dual-protein emulsions before and after HPH treatment were also explored. The results show that with an increase in the oil phase concentration, the droplet size and interface protein adsorption rate of the untreated dual-protein emulsions increased, while the absolute value of the zeta potential decreased. When the oil phase concentration was 10% (*w*/*w*), HPH treatment could significantly reduce the droplet size of the dual-protein emulsion, increase the interface protein adsorption rate, and improve the elasticity of the emulsion. Compared with other oil phase concentrations, the physical stability of the dual-protein emulsion containing a 10% (*w*/*w*) oil phase concentration was the best, so the in vitro digestion characteristics and storage stability of the emulsions were studied. Compared with the control group, the droplet size of the HPH-modified dual-protein emulsion was significantly reduced after gastrointestinal digestion, and the in vitro digestibility and release of free amino groups both significantly increased. The storage stability results show that the HPH-modified dual-protein emulsion showed good stability under different storage methods, and the storage stability of the steam-sterilized dual-protein emulsion stored at room temperature was the best. These results provide a theoretical basis for the development of new nutritional and healthy dual-protein liquid products.

Keywords: dual protein; high pressure homogenization; emulsifying properties; in vitro digestion; storage stability

Citation: Wu, M.; He, X.; Feng, D.; Li, H.; Han, D.; Li, Q.; Zhao, B.; Li, N.; Liu, T.; Wang, J. The Emulsifying Properties, In Vitro Digestion Characteristics and Storage Stability of High-Pressure-Homogenization-Modified Dual-Protein-Based Emulsions. *Foods* **2023**, *12*, 4141. https://doi.org/10.3390/foods12224141

Academic Editor: Chin Ping Tan

Received: 9 October 2023
Revised: 9 November 2023
Accepted: 14 November 2023
Published: 15 November 2023

1. Introduction

The oil-in-water (O/W) emulsion system is a typical form of foods such as milk, mayonnaise and milk beverages [1]. In an O/W emulsion, the oil exists as a dispersed phase in the form of small droplets suspended in water (continuous phase) [2]. The choice of natural proteins as emulsifiers to prepare O/W emulsions has been attracting the interest of researchers [3–5]. As a kind of amphiphilic interfacial active molecule, protein can rapidly absorb at the O/W interface, which changes the molecule's structure; the hydrophobic and hydrophilic groups combine with the oil phase and water phase separately, reducing the surface tension and forming a viscoelastic interface film [6].

In recent years, in order to solve the shortage of protein resources caused by global population growth and promote environmental sustainability, improving the development and utilization of high-quality plant protein resources and researching the partial replacement of animal protein with plant protein in formulated products have gradually become hot research topics [7,8]. A "dual-protein", referring to an edible protein source obtained from natural, high-quality plant protein such as soy protein and animal protein such as whey protein, which was discovered by studying the dose–effect relationship and the mechanisms of synergism and interactions of animal–plant mixed proteins, has a complete range of amino acids and is rich in essential amino acids. Its high nutritional value and bioavailability play important roles in maintaining physical health and promoting body growth. Preliminary findings demonstrated that a dual protein had significant nutritional quality improvement in hematopoietic reconstruction, immune function, muscle level, lipid metabolism, and osteoporosis [9–11]. Whey protein isolate (WPI) and soy protein isolate (SPI) are the main animal–plant protein ingredients in dual proteins; both have excellent functional processing properties such as emulsification, gelation and foaming, which are widely used in food processing [12,13]. However, commercial SPI always exhibits inferior solubility and emulsifying properties because of its partial denaturation during processing and cannot form a stable O/W emulsion structure [14]. Therefore, it is necessary to study suitable modification methods for improving the emulsifying properties of dual proteins, which has profound significance for improving their application in food processing.

High-pressure homogenization (HPH) is an emerging processing technology that has been widely used in food processing. Originally, HPH was only used to homogenize complex liquid products such as fruit and vegetable juices to enhance their rheological properties and physical stability as a standardization step [15,16]. The working principle can be generally summarized as follows: when a liquid material flows quickly through a homogeneous valve gap with a special internal structure, it is subjected to mechanical forces such as high-speed shear, shock, and the cavitation phenomenon, which crush and break the large liquid particles into small ones [17,18]. HPH is widely used to prepare protein-stabilized emulsions, and suitable homogenization pressure could improve the stability of a protein emulsion by reducing the droplet size and enhancing the rheological and interfacial properties of the emulsion. Ma et al. found that HPH promoted the adsorption of cod protein molecules at the O/W interface, and the elastic modulus (G′) of the emulsion interface film significantly increased [19]. Fernandez-Avila et al. observed that 100 MPa and 200 MPa treatments could allow more SPI to be loaded onto the O/W interface, thereby improving the emulsion's oxidative stability [20]. Melchior et al. found that the oil-holding capacity of a pea protein emulsion improved as the HPH intensity increased, and the in vitro digestibility performed best with a 70 MPa treatment, comparable to original pea protein [21].

HPH-modified protein emulsions have been extensively studied, while dual-protein emulsion modification via HPH has not been deeply researched so far. In this study, the structure and interfacial properties of HPH-modified dual-protein emulsions were researched, and their in vitro digestion properties and storage stability were assessed.

2. Materials and Methods

2.1. Materials

SPI (protein content, 90%) was obtained from IFF Inc. (Shanghai, China). WPI (protein content, 95%) was obtained from Agropur Cooperative (Montreal, QC, Canada). Soy oil was purchased from Yihai Karry Co., Ltd. (Shanghai, China). All reagents were of analytical-grade purity.

2.2. Preparation of Protein Solution and HPH Treatment

The protein powder (SPI and WPI) were dispersed in de-ionized water at a total protein concentration of 6% (*w/v*) to achieve SPI, WPI, and dual protein (SPI:WPI = 1:1) solutions, and they were stirred at room temperature (25 °C) for 1.5 h. After that, all the solutions were stored at 4 °C for 12 h to ensure full hydration. After the protein solutions were restored to room temperature, they were filtered using a 0.3 mm sieve and homogenized for three cycles at 60 MPa, using a high-pressure homogenizer (Panda PLUS 2000, GEA Group, Düsseldorf, Germany); each cycle time was approximately 13 s. The high-pressure homogenizer was connected to a condenser to keep the temperature of the protein solutions at 18 ± 1 °C after each HPH treatment cycle. Untreated protein solutions were used as a control.

2.3. Emulsion Preparation

The protein emulsions containing 5%, 10% and 15% (*w*/*w*) oil phase concentrations were prepared as follows: soy oil was mixed with the protein solutions at 1:19, 1:9 and 3:17 mass ratios, using a high-speed homogenizer (Ultra Turrax T25 Digital, IKA-Werke GmbH & Co. KG, Stuttgart, Baden-Württemberg, Germany) at 5000 rpm for 1 min. Then, all the O/W protein emulsions were processed via three cycles of HPH treatment at 50 MPa using a high-pressure homogenizer (Panda PLUS 2000, GEA Group, Düsseldorf, Germany).

2.4. Dual-Protein Emulsion Measurements

2.4.1. Droplet Size and Zeta Potential of Emulsion

All the protein emulsions were diluted to 0.12 mg/mL using de-ionized water. The droplet size and zeta potential were determined using a nanoparticle size analyzer (Zetasizer Nano-ZS90, Malvern Instrument Co., Ltd., Malvern, UK) at room temperature (25 °C), according to a previous report [22]. The zeta potential was indirectly calculated from the electrophoretic mobility using Henry's equation of electric field [23]:

$$\frac{U}{E} = \frac{2\varepsilon\zeta F(\kappa a)}{3\eta} \tag{1}$$

where U/E is the electrophoretic mobility ($m^2s^{-1}V^{-1}$), ζ is the zeta potential (V), ε is the solvent's dielectric permittivity ($kgmV^{-2}s^{-2}$), η is the viscosity ($kgm^{-1}s^{-1}$), and $F(\kappa a)$ is Henry's function (dimensionless), which varies between 1 and 3/2 according to the ratio of the particle size to the Debye length, $1/\kappa$.

2.4.2. Interfacial Protein Absorption Rate of Emulsions

As described by Sun et al. [24], the interfacial protein adsorption rates (AP%) of the emulsions were measured using the Bradford method, and bovine serum albumin (BSA) was used as a standard. All the protein emulsions were centrifuged at 19,890 rpm for 5 h (4 °C). After centrifugation, the emulsion was separated into two layers; then, the cream phase was removed, and the aqueous phase was filtered through a 0.45 μm filter and diluted to 0.06 mg/mL using de-ionized water. The diluted aqueous phase was mixed with Coomassie brilliant blue G-250 at a volume ratio of 1:5 for 2 min, and the absorbance was measured using an ultraviolet–visible spectrophotometer (UV1780, Shimadzu Corporation, Kyoto, Japan) at 595 nm. By substituting the absorbance into the BSA standard curve, the protein concentration of the aqueous phase was calculated. Finally, the AP% was calculated according to the following equation:

$$AP\% = \frac{C_0 - C_1}{C_0} \times 100\% \tag{2}$$

In Equation (2), C_0 and C_1 are the initial protein concentration of the emulsion and the protein concentration of the aqueous phase, respectively.

2.4.3. Physical Stability of Emulsions

A Tubiscan stability analyzer (Tubiscan Tower, Formulaction, Toulouse, France) was used to measure the physical stability of the protein emulsions. According to the previous method reported by Shi et al. [25], a back-scattered light was set to measure at 15 min intervals for 5.75 h, and all the tests were performed at room temperature (25 °C). The changes in the backscattering profiles and Turbiscan stability index (TSI) were calculated using Towersoft 2.0.0.9.

2.4.4. Micro-Rheological Properties of Emulsions

Micro-rheology of the protein emulsions was performed using an optical microrheometer (Rheolaser Master, Formulaction, Toulouse, France). According to the previous method described by Zhu et al. [26], 20 mL of a protein emulsion was added into a glass tube and then placed in the Rheolaser chamber. The emulsion was measured every 5 s for a total of 20 min at room temperature (25 °C). The elasticity index (EI), macro-viscosity index (MVI) and solid-liquid balance (SLB) of the protein emulsion were obtained using a Rheosoft Master 1.4.0.0.

2.5. In Vitro Digestion Behavior Analysis of Dual-Protein Emulsions

2.5.1. In Vitro Digestion Treatment

According to the previous method described by Ding et al. [27], a simulation of gastric–intestinal digestion was designed to conduct in vitro static digestion.

Gastric phase: 2 g of NaCl and 7 mL of HCl (37%) were dissolved in 1 L of de-ionized water, and then 3.2 g of pepsin was added to the solution. The prepared simulated gastric fluid (SGF) was mixed with the emulsion at a volume ratio of 1:1, and the pH of the mixture was adjusted to 2.0 using 1.0 M of HCl and 1.0 M of NaOH. Then, the mixture was incubated at 37 °C for 2 h with continuous agitation in a thermostatic water bath (DF-101S, Lichen Experimental Equipment Manufacturing Co., Ltd., Shanghai, China).

Intestine phase: The pH of each sample after gastric digestion was adjusted to 7.0; then, 1.5 mL of salt solution (10 mM of $CaCl_2$ and 150 mM of NaCl), 3.5 mL of porcine bile extract solution (54 mg/mL in a phosphate buffer, pH 7.0) and 2.5 mL of trypsin solution (24 mg/mL in a phosphate buffer, pH 7.0) were added to the sample in order. After that, the pH of the mixture was adjusted to 7.0 using 1.0 M of HCl and 1.0 M of NaOH. Then, the mixture was incubated at 37 °C for 2 h with continuous agitation in a thermostatic water bath (DF-101S, Lichen Experimental Equipment Manufacturing Co., Ltd., Shanghai, China).

2.5.2. The Droplet Size and Zeta Potential of the Emulsions during In Vitro Digestion

Digestive samples were taken at 30, 60, 90, and 120 min during gastric digestion and intestinal digestion, respectively, which were treated based on enzyme inactivation. All the samples were diluted to 0.12 mg/mL using de-ionized water, and the droplet size and zeta potential were measured using a nanoparticle size analyzer (Zetasizer Nano-ZS90, Malvern Instrument Co., Ltd., Malvern, UK) at room temperature (25 °C).

2.5.3. The Free Amino Groups of the Emulsions during In Vitro Digestion

According to Church et al. [28], the free amino groups (-NH_2) of the emulsions during in vitro digestion were measured via the ortho-phthalaldehyde (OPA) method, and DL-serine was used as a standard. The OPA solution was made by combining the following reagents and diluted to a final volume of 1 L with de-ionized water: 38.1 g of sodium tetraborate, 1 g of sodium dodecyl sulfate (SDS), 0.8 g of OPA (dissolved in 20 mL of absolute ethanol), and 0.88 g of dithiothreitol (DTT). This reagent was prepared freshly every day. The enzyme inactive samples were taken at 30, 60, 90, and 120 min during gastric digestion and intestinal digestion, respectively, and then they were centrifuged at 19,890 rpm for 3 h (4 °C). After centrifugation, the cream phase was removed, and the aqueous phase was diluted to 0.6 mg/mL using de-ionized water. Then, 400 μL of the diluted aqueous phase was mixed with 3 mL of the OPA solution for 2 min, and the absorbance was measured using an ultraviolet–visible spectrophotometer (UV1780, Shimadzu Corporation, Kyoto, Japan) at 340 nm. By substituting the absorbance into the DL-serine standard curve, the -NH_2 of the aqueous phase was calculated.

2.5.4. In Vitro Digestibility of Emulsions

The in vitro digestibility of the emulsions was measured using the Bradford method. The enzyme-inactive samples of all the emulsions were taken at the end of the gastric–intestinal digestion, and they were centrifuged at 19,890 rpm for 3 h (4 °C). After centrifugation, the cream phase was removed, and the aqueous phase was diluted to 0.12 mg/mL using de-ionized water. The diluted aqueous phase was mixed with Coomassie brilliant blue G-250 at a volume ratio of 1:5 for 2 min, and the absorbance was measured using an ultraviolet–visible spectrophotometer (UV1780, Shimadzu Corporation, Kyoto, Japan) at 595 nm. By substituting the absorbance into the BSA standard curve, the protein concentration of the aqueous phase was calculated. Finally, the in vitro digestibility was calculated according to the following equation:

$$\text{In vitro digestion\%} = \frac{C_0 - C_1}{C_0} \times 100\% \tag{3}$$

In Equation (3), C_0 and C_1 are the initial protein concentration of the emulsion and the protein concentration of the aqueous phase after in vitro digestion, respectively.

2.6. Storage Stability of Dual-Protein Emulsions

All the emulsions were treated with 65 °C pasteurization for 30 min, 95 °C boiled sterilization for 10 min, and 121 °C steam sterilization for 20 min, respectively. Untreated emulsions were used as controls. The control and pasteurization samples were stored at 4 °C, and the boiled-sterilization and steam-sterilization samples were stored at room temperature (25 °C). The storage stability of the emulsions was measured by the changes in the appearances of the emulsions at 0, 7, 14, 21, and 28 d.

2.7. Statistical Analysis

All the statistical differences in the data were analyzed using SPSS 26 (IBM Corporation, Armonk, NY, USA) through a one-way analysis of variance (ANOVA), and Origin 2021 was used for data analysis and visualization. The differences were considered significant at $p < 0.05$.

3. Results and Discussion

3.1. Droplet Size and Zeta Potential

The droplet size and zeta potential of a protein emulsion depend on the protein trapped on the surface of the oil, which are important factors affecting the physicochemical properties of emulsion [29]. Figure 1a,b show the droplet size and zeta potential results of the protein emulsions, respectively. It can be seen that as the oil phase concentration rose, the droplet size of the untreated protein emulsion increased significantly, which was because the emulsified oil droplet volume needed to be increased to satisfy the limit of the protein emulsifier absorbed on the surface of the oil. The mean droplet size of the HPH-modified SPI emulsion containing the 10% (*w*/*w*) oil phase concentration decreased obviously from 685 nm to 547 nm, and the absolute value of its zeta potential also increased to the maximum (32 mV). Perhaps because the high shear rate could cut the emulsion droplets into smaller ones during HPH treatment and more negative charges were exposed on the surface of the protein, which increased the electrostatic repulsion between droplets, the stability of the emulsion was enhanced. A similar result demonstrated that HPH played a key role not only in the reduction of the droplet size but also in the particle dispersion, as Song et al. [30] reported before. The same changes took place in the HPH-modified dual-protein emulsion containing a 10% (*w*/*w*) oil phase concentration: the droplet size decreased from 485 nm to 425 nm, and the absolute value of the zeta potential increased to 34.2 mV. The mean droplet size of the HPH-modified WPI emulsion containing a 10% (*w*/*w*) oil phase concentration had no significant difference compared to the control, whereas the absolute value of the zeta potential increased significantly from 35.4 mV to 36.9 mV, indicating that HPH was beneficial in improving the dispersion stability of the

WPI emulsion. As the oil phase concentration rose to 15% (*w*/*w*), the mean droplet size of the HPH-modified SPI and dual-protein emulsions both obviously increased to 1085.4 nm and 748.1 nm separately, and the absolute values of the zeta potential decreased, suggesting that excessive oil phase aggregation might have caused the coalescence and flocculation of the emulsion.

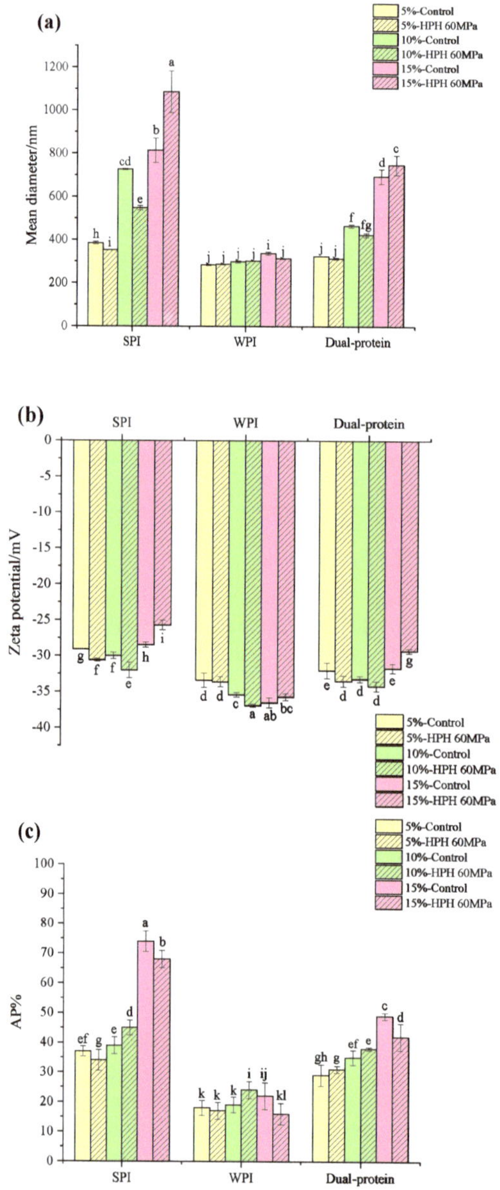

Figure 1. The droplet size (**a**), zeta potential (**b**), and interfacial protein absorption rate (**c**) of protein emulsions containing different oil phase concentrations. Different lowercase letters in Figures (**a**–**c**) indicate significant differences in the different protein emulsions ($p < 0.05$).

3.2. Interfacial Protein Absorption Rate

The AP% at the O/W interface is also one of the most important indexes for measuring the stability of an emulsion [31]. Figure 1c shows the AP% of the protein emulsions. As the oil phase concentration rose, the AP% at the interface of the untreated protein emulsion increased gradually. The AP% of the dual-protein emulsion was lower than the SPI emulsion, but it was much higher than the WPI emulsion, indicating that the existence of SPI could improve the absorption of WPI molecules at the O/W interface. All the AP% values of the HPH-modified protein emulsions improved at the 10% (w/w) oil phase concentration, which might be attributed to the improvement of the surface hydrophobicity of protein after HPH treatment, promoting the interface's absorption capacity at this oil phase concentration [32]. The AP% values of the HPH-modified SPI and WPI emulsions only obviously increased compared with the control ones at the 10% (w/w) oil phase concentration, while the AP% of the HPH-modified dual-protein emulsion increased at both the 5% and 10% (w/w) oil phase concentrations, indicating that HPH had more positive effect on enhancing the interfacial absorption property of the dual-protein emulsion. However, as the oil phase concentration rose to 15% (w/w), compared with the control ones, all the AP% values of HPH modified protein emulsions decreased, which might be because the interface film was loaded its maximum absorption capacity, and the aggregation of unloaded oil droplets inhabited the absorption of protein molecules at the O/W interface.

3.3. Physical Stability Properties

The Turbiscan stability analyzer is based on the principle of multiple heavy light scattering, carrying out top-down scanning using transmitted light and backscattered light sources passing through the static sample after a certain period of time to characterize changes in the liquid sample according to the collected spectral line data and analyzing the mechanism of system instability in real time [33]. Figure 2a–c show the backscattering changes of the SPI, WPI, and dual-protein emulsions, respectively. Clearly, the backscattering of the untreated SPI emulsion showed the condition change from floating to sediment as the oil phase concentration rose, whereas the untreated WPI and dual-protein emulsions both showed the floating condition at three oil phase concentrations, and the backscattering of the right side decreased as the oil phase concentration rose, reflecting that the floating phenomenon of the emulsion declined because the increase in the emulsion's particle size improved the flow resistance of the emulsion, according to what Hebishy et al. [34] reported previously. The backscattering of the untreated dual-protein emulsion was significantly lower than the untreated WPI emulsion at the same oil phase concentration, indicating that the stability of the dual-protein emulsion was better than the WPI emulsion. At the 10% (w/w) oil phase concentration, the backscattering of the SPI emulsion reduced after HPH treatment, and the same phenomenon could be found in the WPI emulsion, indicating that HPH could improve the physical stability of the SPI and WPI emulsions at this oil phase concentration. However, after HPH treatment, the backscattering of the right side of the dual-protein emulsions slightly increased at three oil phase concentrations; as reported previously, this might be caused by the supramolecular aggregates formed, which may reduce the rate of diffusion and rearrangement on oil–water interface, thus preventing the interface film from forming in time and resulting in the aggravation of floating [35].

The Turbiscan Stability Index (TSI) can reflect the kinetic instability of an emulsion system within a certain time. The smaller the TSI value, the better the system's physical stability. It was found that the TSI of all the protein emulsions increased gradually with time, and the TSI values of the protein emulsions at the last scanning time are shown in Table 1. The TSI of the untreated SPI emulsion reached a maximum at the 15% (w/w) oil phase concentration, and the TSI of the untreated dual-protein emulsion was lower than the untreated WPI emulsion at the same oil phase concentration. The HPH-modified SPI emulsion had the lowest TSI at the 10% (w/w) oil phase concentration, 0.1, while the TSI values of the HPH-modified WPI and dual-protein emulsions both increased. It is worth noting that the TSI of the HPH-modified dual-protein emulsion was still lower than the HPH-modified WPI emulsion at the same oil phase concentration, suggesting that the dual-protein emulsion was more stable than the WPI emulsion after HPH treatment.

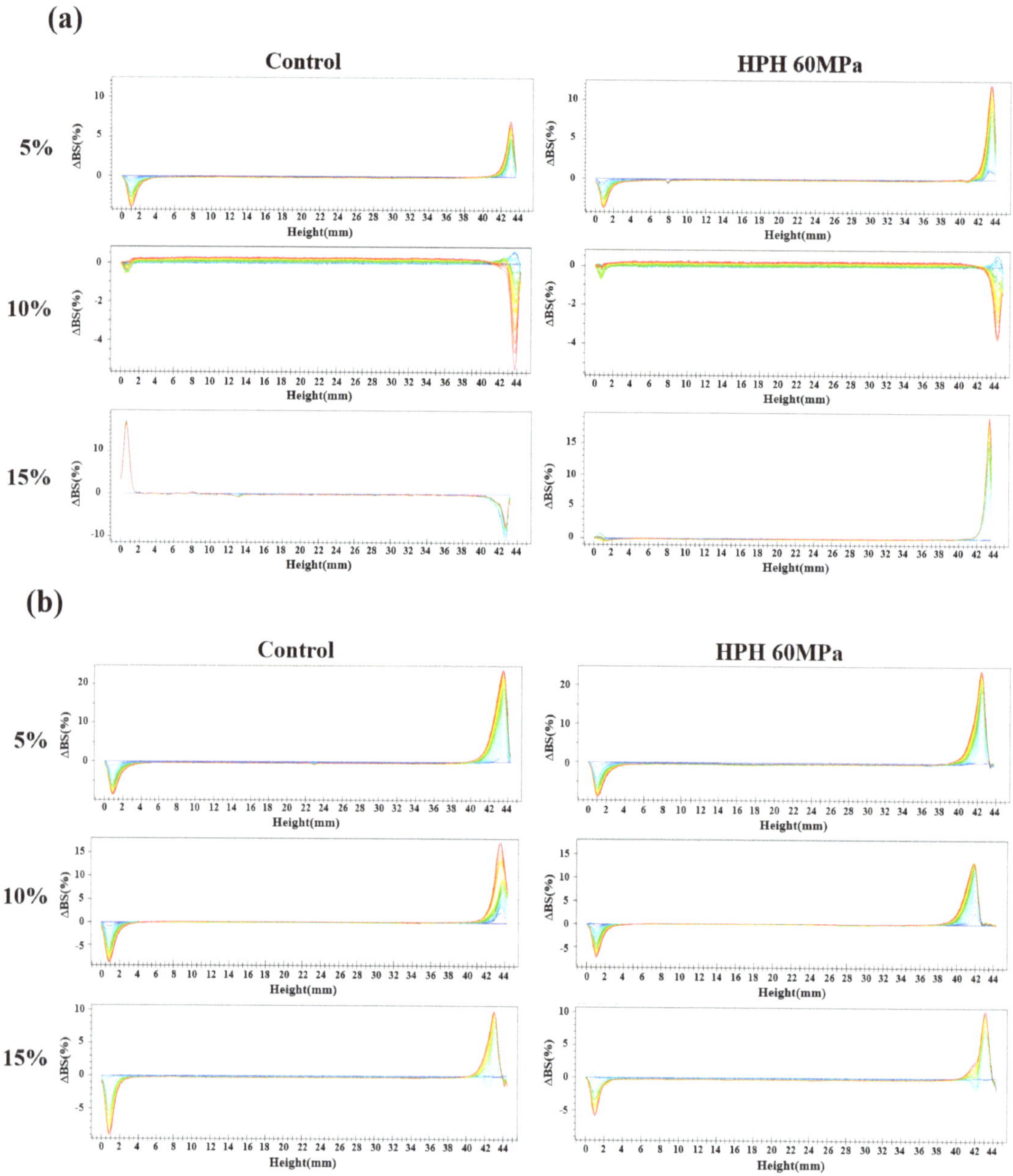

Figure 2. *Cont.*

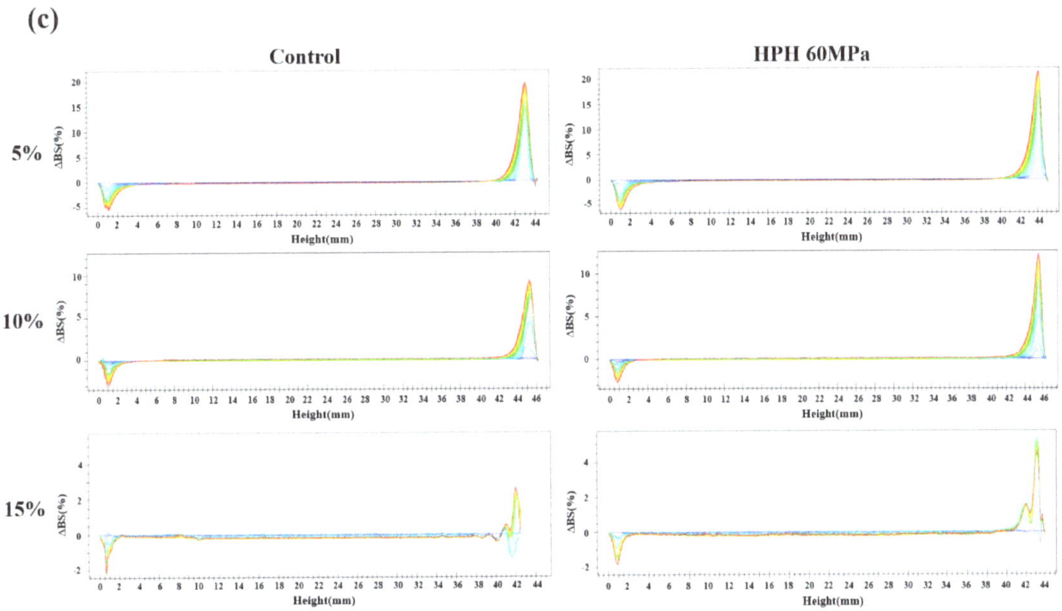

Figure 2. The changes in the backscattering profiles of SPI (**a**), WPI (**b**) and dual-protein (**c**) emulsions containing different oil phase concentrations. The color of the lines in Figures (**a–c**) gradually changes from blue to green and eventually to red as time goes on.

Table 1. The TSI values of protein emulsions containing different oil phase concentrations.

Oil Phase Concentration (%, *w*/*w*)	Control			HPH 60 MPa		
	SPI	WPI	Dual Protein	SPI	WPI	Dual Protein
5	0.2	0.9	0.7	0.3	1.2	0.8
10	0.2	0.4	0.2	0.1	0.6	0.3
15	0.5	0.6	0.4	0.8	0.7	0.5

3.4. Micro-Rheology of Emulsions

Micro-rheology can usually be used to trace the motion of colloidal particles at a scale of 0.1–10 μm. By detecting the Brownian motion of droplets, the displacement of particles can be accurately detected, and the micro-rheological behavior of liquids can be analyzed [36]. A Rheolaser Master was used to detect structural and viscoelastic changes in the emulsion system, using the mean square displacement (MSD) of the emulsion droplet as a formula for the de-correlation time. Figure 3a–c show the MSD-time curves of the SPI, WPI and dual-protein emulsions. It can be seen that as the oil phase concentration increased, the MSD curve of the untreated protein emulsion became narrower and denser, indicating that the structure of the emulsion became tighter. The linear relationship between the MSD curve and the de-correlation time of the HPH-modified SPI emulsion was reduced at the 5% and 10% (*w*/*w*) oil phase concentrations, but it returned to a linear relationship again at the 15% (*w*/*w*) oil phase concentration, which mainly reflected that the emulsion's viscosity became obvious as the oil phase concentration rose. This same phenomenon could be found in the HPH-modified dual-protein emulsion. Regarding the HPH-modified WPI emulsion, it showed a more significant linear relationship at the 10% (*w*/*w*) oil phase concentration, suggesting that the emulsion had strong viscosity at this oil phase concentration.

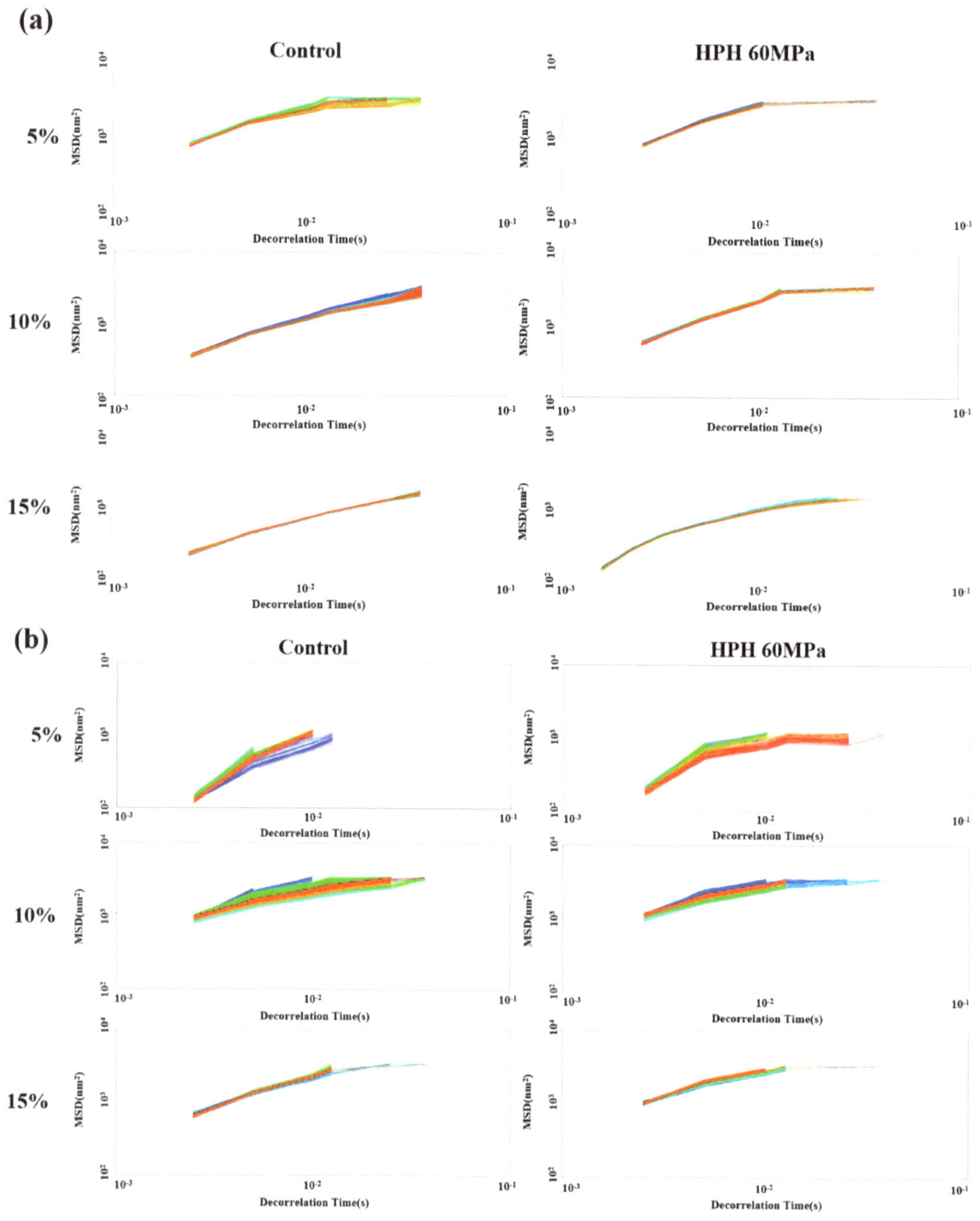

Figure 3. *Cont.*

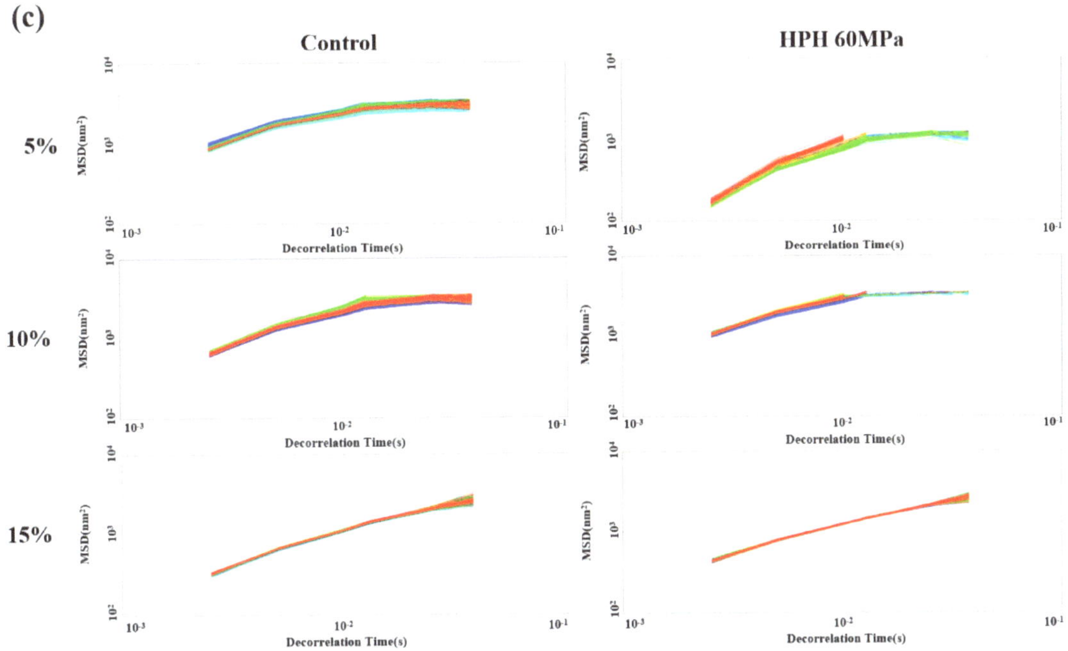

Figure 3. The MSD-time curves of SPI (**a**), WPI (**b**) and dual-protein (**c**) emulsions containing different oil phase concentrations. The color of the lines in Figures (**a–c**) gradually changes from blue to green and eventually to red as time goes on.

The elasticity index (EI) and macroscopic viscosity index (MVI) can reflect the elastic and viscous characteristics of an emulsion, respectively. Figure 4a,b show the EI and MVI of protein emulsions calculated using Rheosoft Master 1.4.0.0 software, respectively. It can be clearly seen that the EI values of the untreated SPI and dual-protein emulsions increased as the oil phase concentration rose, and the EI increased further at the 5% and 10% (*w*/*w*) oil phase concentrations after HPH treatment, indicating that the elasticity of the emulsions was enhanced. On the contrary, the EI of the HPH-modified WPI emulsion declined at three oil phase concentrations, and the elasticity of the emulsion reduced. In addition, the MVI of the protein emulsion decreased after HPH treatment, but the MVI of the HPH-modified SPI emulsion containing the 15% (*w*/*w*) oil phase concentration abnormally increased, indicating that there was abundant oil phase which overflowed, the aggregation of massive oil droplets led to serious damage to the structure of the emulsion, and the viscosity of the emulsion significantly increased after HPH treatment.

The SLB is the slope value of the MSD curve plateau in micro-rheology, which can reflect the solid–liquid equilibrium state of a system. When the SLB < 0.5, the droplet movement speed of the system is slow, which mainly shows elastic solid behavior. When SLB $= 0.5$, the system is in a solid–liquid equilibrium state. When $0.5 <$ SLB < 1, viscous liquid behavior is mainly manifested. When SLB > 1, sediment occurs [37]. Figure 4c shows the SLB of the protein emulsions, calculated using Rheosoft Master 1.4.0.0 software. The SLB values of the untreated SPI and WPI emulsions were both between 0.5 and 1, while the SLB of the untreated dual-protein emulsion containing the 5% (*w*/*w*) oil phase concentration was slightly lower than 0.5, showing mild solid behavior. After HPH treatment, the SLB of the SPI emulsion containing the 10% (*w*/*w*) oil phase concentration and the WPI emulsion containing the 5% (*w*/*w*) oil phase concentration both increased to beyond 1, and a slight sediment existed in the emulsions; additionally the SLB of the SPI emulsion containing the 15% (*w*/*w*) oil phase concentration and the WPI emulsion containing 10% (*w*/*w*) decreased obviously to 0.41 and 0.36, respectively, and the structures of the emulsions tended to be more solid. However, the SLB of the HPH-modified dual-protein emulsions were maintained between 0.5 and 1 at the three oil phase concentrations, indicating that the dual-protein emulsion had significant viscous liquid characteristics compared to a single-protein emulsion after an HPH treatment.

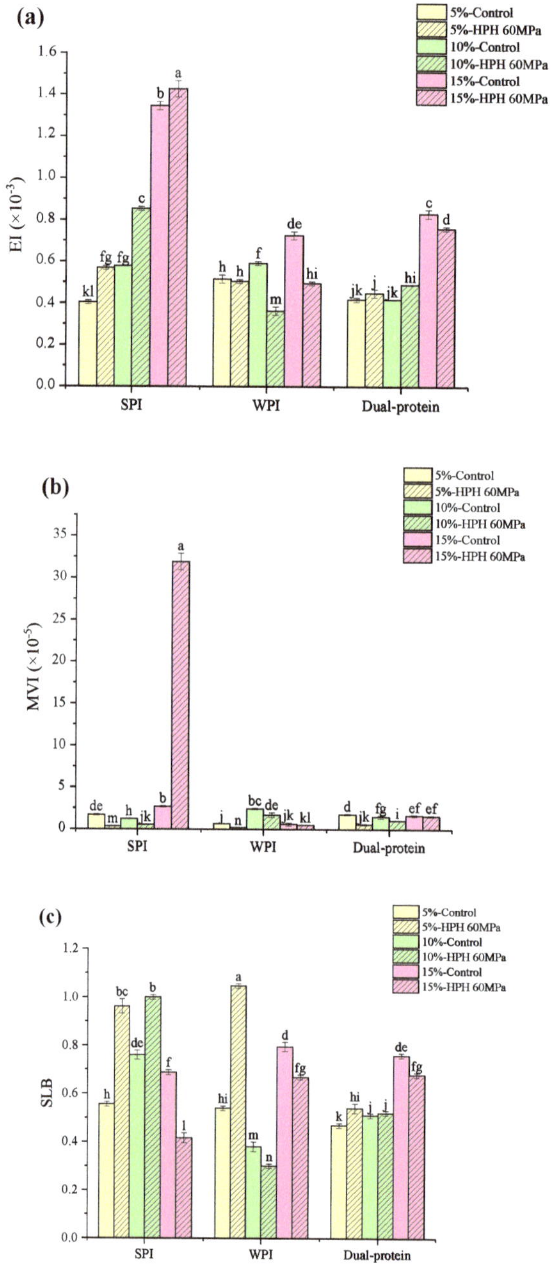

Figure 4. The EI (**a**), MVI (**b**) and SLB (**c**) of protein emulsions containing different oil phase concentrations. Different lowercase letters in Figures (**a**–**c**) indicate significant differences in the different protein emulsions ($p < 0.05$).

In summary, the HPH-modified dual-protein emulsion containing the 10% (*w*/*w*) oil phase concentration had the better diffusion stability, the AP% at the O/W interface increased, and the emulsion's elasticity was enhanced, showing a more significant solid–liquid equilibrium than a single-protein emulsion. Therefore, the HPH-modified dual-protein emulsion containing the 10% (*w*/*w*) oil phase concentration was selected as the follow-up research object.

3.5. In Vitro Digestion Properties of Emulsion

3.5.1. Droplet Size and Zeta Potential during In Vitro Digestion

Changes in the droplet size and zeta potential during the in vitro digestion of the SPI, WPI and dual-protein emulsions containing a 10% (*w*/*w*) oil phase concentration are shown in Figure 5a–c, respectively. It can be observed that the zeta potentials of all the protein emulsions decreased significantly during gastric digestion, which was due to the high concentration of salt ions in the gastric digestive fluid and because the low-pH environment reduced the strength of the electrostatic repulsion between droplets and accelerated the droplets' aggregation [38]. The droplet size of all the SPI and dual-protein emulsions obviously increased in gastric digestion, and the droplet size of the HPH-modified dual-protein emulsion was larger than the untreated one, which might be attributed to the complex interface protein composition that decreased the stability in the gastric digestion phase. Contrarily, the droplet size of the WPI emulsion did not increase significantly during gastric digestion; perhaps the β-lactoglobulin (β-Lg) in the WPI had a certain resistance to pepsin that was not easy to hydrolyze, so the droplet size did not obviously change [39].

In the intestinal digestion phase, the zeta potentials of all the protein emulsions were significantly higher than in the gastric digestion phase because the pH environment during intestinal digestion (7.0) was much higher than the isoelectric point of the protein, which was beneficial for increasing the net charge on the surfaces of the droplets to improve the electrostatic balance between the droplets [40]. The droplet size of all the SPI and dual-protein emulsions obviously decreased in the initial period of the intestinal digestion phase, while the droplet size of the dual-protein emulsion increased again after 60 min of intestinal digestion, and the same phenomenon could be found in the WPI emulsion; perhaps it was related to the aggregation of large particles, such as free fatty acids and other digestive products [41]. Moreover, the droplet size of the HPH-modified dual-protein emulsion in the terminal period of intestinal digestion was lower than the untreated one, indicating that the degree of digestion and the decomposition of the dual-protein emulsion improved after HPH treatment.

3.5.2. The Release of Free Amino Groups during In Vitro Digestion

The release of free amino groups (-NH_2) could reflect the degree of protein digestion in the gastrointestinal tract. The higher the concentration of released -NH_2, the greater the degree of protein digestion and hydrolysis [42]. Figure 5d shows the release of free amino groups from the protein emulsions during in vitro digestion. In gastric digestion, the rising trends of the release of free amino groups of all the protein emulsions were slow; however, the release of free amino groups rose significantly at 30 min of intestinal digestion and kept rising in subsequent intestinal digestion. The release of free amino groups in the HPH-modified SPI emulsion had no significant difference compared to the untreated one, but the release of free amino groups of the WPI and dual-protein emulsions increased further after HPH treatment. In particular, the increasing trend of the HPH-modified dual-protein emulsion was significantly higher than that of the HPH-modified WPI emulsion compared to the control, indicating that HPH had a better effect on promoting the digestion and hydrolysis of the dual-protein emulsion.

3.5.3. In Vitro Digestibility

The in vitro digestibility of the emulsions is shown in Figure 5e. It can be seen that after HPH treatment, the in vitro digestibility of the SPI and dual-protein emulsions significantly increased from 34% to 49% and from 44% to 55%, respectively. These results might be attributed to the disruptions in non-covalent and covalent bonds that caused the structure of protein to unfold after HPH treatment, making it easier for digestive enzymes to attack and combine with enzyme digestion sites. However, the in vitro digestibility of the HPH-modified WPI emulsion did not increase, which could be related to disulfide-bond-mediated protein aggregation that reduced the accessibility of digestive enzymes [43].

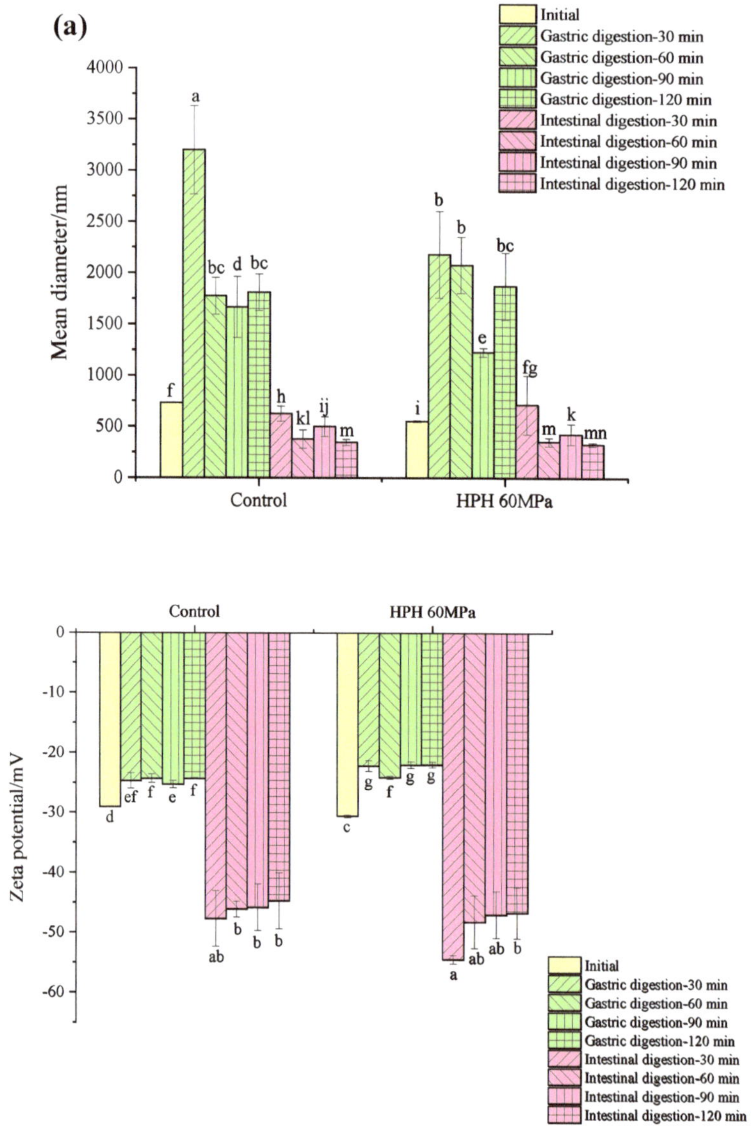

Figure 5. *Cont.*

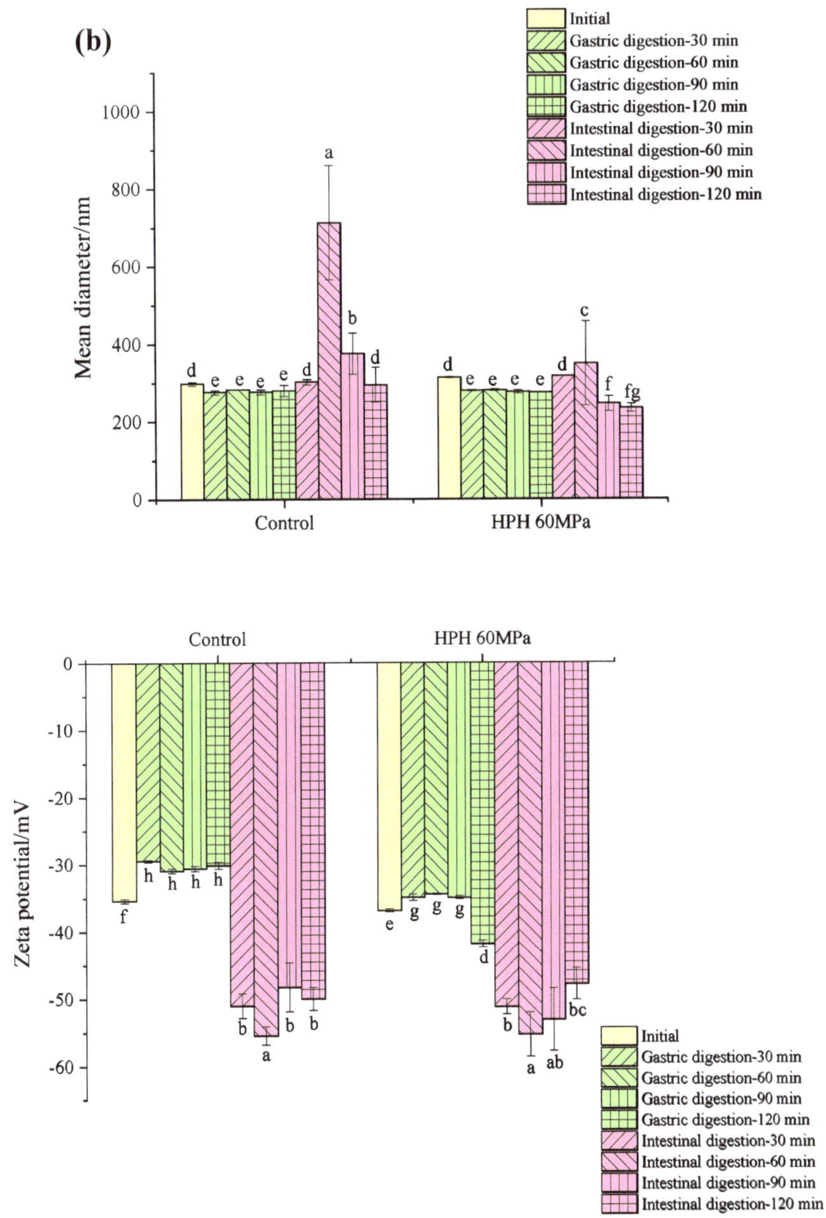

Figure 5. *Cont.*

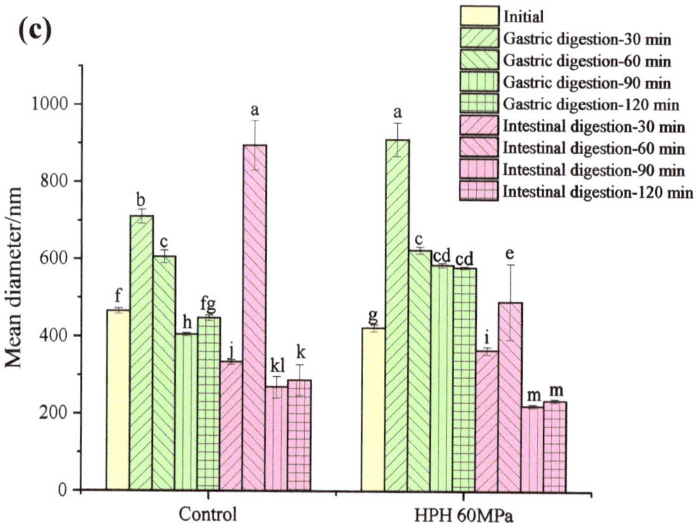

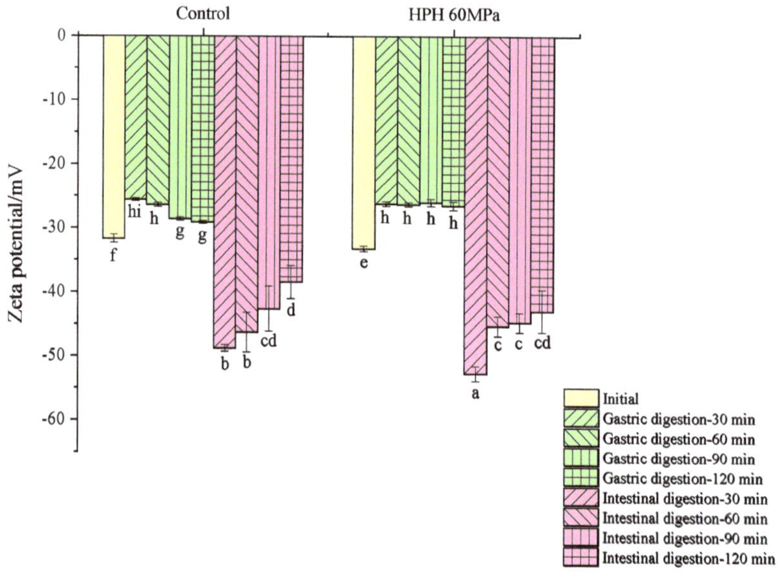

Figure 5. *Cont.*

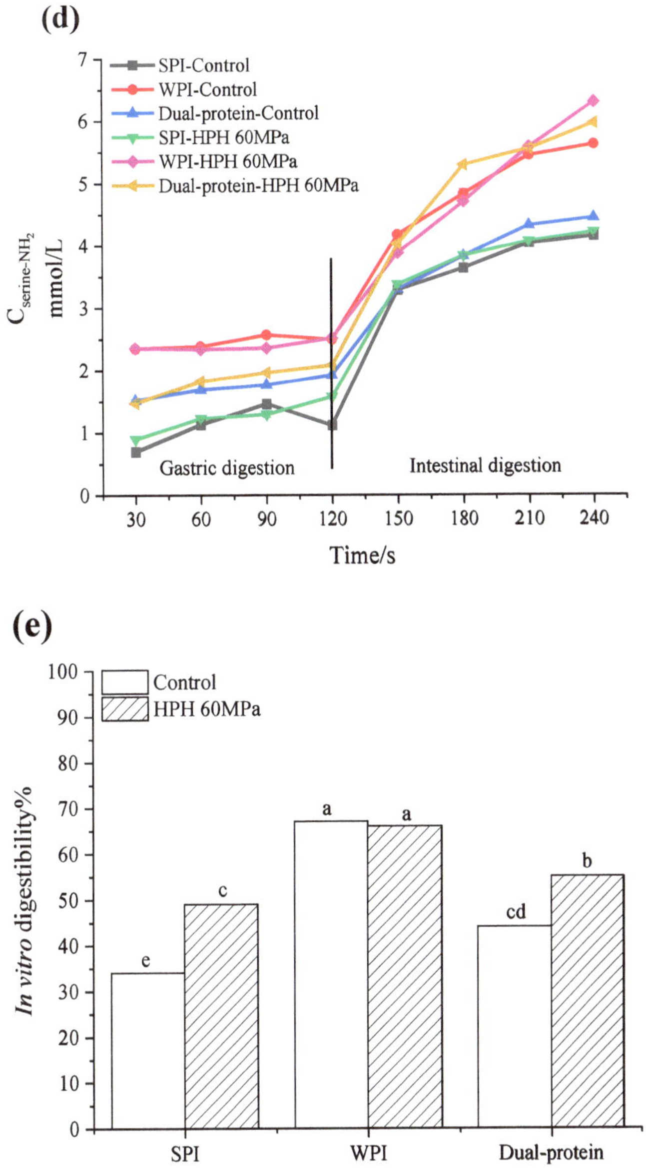

Figure 5. The particle size and zeta potential of SPI (**a**), WPI (**b**) and dual-protein (**c**) emulsions containing a 10% (*w*/*w*) oil phase during in vitro digestion, as well as the free amino groups (-NH_2) (**d**) and in vitro digestibility (**e**) of protein emulsions containing a 10% (*w*/*w*) oil phase. Different lowercase letters in Figures (**a–c**,**e**) indicate significant differences in the different protein emulsions.

3.6. Storage Stability

Changes in the appearances of the protein emulsions under different sterilization and storage temperatures are shown in Figure 6a,b. In low-temperature storage (4 °C), it can be seen that there was no significant change in any of the samples within 14 d. As storage time increased, a slight oil–water stratification appeared on the untreated SPI emulsion in the control group at both 21 and 28 d. However, in the pasteurization group, the stratification phenomenon in the untreated SPI emulsion only happened at 28 d, therefore indicating that pasteurization combined with low-temperature storage could significantly improve the storage stability of an untreated SPI emulsion. Beyond that, the other protein samples maintained a uniform state during 28 d of storage.

Figure 6. *Cont.*

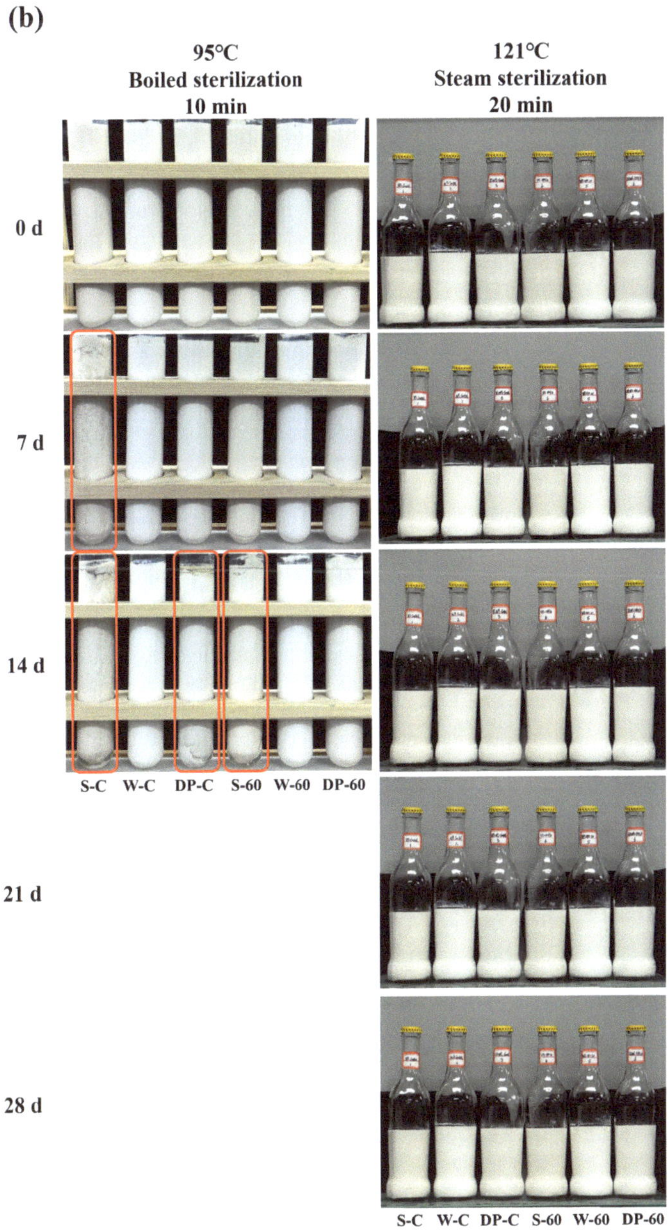

Figure 6. The appearances of protein emulsions containing different oil phase concentrations under different sterilization methods during low-temperature (4 °C) storage (**a**) and room-temperature (25 °C) storage (**b**). The "S-C", "W-C", "DP-C", "S-60", "W-60", and "DP-60" in Figure (**a**,**b**) indicate "SPI-control", "WPI-control", "Dual-protein-control", "SPI-HPH 60 MPa", "WPI-HPH 60 MPa", and "Dual-protein-HPH 60 MPa", respectively.

In room-temperature (25 °C) storage, obvious flocculation appeared in the untreated SPI emulsion at 7 d in the boiled-sterilization group, and this phenomenon also appeared in the untreated dual-protein emulsion and the HPH-modified SPI emulsion in the same group at 14 d; it might be attributed to the microorganisms remaining in the emulsion growing and reproducing, which caused the emulsion flocculation. It was worth noting no flocculation happened in the HPH-modified dual-protein emulsion after boiled sterilization within 14 d, suggesting that HPH could improve the antibacterial effect in a dual-protein emulsion to a certain extent. Since the microorganism contamination in the boiled-sterilization emulsions gradually increased after 14 d at room-temperature storage, which we could no longer carry on observing, the appearance changes of the emulsions are shown for only 14 d. Oppositely, the appearances of the emulsions after steam sterilization did not change significantly within 28 d of room-temperature storage. By comparison, it was found that the HPH-modified dual-protein emulsion had the best stability after steam sterilization during the period of room-temperature storage.

4. Conclusions

This study researched the emulsifying properties, in vitro digestion characteristics, and storage stability of HPH-modified dual-protein emulsions. The research results show that increasing the oil phase concentration increased the droplet size and the absorbed protein at the O/W interface of the dual-protein emulsion; meanwhile, it also improved the viscoelasticity of the dual-protein emulsion. It could be found that when the oil phase concentration was 10% (*w*/*w*), the diffusion stability and AP% of the HPH-modified dual-protein emulsion were both significantly better than the control, and the emulsion elasticity was enhanced, showing a more significant viscous liquid behavior than a single-protein emulsion. The in vitro digestion results indicate that HPH treatment obviously improved the in vitro digestibility of a dual-protein emulsion containing a 10% (*w*/*w*) oil phase concentration, promoting the release of free amino groups during in vitro digestion, and the degree of digestion and the decomposition of the emulsion in the gastrointestinal tract significantly improved. In addition, the storage period of the dual-protein emulsion was effectively extended after HPH treatment, demonstrating better storage stability than a single-protein emulsion after steam sterilization and room-temperature storage. This study has great significance for the in-depth development and utilization of HPH in liquid dual-protein products, and it also enriches the research content on the application of non-thermal processing, especially HPH technology, in animal–plant mixed proteins, which is helpful for using them in the food industry and much more.

Author Contributions: Conceptualization, M.W. and X.H.; data curation, M.W.; formal analysis, M.W. and D.H.; funding acquisition, J.W.; investigation, M.W.; methodology, M.W., X.H., H.L. and N.L.; project administration, X.H., H.L. and J.W.; resources, X.H., H.L. and J.W.; supervision: X.H., H.L., D.F., Q.L. and J.W.; validation, B.Z. and T.L. visualization: M.W.; writing—original draft: M.W. and X.H.; writing—review and editing: M.W., X.H. and J.W. All authors have read and agreed to the published version of the manuscript.

Funding: This study was supported by the Innovation Engineering Project of the Chinese Academy of Agricultural Sciences (CAAS-ASTIP-2023-IFND).

Data Availability Statement: The data presented in this study are available upon request from the corresponding author.

Conflicts of Interest: The authors declare no conflict of interest.

References

1. McClements, D.J.; Rao, J.J. Food-grade nanoemulsions: Formulation, fabrication, properties, performance, biological fate, and potential toxicity. *Crit. Rev. Food Sci. Nutr.* **2011**, *51*, 285–330. [CrossRef] [PubMed]
2. Sivapratha, S.; Sarkar, P. Multiple layers and conjugate materials for food emulsion stabilization. *Crit. Rev. Food Sci. Nutr.* **2018**, *58*, 877–892. [CrossRef] [PubMed]
3. Dickinson, E. Interfacial structure and stability of food emulsions as affected by protein-polysaccharide interactions. *Soft Matter* **2008**, *4*, 932–942. [CrossRef]
4. Rutkevicius, M.; Allred, S.; Velev, O.D.; Velikov, K.P. Stabilization of oil continuous emulsions with colloidal particles from water-insoluble plant proteins. *Food Hydrocoll.* **2018**, *82*, 89–95. [CrossRef]

5. Yan, X.J.; Ma, C.C.; Cui, F.Z.; McClements, D.J.; Liu, X.B.; Liu, F.G. Protein-stabilized pickering emulsions: Formation, stability, properties, and applications in foods. *Trends Food Sci. Technol.* **2020**, *103*, 293–303. [CrossRef]
6. Bos, M.A.; Van Vliet, T. Interfacial rheological properties of adsorbed protein layers and surfactants: A review. *Adv. Colloid Interface Sci.* **2001**, *91*, 437–471. [CrossRef] [PubMed]
7. Henchion, M.; Hayes, M.; Mullen, A.M.; Fenelon, M.; Tiwari, B. Future protein supply and demand: Strategies and factors influencing a sustainable equilibrium. *Foods* **2017**, *6*, 53. [CrossRef]
8. Alves, A.C.; Tavares, G.M. Mixing animal and plant proteins: Is this a way to improve protein techno-functionalities? *Food Hydrocoll.* **2019**, *97*, 105171. [CrossRef]
9. Ren, G.X.; Zhang, J.P.; Li, M.H.; Yi, S.Q.; Wang, J. Protein blend ingestion before allogeneic stem cell transplantation improves protein-energy malnutrition in patients with leukemia. *Nutr. Res.* **2017**, *46*, 68–77. [CrossRef]
10. Zhang, J.J.; Zhang, Q.W.; Liu, H.; Liu, X.Y.; Yu, Y.H.; Han, D.; He, X.Y.; Zeng, P.; Wang, J. Soy-whey dual-protein alleviates osteoporosis of ovariectomized rats via regulating bone fat metabolism through gut-liver-bone axis. *Nutrition* **2022**, *103–104*, 111723. [CrossRef]
11. Huang, Y.C.; Zhang, K.; Zhang, L.; Qiu, J.H.; Fu, L.; Ying, T.Y.; Wang, J.; Qin, R.; Zhang, J.J.; Dong, X.W.; et al. Dosage of dual-protein nutrition differentially impacts the formation of atherosclerosis in apoE-/-mice. *Nutrients* **2022**, *14*, 855. [CrossRef] [PubMed]
12. Nishinari, K.; Fang, Y.; Guo, S.; Phillips, G.O. Soy proteins: A review on composition, aggregation and emulsification. *Food Hydrocoll.* **2014**, *39*, 301–318. [CrossRef]
13. Wagner, J.; Biliaderis, C.G.; Moschakis, T. Whey proteins: Musings on denaturation, aggregate formation and gelation. *Crit. Rev. Food Sci. Nutr.* **2020**, *60*, 3793–3806. [CrossRef] [PubMed]
14. Tang, C.H. Nanostructured soy proteins: Fabrication and applications as delivery systems for bioactives (a review). *Food Hydrocoll.* **2019**, *91*, 92–116. [CrossRef]
15. Szczepanska, J.; Skapska, S.; Polaska, M.; Marszalek, K. High pressure homogenization with a cooling circulating system: The effect on physiochemical and rheological properties, enzymes, and carotenoid profile of carrot juice. *Food Chem.* **2022**, *370*, 131023. [CrossRef]
16. Kruszewski, B.; Domian, E.; Nowacka, M. Influence of high-pressure homogenization on the physicochemical properties and betalain pigments of red beetroot (*Beta vulgaris* L.) juice. *Molecules* **2023**, *28*, 2018. [CrossRef]
17. Dumay, E.; Chevalier-Lucia, D.; Picart-Palmade, L.; Benzaria, A.; Gracia-Julia, A.; Blayo, C. Technological aspects and potential applications of (ultra) high-pressure homogenisation. *Trends Food Sci. Technol.* **2013**, *31*, 13–26. [CrossRef]
18. Sahil; Madhunita, M.; Prabhakar, P.K.; Kumar, N. Dynamic high pressure treatments: Current advances on mechanistic-cumtransport phenomena approaches and plant protein functionalization. *Crit. Rev. Food Sci. Nutr.* **2022**, *2022*, 2125930. [CrossRef]
19. Ma, W.C.; Wang, J.M.; Wu, D.; Xu, X.B.; Wu, C.; Du, M. Physicochemical properties and oil/water interfacial adsorption behavior of cod proteins as affected by high-pressure homogenization. *Food Hydrocoll.* **2019**, *100*, 105429. [CrossRef]
20. Fernandez-Avila, C.; Trujillo, A.J. Ultra-high pressure homogenization improves oxidative stability and interfacial properties of soy protein isolate-stabilized emulsions. *Food Chem.* **2016**, *209*, 104–113. [CrossRef] [PubMed]
21. Melchior, S.; Moretton, M.; Calligaris, S.; Manzocco, L.; Nicoli, M.C. High pressure homogenization shapes the techno-functionalities and digestibility of pea proteins. *Food Bioprod. Process.* **2022**, *131*, 77–85. [CrossRef]
22. Dybowska, B.E.; Krupa-Kozak, U. Stability of oil-in-water emulsions as influenced by thermal treatment of whey protein dispersions or emulsions. *Int. J. Dairy Technol.* **2020**, *73*, 513–520. [CrossRef]
23. Lunardi, C.N.; Gomes, A.J.; Rocha, F.S.; Tommaso, J.D.; Patience, G.S. Experimental methods in chemical engineering: Zeta potential. *Can. J. Chem. Eng.* **2020**, *99*, 627–639. [CrossRef]
24. Sun, S.; Li, S.H.; Yan, H.J.; Zou, H.N.; Yu, C.P. The conformation and physico-chemical properties of pH-treated golden pompano protein on the oil/water interfacial properties and emulsion stability. *Int. J. Food Sci. Technol.* **2022**, *57*, 5611–5620. [CrossRef]
25. Shi, J.Y.; Xiao, J.X.; Liu, L.; Dong, X.Y. Ultrasonic assisted oil-in-water emulsions stabilized by flaxseed protein isolate: Influence of different oils. *J. Dispers. Sci. Technol.* **2021**, *43*, 1789–1800. [CrossRef]
26. Zhu, Q.; Qiu, S.; Zhang, H.; Cheng, Y.; Yin, L. Physical stability, microstructure and micro-rheological properties of water-in-oil-in-water (W/O/W) emulsions stabilized by porcine gelatin. *Food Chem.* **2018**, *253*, 63–70. [CrossRef] [PubMed]
27. Ding, M.Z.; Liu, L.J.; Zhang, T.; Tao, N.P.; Wang, X.C.; Zhong, J. Effect of interfacial layer number on the storage stability and in vitro digestion of fish oil-loaded multilayer emulsions consisting of gelatin particle and polysaccharides. *Food Chem.* **2021**, *336*, 127686. [CrossRef] [PubMed]
28. Church, F.C.; Swaisgood, H.E.; Porter, D.H.; Catignani, G.L. Spectrophotometric assay using o-phthaldialdehyde for determination of proteolysis in milk and isolated milk proteins. *J. Dairy Sci.* **1983**, *66*, 1219–1227. [CrossRef]
29. Gupta, A.; Badruddoza, A.Z.M.; Doyle, P.S. A general route for nanoemulsion synthesis using low-energy methods at constant temperature. *Langmuir ACS J. Surf. Colloids* **2017**, *33*, 7118–7123. [CrossRef]
30. Song, X.Z.; Zhou, C.J.; Fu, F.; Chen, Z.L.; Wu, Q.L. Effect of high-pressure homogenization on particle size and film properties of soy protein isolate. *Ind. Crops Prod.* **2013**, *43*, 538–544. [CrossRef]
31. Liu, F.; Tang, C.H. Soy glycinin as food-grade pickering stabilizers: Part. I. structural characteristics, emulsifying properties and adsorption/arrangement at interface. *Food Hydrocoll.* **2016**, *60*, 606–619. [CrossRef]

32. Zhang, Y.H.; Zhou, F.B.; Shen, P.H.; Zhao, Q.Z.; Zhao, M.M. Influence of thermal treatment on oil-water interfacial properties and emulsion stabilization prepared by sono-assembled soy peptide nanoparticles. *Food Hydrocoll.* **2020**, *103*, 105646. [CrossRef]
33. Christian, C.; Elena, T.; Donato, C.; Donatella, P.; Massimo, F. Turbiscan Lab® expert analysis of the stability of ethosomes® and ultradeformable liposomes containing a bilayer fluidizing agent. *Colloids Surf. B Biointerfaces* **2019**, *72*, 155–160. [CrossRef]
34. Hebishy, E.; Ferragut, V.; Blasco-Moreno, A.; Trujillo, A.J. Impact of oil phase concentration on physical and oxidative stability of oil-in-water emulsions stabilized by sodium caseinate and ultra-high pressure homogenization. *J. Dispers. Sci. Technol.* **2019**, *42*, 46–57. [CrossRef]
35. Yang, J.Q.; Liu, G.Y.; Zeng, H.B.; Chen, L.Y. Effects of high pressure homogenization on faba bean protein aggregation in relation to solubility and interfacial properties. *Food Hydrocoll.* **2018**, *83*, 275–286. [CrossRef]
36. Meleties, M.; Martineau, R.L.; Gupta, M.K.; Montclare, J.K. Particle-based microrheology as a tool for characterizing protein based materials. *ACS Biomater. Sci. Eng.* **2022**, *8*, 2747–2763. [CrossRef] [PubMed]
37. Tisserand, C.; Fleury, M.; Brunel, L.; Bru, P.; Meunier, G. Analytical series soft matter analysis by means of microrheology. *JCT CoatingsTech* **2012**, *9*, 46–51.
38. Li, Q.; He, S.H.; Xu, W.L.; Peng, F.S.; Gu, C.; Wang, R.C.; Ma, Y. Formation, stability and in vitro digestion of beta-carotene in oil-in-water milk fat globule membrane protein emulsions. *Food Biophys.* **2018**, *13*, 198–207. [CrossRef]
39. Ozel, B.; Zhang, Z.Y.; He, L.L.; McClements, D.J. Digestion of animal-and plant-based proteins encapsulated in κ-carrageenan/protein beads under simulated gastrointestinal conditions. *Food Res. Int.* **2020**, *137*, 1096642. [CrossRef]
40. Zhang, R.J.; Zhang, Z.P.; Zhang, H.; Decker, E.A.; McClements, D.J. Influence of lipid type on gastrointestinal fate of oil-in-water emulsions: In vitro digestion study. *Food Res. Int.* **2015**, *75*, 71–78. [CrossRef]
41. Yao, M.F.; Xiao, H.; McClements, D.J. Delivery of lipophilic bioactives: Assembly, disassembly, and reassembly of lipid nanoparticles. *Annu. Rev. Food Sci. Technol.* **2014**, *5*, 53–81. [CrossRef] [PubMed]
42. Nielsen, P.M.; Petersen, D.; Dambmann, C. Improved method for determining food protein degree of hydrolysis. *J. Food Sci.* **2001**, *66*, 642–646. [CrossRef]
43. Sun, M.J.; Mu, T.H.; Sun, H.N.; Zhang, M. Digestibility and structural properties of thermal and high hydrostatic pressure treated sweet potato (*Ipomoea batatas* L.) protein. *Plants Foods Hum. Nutr.* **2014**, *69*, 270–275. [CrossRef] [PubMed]

foods

MDPI

Article

Biosynthesis of Quercetin-Loaded Melanin Nanoparticles for Improved Antioxidant Activity, Photothermal Antimicrobial, and NIR/pH Dual-Responsive Drug Release

Di Zhang [1], Xianrui Chen [1], Nitong Bu [1], Liying Huang [1], Huanglong Lin [1], Lizhen Zhou [1], Ruojun Mu [1], Lin Wang [2,*] and Jie Pang [1,*]

[1] College of Food Science, Fujian Agriculture and Forestry University, Fuzhou 350002, China; mu@fafu.edu.cn (R.M.)

[2] Department of Engineering Mechanics, Tsinghua University, Beijing 100084, China

* Correspondence: linwang0528@mail.tsinghua.edu.cn (L.W.); pang3721941@fafu.edu.cn (J.P.)

Abstract: Quercetin (QCT) is a promising dose-dependent nutraceutical that usually suffers from poor water solubility and low bioavailability issues. In this work, a novel QCT-loaded nanoscale delivery system was constructed based on the oxidative self-polymerization of melanin (Q@MNPs). The FT-IR, XRD, and Zeta potential analyses confirmed that QCT was successfully absorbed on the melanin nanoparticles (MNPs) via $\Pi-\Pi$ and hydrogen bonding interactions. The encapsulation efficiency and particle size of Q@MNPs were 43.78% and 26.68 nm, respectively. Q@MNPs improved the thermal stability of QCT and the antioxidant properties in comparison to MNPs. Meanwhile, Q@MNPs presented fantastic photothermal conversion capacity and stability triggered by the NIR laser, which significantly enhanced the antibacterial capability with a sterilization rate of more than 98% against *E. coli* and *S. aureus*. More importantly, Q@MNPs exhibited NIR/pH dual-responsive drug release behavior and good biocompatibility (at concentrations of < 100 μg/mL). Thus, Q@MNPs show promising prospects for flavonoid delivery.

Keywords: melanin nanoparticles; quercetin; antioxidant activity; photothermal antimicrobial; stimuli drug release

Citation: Zhang, D.; Chen, X.; Bu, N.; Huang, L.; Lin, H.; Zhou, L.; Mu, R.; Wang, L.; Pang, J. Biosynthesis of Quercetin-Loaded Melanin Nanoparticles for Improved Antioxidant Activity, Photothermal Antimicrobial, and NIR/pH Dual-Responsive Drug Release. *Foods* **2023**, *12*, 4232. https://doi.org/10.3390/foods12234232

Academic Editor: Andrea Gomez-Zavaglia

Received: 3 November 2023
Revised: 18 November 2023
Accepted: 21 November 2023
Published: 23 November 2023

1. Introduction

Food spoilage and deterioration caused by bacterial infection are still a challenge in the world. There are considerable concerns about developing and designing antimicrobial materials for food preservation. Finding safe antibacterial materials to reduce environmental pollution is a current trend. Quercetin (QCT) is an abundant natural flavonoid found in many common fruits and vegetables with many bioactivities, including antioxidant, antibacterial, and anticancer activities, and it is used in the food and medical fields. Arcan et al. incorporated QCT into zein films as bioactive packaging materials [1]. George et al. found that QCT has anticancer activities in cancerous cells [2]. However, the poor water solubility and instability of QCT during processing limits its application. With the development of nanotechnology, many advanced research paths have opened up in different fields. Nanotechnology is a constantly evolving and innovative technology that can be used to develop nanoparticles (NPs), which are widely used for encapsulating functional compounds, such as essential oils, antimicrobial agents, and natural extracts [3,4]. Moreover, it can improve the stability of functional compounds, control delivery, and enhance bioactivity. Nanoparticles have been used for the delivery of QCT. To improve its bioavailability, organic solvents like dimethyl sulfoxide (DMSO) have been used to produce nanoparticles loaded with QCT [5], whereas the safety of organic solvents is a potential issue. Some methods use chemical crosslinking to prepare nanoparticles loaded with QCT [6]. However, the crosslinking agents used to prepare chemically crosslinked nanoparticles have toxicity concerns. Hence, natural polymer-based nanomaterials, including proteins and polysaccharides, were further employed to fabricate nanoparticles to encapsulate QCT. For example, zein nanoparticles have been reported to encapsulate QCT, although the stability of zein nanoparticles is far from satisfactory. Therefore, researchers attempted to add polysaccharides to enhance the stability of the nanoparticles. Zou et al. encapsulated QCT in chitosan-coated zein nanoparticles to enhance its water dispersibility, antioxidant activity, and bio-accessibility [7]. However, it is still a challenge to create nanoparticles with controlled and stimuli-responsive abilities to release QCT.

Among these nanoparticles, melanin-like nanoparticles (MNPs) have received much interest because of their biocompatibility, biodegradability, free radical scavenging ability, and photothermal properties with strong absorption of near-infrared (NIR) laser and conversion of light energy to heat energy [8]. Therefore, it is widely used in food, energy, and medical fields. MNPs can be produced from natural microorganisms, animals, and plants [9]. However, the practical applications of natural melanin are limited due to its poor purity and extraction rate [10]. Another alternative method is the use of a green synthetic for dopamine oxidation. The dopamine hydrochloride can dissolve and oxidize in an alkaline aqueous solution to form MNPs, which demonstrate comparable physical and chemical properties compared to those of natural MNPs [11]. It has been proven that MNPs can be directly modified using aromatic molecules and further applied as surface modification agents for nanoparticles, filler materials for light absorption, and drug-controllable and -stimulating delivery platforms [12]. Specifically, there are many conjugated polymeric structures and reactive functional groups, including quinone groups, phenolic hydroxyl, hydroxyl groups, and amino groups in MNPs, that enable it to easily modify the surface and combine with the aromatic structures of substances via $\Pi-\Pi$ stacking and hydrogen bonding interactions [13]. In previous studies, Xing et al. formed core-shell lignin-melanin NPs (LMNP) using in situ polymerization of dopamine on the surface of lignin nanoparticles [14]. Liu et al. synthesized silver/melanin-like nanoparticles (Ag-SMNPs) to fabricate cotton fabric with multifunctional UV-resistant, antibacterial, and photothermal properties [15]. Recently, Wang et al. extracted natural melanin to produce MNPs and encapsulated QCT on the surface of MNPs, which exhibited excellent antibacterial and anti-tumor properties [16], improved the bioavailability of QCT, and expanded the application of natural MNPs. However, related research on the photothermal properties and drug release behavior of QCT-loaded MNPs from the green synthesis of dopamine oxidation is lacking.

In this work, a novel functional nano-carrier of dopamine hydrochloride-derived MNPs was fabricated and loaded with QCT to obtain Q@MNPs (Figure 1a). Fourier-transform infrared spectroscopy (FT-IR), X-ray diffraction analysis (XRD), thermo-gravimetric analysis (TGA), transmission electron microscopy (TEM), and Zeta potential values were used to characterize Q@MNPs. The photothermal performance and stability of Q@MNPs were evaluated using photothermal experiments. The antioxidant activities of Q@MNPs, and antibacterial activities of Q@MNPs combining photothermal effects against *S. aureus* and *E. coli* were explored. The pH and NIR dual stimuli release of QCT and the biocompatibility of Q@MNPs were also investigated. The findings of this research can broaden the application of melanin nanoparticles in food processing and provide an appealing method to produce specific drug delivery systems.

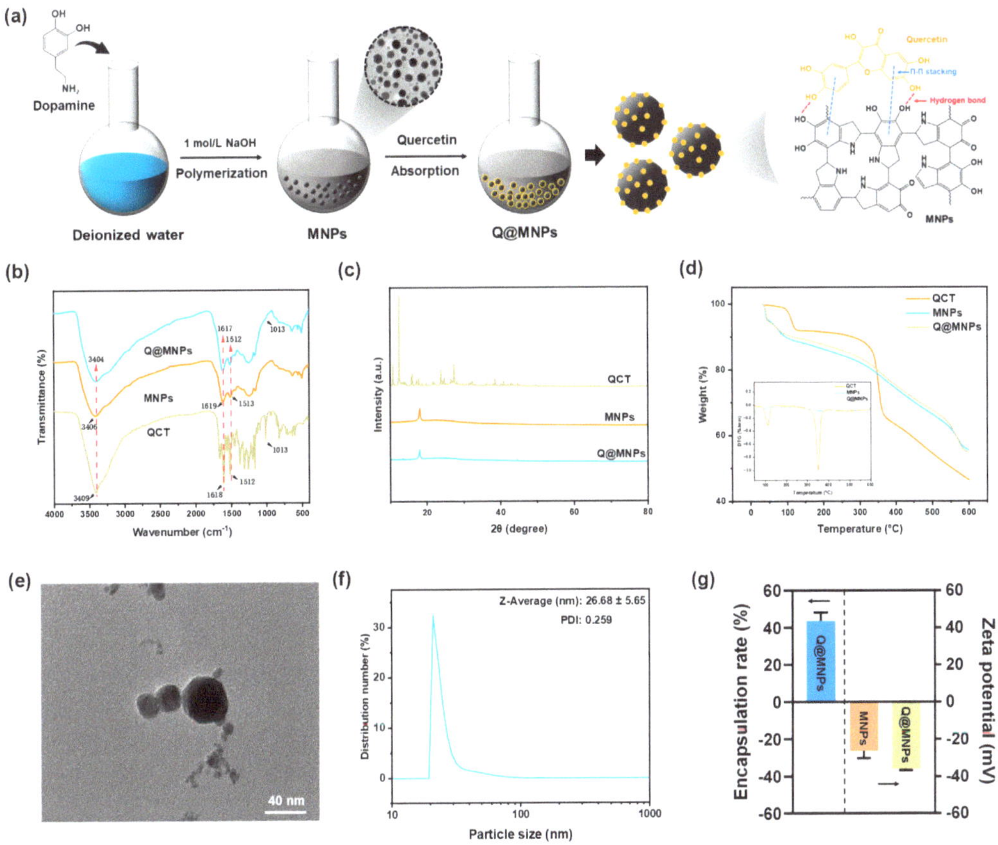

Figure 1. Schematic illustration of the procedure for the preparation of Q@MNPs (**a**); FT-IR spectra of QCT, MNPs, and Q@MNPs (**b**); XRD pattern of QCT, MNPs, and Q@MNPs (**c**); TGA curve of QCT, MNPs, and Q@MNPs (**d**); TEM image of Q@MNPs (scale bar: 40 nm) (**e**); particle size distribution of Q@MNPs (**f**); encapsulation rate and Zeta potential (**g**) of Q@MNPs.

2. Materials and Methods

2.1. Materials

Dopamine hydrochloride (>98%, purity), 2,2-Diphenyl-1-picrylhydrazyl (DPPH), 2,2′-azino-bis (3-ethylbenzothiazoline-6-sulfonic acid) (ABTS), and potassium persulfate ($K_2S_2O_8$) were obtained from Aladdin Biochemical Technology Co. Ltd. (Shanghai, China). Quercetin (97%, purity) was purchased from Maclin Biochemical Technology Co., Ltd. (Shanghai, China). *Escherichia coli* (*E. coli*) and *Staphylococcus aureus* (*S. aureus*) were supplied by the food microbiology laboratory at the College of Food Science, Fujian Agriculture and Forestry University (Fuzhou, China). The ultrapure water was used in this experiment, and all other reagents used were of analytical grade.

2.2. Preparation of MNPs

The MNPs were produced based on a previous method, with slight modifications [12]. In brief, 2 g of dopamine hydrochloride was dissolved in 200 mL of deionized water under magnetic stirring at 600 rpm at 40 °C. After dopamine hydrochloride was completely dissolved, 6 mL of 1 mol/L NaOH was added to the solution. During the reaction, the color of the solution changed from light yellow to brownish-black. The solution was continuously stirred at this temperature in the dark for 24 h. After polymerization, the synthesized

MNPs were obtained via centrifugation at 11,000× g and washed with deionized water. This procedure was repeated several times until the supernatant was pale and neutral. Subsequently, the precipitate was dried in an oven (40 °C) for 24 h to obtain MNP powder.

2.3. Preparation of Q@MNPs

First, QCT (200 mg) was weighed and dissolved in 100 mL of anhydrous ethanol. Then, the prepared MNP powder (100 mg) was added to the solution with magnetic stirring (600 rpm) in the dark at 25 °C for 12 h. After stirring, the Q@MNPs were obtained via centrifugation at 11,000× g. The precipitate was dried in an oven (40 °C) for 12 h to obtain Q@MNPs powder. Eventually, the Q@MNPs powder was stored at 4 °C until further experiments.

Encapsulation Efficiency (EE) of QCT

The above-mentioned supernatant of Q@MNPs was measured to determine the absorbance of free QCT spectrophotometrically at 373 nm using a UV-vis spectrophotometer (UV-2600, Shimadzu, Japan). The standard curve of QCT dissolved in anhydrous ethanol was measured ($y = 0.0707\ x + 0.0365,\ R^2 = 0.9992$). The amount of free QCT was evaluated using a standard curve. The encapsulation efficiency (EE) of QCT was calculated as follows:

$$EE\ (\%) = \frac{m_t - m_f}{m_t} \times 100$$

where m_t was the total amount of QCT and m_f was the free amount of QCT.

2.4. Characterization of Q@MNPs

2.4.1. Fourier-transform Infrared Spectroscopy (FT-IR)

The FT-IR spectra of QCT, MNPs, and Q@MNPs were measured using a Bruker Vertex 70 FT-IR spectrometer (Karlsruhe, Germany) using the KBr tableting method, and the scanning range was from 4000 to 400 cm^{-1} with a resolution of 4 cm^{-1}.

2.4.2. X-ray Diffraction Analysis (XRD)

The XRD spectra of QCT, MNPs, and Q@MNPs were obtained using an X-ray diffractometer (D8, Bruker, Germany) with a diffraction angle of 10–80° and at a scanning speed of 5°/min.

2.4.3. Thermo-Gravimetric Analysis (TGA)

The thermal stability of QCT, MNPs, and Q@MNPs was investigated using a thermal analyzer (Q500 TGA, Newcastle, DE, USA) heated at a rate of 10 °C/min with a scan range from 30 to 600 °C under a nitrogen atmosphere.

2.4.4. Transmission Electron Microscopy (TEM)

The morphology of Q@MNPs was characterized using TEM (JEOL JEM 2100F, Japan Electronics Co., Ltd., Tokyo, Japan).

2.4.5. Zeta Potential

Based on the method of Wang et al. [17], 10 mg of MNPs and QCT@MNPs powder was dispersed in 10 mL of deionized water and ultrasonicated for 10 min. Then, the Zeta potential of the MNPs and Q@MNPs solutions was examined using a Litesizer 500 (Anton Parr, Graz, Austria) at 25 °C, and three parallel measurements were performed.

2.4.6. Particle Size of Q@MNPs

A Litesizer 500 (Anton Parr, Graz, Austria) was used to measure the particle size and polydispersity index (PDI) of Q@MNPs at 25 °C. The sample was dispersed and diluted with deionized water before observation.

2.5. Photothermal Performance

The photothermal performance of Q@MNPs was tested using the method described by Liang et al., with some modifications [8]. In brief, the Q@MNPs powder was weighed (0.1 g) and pressed into a disc with a diameter of 1.3 cm. Then, the sample was irradiated by an 808 nm laser (0.25 W/cm^2) for 10 min. At the same time, the thermograph and the infrared camera were used to record the temperature on the surface of the samples. To evaluate the photothermal stability of Q@MNPs, the sample was exposed to an 808 nm laser, irradiated, and cooled for 10 min. The heating–cooling curves were obtained for Q@MNPs over 4 cycles.

2.6. Antioxidant Activity of Q@MNPs

2.6.1. DPPH Radical Scavenging Activity

Following the method described by Zhao et al. [18], the DPPH radical scavenging activity of Q@MNPs was measured. In brief, different quantities of Q@MNPs were weighed (0.5, 1, 2, 4, and 8 mg) and then immersed in 10 mL DPPH anhydrous ethanol solution to ensure that the concentrations of Q@MNPs solutions were 0.05, 0.1, 0.2, 0.4, and 0.8 mg/mL (w/v), respectively. Then, the solution was allowed to react in the dark at room temperature for 15 min. After the reaction, the absorbance of the solutions was measured at 517 nm using a UV-vis spectrophotometer (UV–2600, Shimadzu, Japan). The DPPH radical scavenging ability of different concentrations of Q@MNPs was determined using the following equation:

$$Scavenging\ ability\ of\ DPPH\ (\%) = \frac{A_0 - A_i}{A_0} \times 100$$

where A_0 and A_i are the absorbance of the blank and samples, respectively.

2.6.2. ABTS Scavenging Activity

The ABTS radical scavenging rate of Q@MNPs was measured according to a method described in a previous study, with slight modifications [19]. Briefly, an equal volume of 7 mM ABTS and 4.9 mM potassium persulfate was mixed and stored at 25 °C in the dark for 16 h. Subsequently, the ABTS/potassium persulfate mixture was diluted with anhydrous ethanol to obtain the absorbance of ABTS solution, 0.7 ± 0.02, at 734 nm. After that, different quantities of Q@MNPs were weighed and then immersed in 10 mL ABTS solution to obtain different concentrations of Q@MNPs solutions (0.05, 0.1, 0.2, 0.4, and 0.8 mg/mL). The solution was allowed to react in the dark at 25 °C for 5 min. After the reaction, the absorbance of the solutions was determined at 734 nm using a UV-vis spectrophotometer (UV-2600, Shimadzu, Japan). The ABTS radical scavenging activity of different concentrations of Q@MNPs was determined as follows:

$$Scavenging\ ability\ of\ ABTS\ (\%) = \frac{A_0^1 - A_i^1}{A_0^1} \times 100$$

where A_0^1 is the absorbance of the blank and A_i^1 is the absorbance of the samples.

2.7. Antibacterial Activity

E. coli and *S. aureus* were applied to test the antibacterial activity of Q@MNPs using a previously reported method, with slight modifications [20]. The quantitative of QCT, MNPs, and Q@MNPs were ultrasonically dispersed in a phosphate-buffered saline (PBS) at pH 7.2 to make QCT, MNPs, and Q@MNPs solutions at a concentration of 200 μg/mL (m/v). This test was divided into four groups for comparison: control (PBS), QCT, MNPs, and Q@MNPs. Each group was divided into two treatment groups, including near–infrared light irradiation and non–irradiation. Subsequently, 100 μL of the bacterial suspension (1×10^6 CFU/mL) was added to a 24-well plate. Subsequently, 900 μL of PBS, QCT, MNPs, and Q@MNPs solutions were mixed with the bacterial suspension, and the mixture was incubated at 37 °C for 1 h. For each group that needed contact, near–infrared light irradiation was applied using an 808 nm laser with 1.0 W/cm^2 for 10 min. Then, 100 μL of the treated bacterial suspension was taken out and inoculated on LB agar plates at 37 °C for 16 h. Eventually, the number of bacteria was counted to show the antibacterial performance of Q@MNPs. The experiments of each sample were performed in triplicate. The number of bacterial colonies in each sample was estimated using the following equation:

$$Bacterial\ survival\ (\%) = \frac{N_i}{N_0} \times 100$$

where N_i is the number of bacterial colonies in the QCT, MNPs, and Q@MNPs groups without or with NIR, respectively; N_0 is the number of bacterial colonies in the control group.

2.8. In Vitro QCT Release

The QCT release experiment was conducted using a previously described method, with some modifications [21], and the Q@MNPs were divided into two groups at different pH (4.5 and 7.2). Each group was divided into two treatment groups, including near–infrared light (808 nm, 1.0 W/cm^2) irradiation for 10 min and non–irradiation before the first extraction. The Q@MNPs (30 mg) were weighed and dispersed in 40 mL of PBS solution. Subsequently, it was placed in a constant-temperature shaker at 80 rpm and 37 °C. At the predetermined time, 4 mL of the mixed solution was collected, and then an equivalent volume of fresh PBS solution was replaced. The extract solution was centrifuged at 5000× g for 5 min, and the absorbance of the supernatant was measured at 373 nm using a UV-vis spectrophotometer (UV-2600, Shimadzu, Japan). The cumulative release rate of QCT was calculated as follows:

$$Cumulative\ release\ (\%) = \frac{V_0 \times C_n + V_1 \times \sum C_{n-1}}{m_0} \times 100$$

where m_0 is the weight of QCT in the sample, and V_0 and V_1 are the total volume and extraction volume of the PBS solutions at the predetermined time, respectively. The C_n and C_{n-1} represent the concentrations of QCT in the n and $n-1$ releases of PBS solutions, respectively.

2.9. In Vitro Cytotoxicity Assay

According to the reference method with some modifications [22], the relative cell cytotoxicity of Q@MNPs was evaluated via CCK–8 assay using L929 cells. Quantitative nanoparticles were sterilized using ultraviolet irradiation for 30 min and ultrasonically dispersed in MEM medium solution for 5 min to prepare different concentrations of Q@MNPs solutions. The 100 μL of fresh MEM medium solution containing different concentrations of Q@MNPs (50 μg/mL, 100 μg/mL, 200 μg/mL, and 500 μg/mL) was added to 96-well plates, and 100 μL of fresh MEM culture medium solution was added to the 96-well plates as the control group. Then, L929 cells were added to the plates at a density of 8 × 10^3 cells/well and cultured for 24 h (5% CO_2, 37 °C). Subsequently, the culture medium was removed and cleaned with PBS solution three times, and 150 μL of MEM medium solution containing 10% CCK−8 was added and incubated in a constant-temperature incubator (5% CO_2, 37 °C). After 2 h, 100 μL of the supernatant from each well was transferred to a new plate. Finally, the absorbance at 450 nm was recorded using a microtiter reader (SPARK 10M, TECAN, Männedorf, Austria). Each experiment was performed in triplicate.

2.10. Statistical Analysis

All results are expressed as the mean ± standard deviation (SD). GraphPad Prism 9 (GraphPad Software, San Diego, CA, USA) was used to analyze the significance of the differences ($p < 0.05$) using the one-way analysis of variance (ANOVA).

3. Results and Discussion

3.1. FT-IR Analysis

FT-IR spectrometry was used to study the structural changes in MNPs after binding to QCT molecules, as presented in Figure 1b. For the characteristic bands of QCT, 3409 cm^{-1} was assigned to O–H stretching, and the band situated at 1664 cm^{-1} was assigned to C=O stretching vibration. The characteristic bands that appeared at 1611 cm^{-1} and 1521 cm^{-1} were the stretching vibrations of the benzene ring skeleton [23]. And the characteristic band at 1014 cm^{-1} corresponded to C–O and C–C–O stretching vibrations [24]. In the spectra of MNPs, the bands at 1619 cm^{-1} and 1513 cm^{-1} were linked to the C=C bond of the benzene rings and C–N in indolequinone, respectively [25]. The band at 1247 cm^{-1} was linked to the C–O–H bending and stretching of phenolics [26]. Moreover, when QCT was combined with MNPs, the peaks of the Q@MNPs changed. The peak at 3416 cm^{-1} shifted to 3404 cm^{-1}, and the intensity of the peaks at 1617 and 1512 cm^{-1} increased. Furthermore, the Q@MNPs showed a new peak at 1013 cm^{-1}, which was the characteristic band of QCT. This can be ascribed to the hydrogen bonding and $\Pi-\Pi$ stacking of the aromatic structure among the MNPs and QCT [12], respectively.

3.2. XRD Analysis

The XRD patterns of QCT, MNPs, and QCT@MNPs are shown in Figure 1c. The characteristic crystalline peaks of QCT were observed at 10.78°, 12.44°, 16.12°, 23.84°, and 27.38°, which indicated the crystalline nature of QCT [27]. For MNPs, the characteristic peak appeared at 18.06°, which is attributed to the amorphous structure of MNPs [28]. However, when the MNPs were combined with QCT, the intensity of the peak decreased, and the crystal characteristics of QCT nearly disappeared, which indicated that the encapsulated QCT in Q@MNPs was in an amorphous form. It is possible that the original crystal structure of QCT was disrupted, and $\Pi-\Pi$ stacking and hydrogen bonds were formed between the MNPs and QCT, which confirmed the results of the FT-IR analysis. A similar encapsulated QCT transformed into an amorphous form was reported in the literature. Zou et al. [7] encapsulated QCT in biopolymer-coated zein nanoparticles, and the diffraction pattern was similar to that of the plain nanoparticles, which showed that the encapsulated QCT was in an amorphous form. In addition, Gan et al. also found that after embedding plant sterols in sodium caseinate/pectin-based nanoparticles, plant sterols transitioned from an ordered crystalline state to an amorphous form [29].

3.3. TGA Analysis

The thermal stability of Q@MNPs was determined using TGA analysis (Figure 1d). The first stage of the weight loss of QCT, MNPs, and Q@MNPs occurred below 120 °C, which was attributed to the evaporation of water [30]. The second stage ranged from 120 to 370 °C, and a sharp decline in QCT appeared. This was attributed to pyrolysis, resulting in the polymer segmentation and degradation of the QCT [31]. From the DTG pattern of the QCT, there was an obvious peak at about 351 °C, showing a point of temperature where a large-scale weight loss occurred. According to the initial weight of the sample, the mass of QCT was lost about 34.36%, whereas those of MNPs and Q@MNPs were 23.12% and 22.15%, respectively. However, the thermal stability of Q@MNPs significantly improved ($p < 0.05$) when QCT was combined with MNPs. The final stage ranged from 370 to 600 °C, which was ascribed to the carbonization of the materials [32]. The weight of QCT, MNPs, and Q@MNPs finally lost about 53.55%, 44.51%, and 45.3%, respectively. Notably, the thermal decomposition patterns of the MNPs were not affected by QCT. However, the TGA curve of Q@MNPs was slightly lower than that of MNPs at the later period of carbonization, which can be explained by the thermal decomposition of QCT in the Q@MNPs, which caused slight weight loss [12]. In general, from the DTG curve, the overall curve trends of Q@MNPs were similar to those of MNPs. It can be concluded that the Q@MNPs improved the thermal stability of QCT.

3.4. TEM Analysis, Particle Size, Encapsulation Efficiency and Zeta Potential

The TEM analysis was used to evaluate the microscopic morphology of Q@MNPs. The size distribution of Q@MNPs was calculated based on the TEM observations. As can be observed in Figure 1e,f, the Q@MNPs exhibited a spherical structure and diameter with an average size of 26.68 ± 5.65 nm. The PDI of Q@MNPs was 0.259, which indicated the uniformity of Q@MNPs in a dispersion [33]. The Zeta potential and encapsulation efficiency of Q@MNPs were estimated. As observed in Figure 1g, the encapsulation efficiency of QCT loaded in Q@MNPs in this study was 43.78 ± 3.61%. It was reported that the stability of nanoparticle solution is dependent on the surface charge of the nanoparticle and higher zeta potential (absolute value) of electrostatic repulsion, and greater zeta potential manifested stronger electrostatic repulsion between nanoparticles [34]. The Zeta potentials of MNPs and Q@MNPs were −28.57 ± 0.46 and −35.87 ± 0.75 mV, respectively. As a result, the Zeta potential of Q@MNPs was greater than that of MNPs, which indicated that the encapsulation of QCT in MNPs could improve the spatial stability of Q@MNPs.

3.5. Photothermal Performance

The UV-vis absorption spectra of the QCT, MNPs, and Q@MNPs are shown in Figure 2a. This indicated that MNPs and Q@MNPs exhibited wide and strong absorption in the NIR region. After being irradiated (808 nm laser, 0.25 W/cm^2) for 10 min (Figure 2b,c), the temperature of QCT, MNPs, and Q@MNPs increased to 64.4 °C, 121.6 °C, and 123.3 °C, respectively. MNPs can transform NIR light into heat and have high photothermal energy conversion efficiency in previous reports [8]. Moreover, there were no significant changes compared with MNPs and Q@MNPs, indicating that the photothermal performance was unaffected by the encapsulation of QCT. Next, in order to further evaluate the photothermal stability of Q@MNPs, heating and cooling cycles were performed four times (Figure 2d). After four cycles, the maximum temperature and change trend of Q@MNPs were not significantly changed ($p < 0.05$), which suggested that the Q@MNPs have good photothermal stability. Therefore, the Q@MNPs exhibited good photothermal properties and photothermal stability, which can be used as photothermal antibacterial nanomaterials in the food industry.

3.6. Antioxidant Activity of Q@MNPs

The DPPH and ABTS radical scavenging experiments were used to evaluate the antioxidant activities of MNPs and Q@MNPs. As can be seen in Figure 3a,b, when the concentration of MNPs increased from 0.05 to 0.8 mg/mL, the DPPH radical scavenging activities significantly improved ($p < 0.05$) from 4.85 ± 0.08 to 70.21 ± 0.74%, and the ABTS radical scavenging activities ranged from 21.89 ± 0.41 to 80.62 ± 0.43%. The antioxidant activity of MNPs is due to the intramolecular non-covalent electrons that can interact with free radicals [35]. As shown in Figure 3b, for Q@MNPs, with the concentration of Q@MNPs increased, the DPPH and ABTS radical scavenging ability significantly increased ($p < 0.05$) from 31.75 ± 0.11 to 86.98 ± 0.71%, 30.51 ± 0.84 to 88.26 ± 0.64%, respectively. Compared with MNPs, the DPPH and ABTS radical scavenging capacity of Q@MNPs at different concentrations was higher than that of MNPs. This can be ascribed to QCT promoting the antioxidant ability of Q@MNPs. The antioxidant ability of QCT was attributed to polyphenolic compounds, and QCT exhibited the ability to scavenge reactive oxygen species and form the equivalent hydrazine through the potential of hydrogen atoms [36]. According to Nalini et al. [37], encapsulating QCT into alginate/chitosan nanoparticles showed higher antioxidant activity than QCT alone. Hence, the Q@MNPs with high antioxidant activity can be used as food antioxidants in food processing.

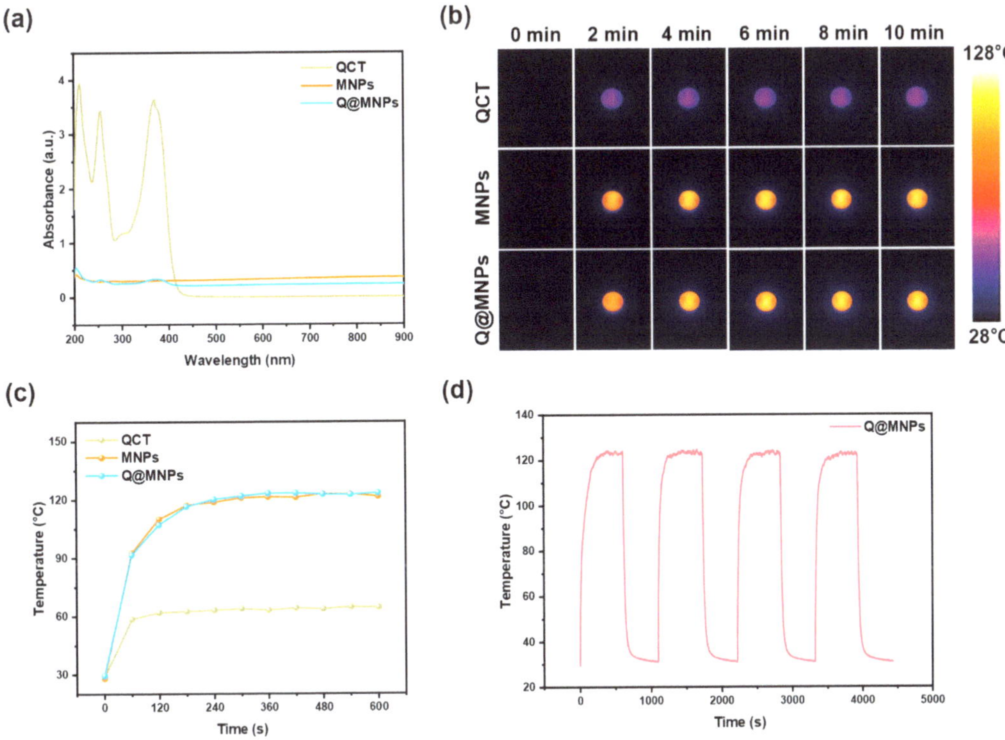

Figure 2. UV-vis absorbance (**a**); the photothermal image of QCT, MNPs, and Q@MNPs exposed to NIR laser at 808 nm (0.25 W/cm^2) for 10 min (**b**); temperature changes in QCT, MNPs, and Q@MNPs irradiated using an 808 nm laser (**c**); and irradiated for four on/off cycles (**d**).

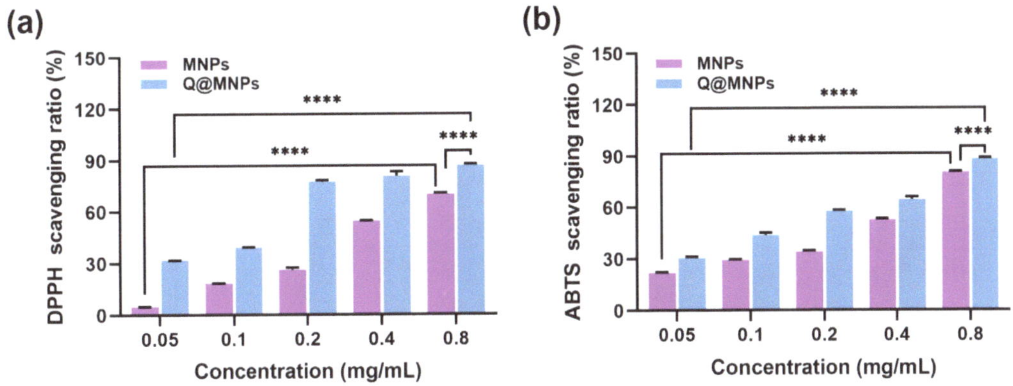

Figure 3. The DPPH radical scavenging ratio (**a**) and ABTS radical scavenging ratio of MNPs and Q@MNPs with increasing concentration from 0.05 to 0.8 mg/mL (**b**). Data presented as mean ± SD, **** $p < 0.0001$.

3.7. Antibacterial Activity

The antibacterial activity of Q@MNPs against *E. coli* and *S. aureus* was next assessed in vitro using the method of plate counting. As can be seen in Figure 4a, after the treatment of each group with *S. aureus*, the survival rates of the QCT and QCT+NIR groups were

84.13%, and 84.42%, respectively, which may be attributed to the weak effects of QCT release [36]. The survival rates of the MNPs and MNPs+NIR groups were 96.08%, and 63.32%, respectively. This suggests that the MNPs may have some antibacterial activities against *S. aureus*. However, the antibacterial properties of *S. aureus* in the MNPs+NIR group significantly ($p < 0.05$) improved. This may be because MNPs are distributed in the cell walls and membrane of bacteria. The high temperature of MNPs produced from irradiation at 808 nm can later damage the bacteria, resulting in the release of the cytoplasmic matrix, thereby affecting its metabolism [38]. The survival rates of the Q@MNPs and Q@MNPs+NIR groups were 75.27% and 0.12%, respectively, suggesting that the photothermal effects of Q@MNPs combined with QCT resulted in good antibacterial activities against *S. aureus*. This could be attributed to the QCT binding to MNPs via Π−Π and hydrogen bonding interactions, and the photothermal effects produced high temperatures and weakened the interactions, resulting in a greater release of QCT [39]. Next, Figure 4b shows the bacteriostatic efficacy of each treatment group for *E. coli*. The survival rates of the QCT, QCT+NIR, MNPs, MNPs+NIR, Q@MNPs, and Q@MNPs+NIR groups were 90.61%, 87.77%, 82.76%, 68.86%, 68.29%, and 1.14%, respectively. Compared with *S. aureus*, MNPs showed stronger antibacterial activity against *E. coli*, which was similar to the report of Roy et al. [40]. Moreover, when Q@MNPs were combined with NIR, they showed excellent antibacterial effects on *E. coli*. Therefore, the excellent photothermal performance of Q@MNPs combined with QCT resulted in excellent antibacterial performance against *S. aureus* and *E. coli*. This can be used as a photothermal and antibacterial material for food processing.

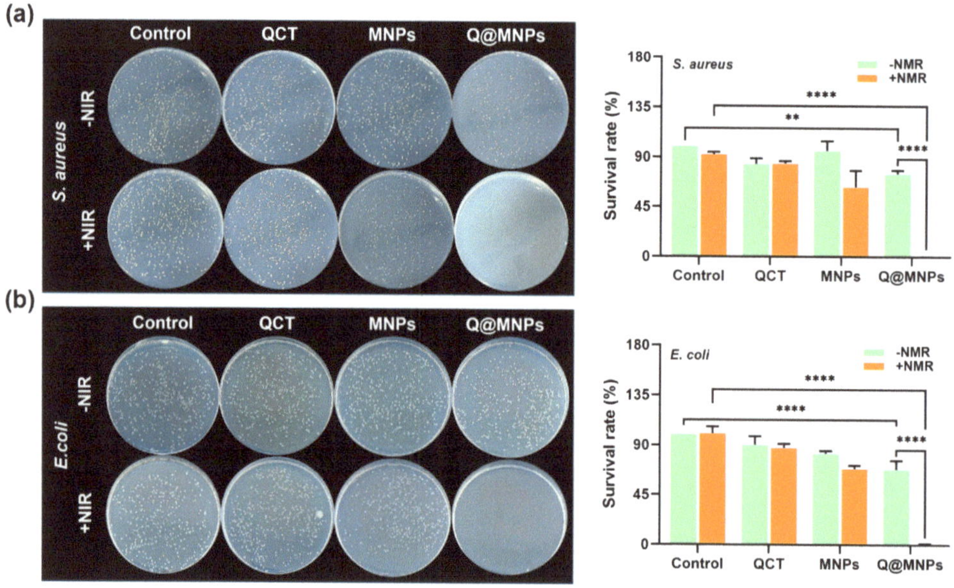

Figure 4. Photographs and survival rate (%) of *S. aureus* (**a**) and *E. coli* (**b**) treated with QCT, MNPs, and Q@MNPs. Data presented as mean ± SD, ** $p < 0.01$, **** $p < 0.0001$.

3.8. In Vitro QCT Release

The QCT in vitro release behavior of Q@MNPs was investigated at pH 4.5 and 7.2 during 24 h. As illustrated in Figure 5a, the release of Q@MNPs exhibited three stages, including burst release, steady release, and release equilibrium. There was a fast release of QCT from Q@MNPs at pH 7.2 in the first 4 h, which reached 22.74%, and then increased to 26.28% at the steady stage (12 h), and eventually reached 27.33% at the equilibrium stage (24 h). The initial fast release can be explained by the QCT crystals located close to the surface of Q@MNPs, resulting in rapid dissolution [4]. It can be observed that the cumulative release of QCT from Q@MNPs at pH 4.5 was lower than at pH 7.2. This is because carboxyl groups in pH 7.2 carried negative charges due to deprotonation, resulting in higher repulsive forces between these groups on the polymer surface, which facilitated the release of QCT from the internal structure of Q@MNPs. In acidic media (pH 4.5), the electrostatic effect was minimized due to the protonation of carboxyl groups, which increased the ability to retain QCT [1]. These results showed that the release of QCT from Q@MNPs may be affected by different pH values [21]. After irradiation with an 808 nm laser, the cumulative release of QCT from Q@MNPs significantly increased ($p < 0.05$) in the first 1 h in PBS solutions of pH 4.5 and 7.2, and the cumulative release of 24 h significantly improved ($p < 0.05$), which can be attributed to the photothermal properties of Q@MNPs weakening the Π−Π and hydrogen bonding interactions between QCT and MNPs, promoting the release of QCT from Q@MNPs [39]. At 24 h, the cumulative release of Q@MNPs at pH 4.5 and 7.2 under irradiating treatments reached 16.31% and 35.36%, respectively. These results indicated that Q@MNPs exhibited fantastic NIR-responsive drug release behavior, which can be applied as photothermal bactericides in food preservation. Therefore, Q@MNPs with NIR/pH dual-responsive have great potential for drug delivery of QCT.

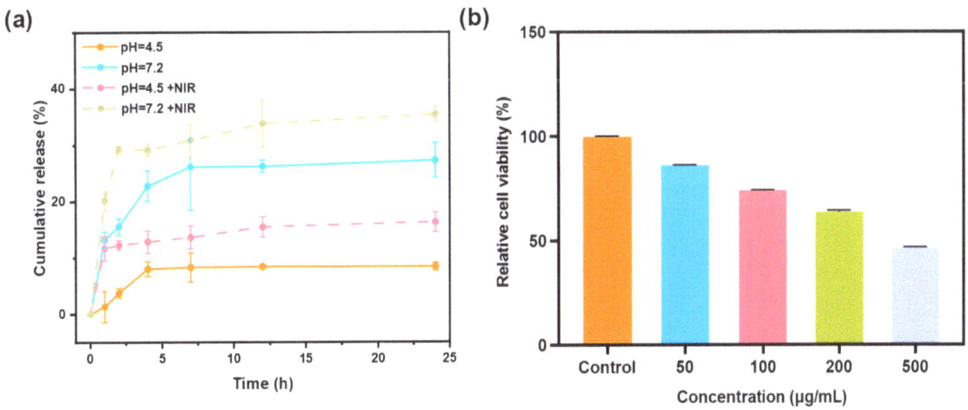

Figure 5. The cumulative release of QCT from Q@MNPs treated at pH = 4.5 (−NIR/+NIR) and pH = 7.2 (−NIR/+NIR) (**a**), and relative cell viability of L929 cells exposed to increasing concentrations of Q@MNPs from 50 to 500 μg/mL (**b**).

3.9. *In Vitro Cytotoxicity Assay*

The relative cell viability of L929 cells incubated with different concentrations of Q@MNPs using the CCK−8 assay is shown in Figure 5b. The relative cell viability of L929 cells was 86.16% and 74.04% at concentrations of 50 μg/mL and 100 μg/mL, respectively. In addition, when the concentration of Q@MNPs increased to 200 and 500 μg/mL, the relative cell viability decreased to 63.72% and 46.57%. According to previous research, the MNPs produced by synthetic methods showed good biocompatibility. Yuan et al. found that PDA60-PEG-NH_2 nanoparticles exposed to L929 cells presented excellent relative cell viability exceeding 90% when the concentration was up to 1.0 mg/mL [41]. Ozlu et al. also stated that MNPs have no significant cytotoxicity to L929 cells [42]. Moreover, compared with the cell cytotoxicity of QCT, George et al. reported that a QCT concentration of 81 μg/mL results in 50% L929 cell death [2]. Therefore, the concentration of QCT increased with the concentration of Q@MNPs from 50 to 500 μg/mL, which may result in the reduced relative cell viability of L929 cells. These results indicated that the Q@MNPs presented good biocompatibility with L929 cells at concentrations below 100 μg/mL [43]. In conclusion, the Q@MNPs at low doses can be applied in the field of food processing.

4. Conclusions

In this work, Q@MNPs were successfully fabricated through the oxidative self-polymerization of dopamine in an alkaline aqueous solution combined with the absorption of QCT via Π−Π stacking and hydrogen bonding interactions. FT-IR, XRD, and Zeta potential analyses indicated that QCT was successfully loaded onto MNPs. The TGA analysis showed the enhanced thermal stability of QCT compounded with MNPs. The photothermal experiments revealed that Q@MNPs had excellent photothermal performance and stability, which could effectively sterilize *E. coli* and *S. aureus* after irradiation with an 808 nm NIR laser. Meanwhile, the encapsulation of QCT also improved the antioxidation performance of MNPs. Moreover, the Q@MNPs showed that the pH and NIR stimulated the sustained release of QCT. The relative cell viability of L929 cells results showed that the Q@MNPs at low doses (<100 μg/mL) presented good biocompatibility. To sum up, the as-prepared Q@MNPs exhibited improved antioxidant activity, photothermal antimicrobial activity, and NIR/pH dual-responsive drug release, which has great potential as a promising functional drug delivery platform in the food industry.

Author Contributions: Conceptualization, D.Z.; methodology, X.C. and N.B.; software, D.Z. and L.H.; validation, D.Z. and X.C.; formal analysis, H.L.; investigation, L.Z.; writing—original draft preparation, D.Z.; writing—review and editing, J.P. and L.W.; visualization, J.P. and L.W.; funding acquisition, R.M., L.W. and J.P. All authors have read and agreed to the published version of the manuscript.

Funding: This work was supported by the Key project of Natural Science Foundation of Fujian Province (2022J02021), Natural Science Foundation from Fujian Provincial Department of Science and Technology (2020J01556), Scientific Research Foundation for Distinguished Young Scientists from Fujian Agriculture and Forestry University (xjq202006) and China Postdoctoral Science Foundation (2022M712180). The authors also thank Guoyu Cao from Department of Food of Minbei Vocational and Technical College for her thoughtful review and editing of this manuscript.

Data Availability Statement: The data used to support the findings of this study can be made available by the corresponding author upon request.

Conflicts of Interest: The authors declare no conflict of interest.

References

1. Araújo, M.; Viveiros, R.; Correia, T.R.; Correia, I.J.; Bonifácio, V.D.B.; Casimiro, T.; Aguiar-Ricardo, A. Natural melanin: A potential pH-responsive drug release device. *Int. J. Pharm.* **2014**, *469*, 140–145. [CrossRef]
2. George, D.; Maheswari, P.U.; Begum, K. Synergic formulation of onion peel quercetin loaded chitosan-cellulose hydrogel with green zinc oxide nanoparticles towards controlled release, biocompatibility, antimicrobial and anticancer activity. *Int. J. Biol. Macromol.* **2019**, *132*, 784–794. [CrossRef] [PubMed]
3. Bu, N.; Zhou, N.; Cao, G.; Mu, R.; Pang, J.; Ma, C.; Wang, L. Konjac glucomannan/carboxymethyl chitosan film embedding gliadin/casein nanoparticles for grape preservation. *Int. J. Biol. Macromol.* **2023**, *249*, 126131. [CrossRef] [PubMed]
4. Eizatahry, A.A.; Eldin, M.S.M. Preparation and characterization of metronidazole loaded chitosan nanoparticles for drug delivery application. *Polym. Adv. Technol.* **2008**, *19*, 1787–1791. [CrossRef]
5. Jiang, F.; Du, C.; Zhao, N.; Jiang, W.; Yu, X.; Du, S.-K. Preparation and characterization of quinoa starch nanoparticles as quercetin carriers. *Food Chem.* **2022**, *369*, 130895. [CrossRef] [PubMed]
6. Momeni, S.; Ramezani, A.M.; Talebi, S.; Nabipour, I. Synthesis of intrinsic fluorescent dopamine/quercetin copolymer nanoparticles and their application as a dual-mode assay for detection of quercetin. *J. Food Compos. Anal.* **2023**, *120*, 105296. [CrossRef]
7. Zou, Y.; Qian, Y.; Rong, X.Y.; Cao, K.X.; McClements, D.J.; Hu, K. Encapsulation of quercetin in biopolymer-coated zein nanoparticles: Formation, stability, antioxidant capacity, and bioaccessibility. *Food Hydrocoll.* **2021**, *120*, 106980. [CrossRef]
8. Liang, Y.; Zhao, Y.; Sun, H.; Dan, J.; Kang, Y.; Zhang, Q.; Su, Z.; Ni, Y.; Shi, S.; Wang, J.; et al. Natural melanin nanoparticle-based photothermal film for edible antibacterial food packaging. *Food Chem.* **2023**, *401*, 134117. [CrossRef] [PubMed]
9. Roy, S.; Rhim, J.-W. Agar-based antioxidant composite films incorporated with melanin nanoparticles. *Food Hydrocoll.* **2019**, *94*, 391–398. [CrossRef]
10. Liu, W.; Yu, Y.; Cheng, W.; Zhou, M.; Cui, L.; Wang, P.; Wang, Q. Melanin-like nanoparticles loaded with Ag NPs for rapid photothermal sterilization and daily protection of textiles. *Colloids Surf. B* **2022**, *219*, 112829. [CrossRef]
11. Shanmuganathan, K.; Cho, J.H.; Iyer, P.; Baranowitz, S.; Ellison, C.J. Thermooxidative Stabilization of Polymers Using Natural and Synthetic Melanins. *MACROMOLECULES* **2011**, *44*, 9499–9507. [CrossRef]

12. Zhao, W.; Liu, Z.; Liang, X.; Wang, S.; Ding, J.; Li, Z.; Wang, L.; Jiang, Y. Preparation and characterization of epigallocatechin-3-gallate loaded melanin nanocomposite (EGCG @MNPs) for improved thermal stability, antioxidant and antibacterial activity. *LWT* **2022**, *154*, 112599. [CrossRef]
13. Zhao, W.; Liang, X.; Wang, X.; Wang, S.; Wang, L.; Jiang, Y. Chitosan based film reinforced with EGCG loaded melanin-like nanocomposite (EGCG@MNPs) for active food packaging. *Carbohydr. Polym.* **2022**, *290*, 119471. [CrossRef] [PubMed]
14. Xing, Q.Q.; Buono, P.; Ruch, D.; Dubois, P.; Wu, L.B.; Wang, W.J. Biodegradable UV-Blocking Films through Core-Shell Lignin-Melanin Nanoparticles in Poly(butylene adipate-co-terephthalate). *ACS Sustain. Chem. Eng.* **2019**, *7*, 4147–4157. [CrossRef]
15. Liu, W.J.; Cheng, W.; Zhou, M.; Xu, B.; Wang, P.; Wang, Q.; Yu, Y.Y. Construction of multifunctional UV-resistant, antibacterial and photothermal cotton fabric via silver/melanin-like nanoparticles. *CELLULOSE* **2022**, *29*, 7477–7494. [CrossRef]
16. Wang, J.Y.; Cheng, H.; Wang, Z.K.; Yang, E.D.; Guo, F.; Wang, W.Y.; Sun, D.D. Human small intestine cancer cell membrane-camouflaged quercetin-melanin for antibacterial and antitumor activity. *J. Biomed. Mater. Res. Part B-Appl. Biomater.* **2021**, *109*, 1534–1551. [CrossRef] [PubMed]
17. Wang, S.; Duan, Z.; Zheng, L.; Yang, Y.; Zheng, X.; Xiao, D.; Ai, B.; Wang, M.; Sheng, Z. Digestive enzyme corona formed in simulated gastrointestinal tract and its impact on EGCG release from banana resistant starch nanoparticles. *Food Hydrocoll.* **2024**, *146*, 109267. [CrossRef]
18. Zhao, S.S.; Jia, R.Y.; Yang, J.; Dai, L.; Ji, N.; Xiong, L.; Sun, Q.J. Development of chitosan/tannic acid/corn starch multifunctional bilayer smart films as pH-responsive actuators and for fruit preservation. *Int. J. Biol. Macromol.* **2022**, *205*, 419–429. [CrossRef]
19. Lee, S.J.; Gwak, M.A.; Chathuranga, K.; Lee, J.S.; Koo, J.; Park, W.H. Multifunctional chitosan/tannic acid composite films with improved anti-UV, antioxidant, and antimicrobial properties for active food packaging. *Food Hydrocoll.* **2023**, *136*, 108249. [CrossRef]
20. Li, Z.; You, S.; Mao, R.; Xiang, Y.; Cai, E.; Deng, H.; Shen, J.; Qi, X. Architecting polyelectrolyte hydrogels with Cu-assisted polydopamine nanoparticles for photothermal antibacterial therapy. *Mater. Today Bio* **2022**, *15*, 100264. [CrossRef]
21. Kumar, D.; Gautam, A.; Paban Kundu, P. Synthesis of acrylamide-g-melanin/itaconic acid-g-psyllium based nanocarrier for capecitabine delivery: In vivo and in vitro anticancer activity. *Int. J. Pharm.* **2023**, *635*, 122735. [CrossRef] [PubMed]
22. Wang, W.; Zhang, W.; Li, L.; Deng, W.; Liu, M.; Hu, J. Biodegradable starch-based packaging films incorporated with polyurethane-encapsulated essential-oil microcapsules for sustained food preservation. *Int. J. Biol. Macromol.* **2023**, *235*, 123889. [CrossRef] [PubMed]
23. Li, X.; Zhou, P.; Luo, Z.; Feng, R.; Wang, L. Hohenbuehelia serotina polysaccharides self-assembled nanoparticles for delivery of quercetin and their anti-proliferative activities during gastrointestinal digestion in vitro. *Int. J. Biol. Macromol.* **2022**, *203*, 244–255. [CrossRef] [PubMed]
24. Ma, L.; Su, C.-R.; Li, S.-Y.; He, S.; Nag, A.; Yuan, Y. Co-delivery of curcumin and quercetin in the bilayer structure based on complex coacervation. *Food Hydrocoll.* **2023**, *144*, 108907. [CrossRef]
25. Wang, Y.; Su, J.; Li, T.; Ma, P.M.; Bai, H.Y.; Xie, Y.; Chen, M.Q.; Dong, W.F. A Novel UV-Shielding and Transparent Polymer Film: When Bioinspired Dopamine-Melanin Hollow Nanoparticles Join Polymers. *ACS Appl. Mater. Interfaces* **2017**, *9*, 36281–36289. [CrossRef] [PubMed]
26. Fu, J.; Chen, Z.; Wang, M.; Liu, S.; Zhang, J.; Zhang, J.; Han, R.; Xu, Q. Adsorption of methylene blue by a high-efficiency adsorbent (polydopamine microspheres): Kinetics, isotherm, thermodynamics and mechanism analysis. *Chem. Eng. J.* **2015**, *259*, 53–61. [CrossRef]
27. Li, S.-F.; Hu, T.-G.; Wu, H. Fabrication of colon-targeted ethyl cellulose/gelatin hybrid nanofibers: Regulation of quercetin release and its anticancer activity. *Int. J. Biol. Macromol.* **2023**, *253*, 127175. [CrossRef] [PubMed]
28. Ghorbani, F.; Zamanian, A.; Behnamghader, A.; Joupari, M.D. A facile method to synthesize mussel-inspired polydopamine nanospheres as an active template for in situ formation of biomimetic hydroxyapatite. *Mater. Sci. Eng. C* **2019**, *94*, 729–739. [CrossRef]
29. Gan, C.F.; Liu, Q.; Zhang, Y.; Shi, T.Y.; He, W.S.; Jia, C.S. A novel phytosterols delivery system based on sodium caseinate-pectin soluble complexes: Improving stability and bioaccessibility. *Food Hydrocoll.* **2022**, *124*, 107295. [CrossRef]
30. Zhang, H.; Feng, H.; Ling, J.; Ouyang, X.-k.; Song, X. Enhancing the stability of zein/fucoidan composite nanoparticles with calcium ions for quercetin delivery. *Int. J. Biol. Macromol.* **2021**, *193*, 2070–2078. [CrossRef]
31. Wang, L.; Li, Y.; Lin, L.; Mu, R.; Pang, J. Novel synthesis of mussel inspired and Fe3+ induced pH-sensitive hydrogels: Adhesion, injectable, shapeable, temperature properties, release behavior and rheological characterization. *Carbohydr. Polym.* **2020**, *236*, 116045. [CrossRef] [PubMed]
32. Zhang, Z.; Belda Marín, C.; Lefebvre, M.; Lefebvre, C.; Terrasson, V.; Guénin, E. The preparation of stable spherical alkali lignin nanoparticles with great thermal stability and no cytotoxicity. *Int. J. Biol. Macromol.* **2022**, *222*, 1830–1839. [CrossRef] [PubMed]
33. Carrasco-Sandoval, J.; Aranda, M.; Henríquez-Aedo, K.; Fernández, M.; López-Rubio, A.; Fabra, M.J. Impact of molecular weight and deacetylation degree of chitosan on the bioaccessibility of quercetin encapsulated in alginate/chitosan-coated zein nanoparticles. *Int. J. Biol. Macromol.* **2023**, *242*, 124876. [CrossRef] [PubMed]
34. Ribeiro, E.F.; de Barros-Alexandrino, T.T.; Assis, O.B.G.; Junior, A.C.; Quiles, A.; Hernando, I.; Nicoletti, V.R. Chitosan and crosslinked chitosan nanoparticles: Synthesis, characterization and their role as Pickering emulsifiers. *Carbohydr. Polym.* **2020**, *250*, 116878. [CrossRef] [PubMed]

35. Roy, S.; Van Hai, L.; Kim, H.C.; Zhai, L.; Kim, J. Preparation and characterization of synthetic melanin-like nanoparticles reinforced chitosan nanocomposite films. *Carbohydr. Polym.* **2020**, *231*, 115729. [CrossRef] [PubMed]
36. Roy, S.; Rhim, J.-W. Fabrication of chitosan-based functional nanocomposite films: Effect of quercetin-loaded chitosan nanoparticles. *Food Hydrocoll.* **2021**, *121*, 107065. [CrossRef]
37. Nalini, T.; Basha, S.K.; Sadiq, A.M.; Kumari, V.S. In vitro cytocompatibility assessment and antibacterial effects of quercetin encapsulated alginate/chitosan nanoparticle. *Int. J. Biol. Macromol.* **2022**, *219*, 304–311. [CrossRef]
38. Sun, M.; Gao, P.; Wang, B.; Li, X.Y.; Shao, D.H.; Xu, Y.; Li, L.J.; Li, Y.H.; Zhu, J.W.; Li, W.L.; et al. Polydopamine-functionalized selenium nanoparticles as an efficient photoresponsive antibacterial platform. *RSC Adv.* **2023**, *13*, 9998–10004. [CrossRef]
39. Zhuang, H.Q.; Su, H.L.; Bi, X.X.; Bai, Y.T.; Chen, L.; Ge, D.T.; Shi, W.; Sun, Y.A. Polydopamine Nanocapsule: A Theranostic Agent for Photoacoustic Imaging and Chemo-Photothermal Synergistic Therapy. *ACS Biomater. Sci. Eng.* **2017**, *3*, 1799–1808. [CrossRef]
40. Roy, S.; Rhim, J.W. Preparation of carrageenan-based functional nanocomposite films incorporated with melanin nanoparticles. *Colloids Surf. B-Biointerfaces* **2019**, *176*, 317–324. [CrossRef]
41. Yuan, X.D.; Zhao, X.; Lin, Y.; Su, Z.H. Polydopamine-Based Nanoparticles for an Antibiofilm Platform: Influence of Size and Surface Charge on Their Penetration and Accumulation in *S. aureus* Biofilms. *Langmuir* **2022**, *38*, 10662–10671. [CrossRef]
42. Ozlu, B.; Kabay, G.; Bocek, I.; Yilmaz, M.; Piskin, A.K.; Shim, B.S.; Mutlu, M. Controlled release of doxorubicin from polyethylene glycol functionalized melanin nanoparticles for breast cancer therapy: Part I. Production and drug release performance of the melanin nanoparticles. *Int. J. Pharm.* **2019**, *570*, 118613. [CrossRef]
43. Erdem, U.; Dogan, D.; Bozer, B.M.; Turkoz, M.B.; Yildirim, G.; Metin, A.U. Fabrication of mechanically advanced polydopamine decorated hydroxyapatite/polyvinyl alcohol bio-composite for biomedical applications: In-vitro physicochemical and biological evaluation. *J. Mech. Behav. Biomed. Mater.* **2022**, *136*, 105517. [CrossRef]

foods

MDPI

Article

Antioxidant Activity, Formulation, Optimization and Characterization of an Oil-in-Water Nanoemulsion Loaded with Lingonberry (*Vaccinium vitis-idaea* L.) Leaves Polyphenol Extract

Siyu Wang [1], Yuan Cheng [1], Jingyi Wang [1], Miao Ding [1] and Ziluan Fan [1,2,*]

[1] School of Forestry, Northeast Forestry University, 26 HeXing Road, XiangFang District, Harbin 150040, China; siyuwang1122@163.com (S.W.); cy_lyz1109@163.com (Y.C.); owo_guyu@163.com (J.W.); 18724457575@163.com (M.D.)

[2] Key Laboratory of Forest Food Resources Utilization, Harbin 150040, China

* Correspondence: fzl_1122@nefu.edu.cn; Tel./Fax: +86-0451-82190514

Abstract: The active ingredients in lingonberry leaves and their beneficial properties to the human body have been well confirmed. In order to improve the stability and antioxidant activity of the active ingredients in lingonberry leaves, the response surface optimization method was used to prepare an oil-in-water nanoemulsion of polyphenol extract from lingonberry leaves. The active components in the extract were analyzed by ultra-performance liquid chromatography with triple quadrupole mass spectrometry (UPLC-TQ-MS), and bioactive compounds such as apigenin, sorbitol, and hesperidin were mainly found. Nanoemulsion droplets of 120 nm in diameter were prepared using ultrasonic emulsification. The optimal nanoemulsion formulation was determined through rigorous testing, and it was determined to be 10% (*w*/*w*) lingonberry extract and 20% (*w*/*w*) medium chain triglyceride (MCT). Additionally, a surfactant mixture was used, which combined soy protein isolate (SPI) and whey protein isolate (WPI) at 4% (*w*/*w*). The preparation method utilized ultrasonic emulsification, applying an ultrasonic power of 360 W for a duration of 300 s. The antioxidant activity (DPPH inhibition rate, ABTS inhibition rate and total reducing power) of the lingonberry nanoemulsion was significantly higher than that of the lingonberry polyphenol (LBP) extract. The nanoemulsion prepared using the optimal formulation had an entrapping efficiency of 73.25% ± 0.73% and a diameter of 114.52 ± 0.015 nm, with a satisfactory particle size of nanoscale and a PDI of 0.119 ± 0.065, demonstrating good stability of the emulsion.

Keywords: nanoemulsion; response surface methodology; antioxidant activity; lingonberry (*Vaccinium vitis-idaea* L.); delivery system

Citation: Wang, S.; Cheng, Y.; Wang, J.; Ding, M.; Fan, Z. Antioxidant Activity, Formulation, Optimization and Characterization of an Oil-in-Water Nanoemulsion Loaded with Lingonberry (*Vaccinium vitis-idaea* L.) Leaves Polyphenol Extract. *Foods* **2023**, *12*, 4256. https://doi.org/10.3390/foods12234256

Academic Editor: Yesim Özogul

Received: 4 September 2023
Revised: 9 November 2023
Accepted: 15 November 2023
Published: 24 November 2023

1. Introduction

Lingonberry (*Vaccinium vitis-idaea* L., family *Ericaceae*) is a natural wild berry with significant nutritional value and a wide range of health-promoting properties; lingonberry is rich in phenolics, which are the main active substances. Lingonberry, often termed a "superfruit" is renowned for its substantial antioxidant properties. Predominantly found in regions of northern and central Europe, Canada, and Asia, including mountainous zones like Northeastern China's Greater Khingan and the Lesser Hinggan Ranges, it is a source of various potent bioactives [1]. While bilberry and lingonberry leaves emerge as primary by-products during berry collection, recent studies have highlighted an intriguing observation [2], i.e., the leaves and stems from the Vaccinium species house a more substantial phenolic content than the berries themselves. The elevated phenolic concentration is congruent with their superior antioxidant capabilities compared to the fruit [3]. Such findings suggest that these by-products could be harnessed as promising reservoirs of bioactive compounds, opening doors for their inclusion in food supplements, nutraceuticals, or functional food products.

Regarding the biochemical makeup of lingonberry leaves, they feature a plethora of compounds, including phenolic acids, anthocyanins, triterpene acids, organic acids, carbohydrates, glycosyl hydroquinones, and proanthocyanidins [4]. This rich profile renders the leaves a more chemically diverse entity than the berry fruits. Notably, recent research has spotlighted flavonoid glycosides as the dominant phenolic entities in lingonberries. These compounds exhibit a myriad of health-promoting attributes, such as astringent, antitussive, diuretic, neuroprotective, antioxidant, and anti-inflammatory activities. Their potential roles in inhibiting cancer cell proliferation further underscore their therapeutic potential [5].

In addition these properties, there have been studies that have indicated that extracts from lingonberry leaves can curtail hepatitis C virus expression and stymie the proliferation of sensitive human promyelocytic leukaemia HL60 cells. Historically and traditionally, lingonberry leaves have been employed to mitigate cholesterol, address gastric complications, treat rheumatic ailments, and combat infections in the bladder and kidneys [6–8]. The various plant parts of the lingonberry are used in the production of pharmaceuticals, cosmeceuticals, and nutritional foods. The fruit is widely consumed and used in the food industry, while the leaves are an excellent source of phenolic compounds, especially functional products containing proanthocyanidins.

However, leaves have always been the raw materials that are not fully developed and utilized by people. Studies have shown that the leaves of lingonberry have higher biological activity than the fruit, making the leaves a promising medicinal resource [9]. They are an ignored renewable resource with huge reserves; the active ingredients in lingonberry leaves have broad application prospects.

Lingonberry polyphenols have antioxidant and free-radical scavenging properties, and their antitumour and anticancer effects should not be underestimated. However, most polyphenols have poor solubility, low stability, easy metabolism and excretion, and extremely low bioavailability, making it difficult for them to exert their effects and limiting their applications in functional foods and drugs.

Nanoemulsions have the characteristics of small particles, large specific surface area, and relatively uniform distribution, which can protect biologically active and unstable substances by encapsulating them in emulsions, and therefore the active substances can maintain their original biological activity for a long time. The small particle size, kinetic stability, and high optical transparency of nanoemulsions compared to conventional emulsions, as well as their ability to adjust the texture of a product, give them an advantage in many technical applications. Currently, nanoemulsions loaded with polyphenols have a bright future with industrial production possibilities in the field of pharmaceuticals and functional foods. Nanoemulsions are generally valued by researchers as a new type of carrier system and are increasingly being used in the fields of biology, medicine and pharmacy, and functional food development. As new active substance carriers, nanoemulsions are mainly formed by dispersion and homogenization followed by mixing water, oil, and emulsifier in appropriate proportions to form thermodynamically stable homogeneous dispersion systems. Nanoemulsions are often set to have specific functional properties, e.g., controlled release, enhanced bioavailability, enhanced potency, synergistic or targeted release, thermodynamic stability, and easy preservation; due to their small particle size and low surface tension, the active substance is more likely to come into direct contact with the gastrointestinal epithelium and promote absorption. Several studies have successfully constructed the delivery system of polyphenol nanoemulsions, improving stability [10,11], biological accessibility [12–14], antioxidant [15,16], and other functional activities of polyphenols. Nanoemulsions have been widely used to improve the bioavailability of insoluble drugs.

It is imperative to note that nanoemulsions are not equilibrium systems. They are non-thermodynamically stable and do not form spontaneously. Their formation is influenced either by their inherent chemical potential within the emulsion system or by external forces. Emulsification methods, which are pivotal to their formation, can be broadly classified based on the energy source utilized. There are high-energy methods that rely on external energy and low-energy methods that harness the system's inherent potential.

Among the high-energy methods, techniques like ultrasound, micronization, and high-pressure homogenization are prevalent. Notably, ultrasound technology stands out due to its efficiency and cost-effectiveness and can deliver nanoemulsions with smaller droplet size, reduced polydispersity index (PDI), and superior stability, all while using fewer surfactants. Ultrasound technology employs a frequency of at least 20 kHz [17]. The ultrasonic vibrations produce pronounced shear and pressure gradients in the emulsion. This facilitates the breakdown of larger droplets, enabling continuous production of the nanoemulsion. Notably, the energy consumption and associated costs are considerably lower than those in high-pressure homogenization.

To conclude, as the quest for optimal drug delivery systems persists, nanoemulsions are progressively recognized for their multifaceted benefits, particularly in the pharmaceutical and functional food industries. Their potential in enhancing the delivery and efficacy of compounds like LBPs is undeniable [18]. The total experimental content of this study is shown in Figure 1.

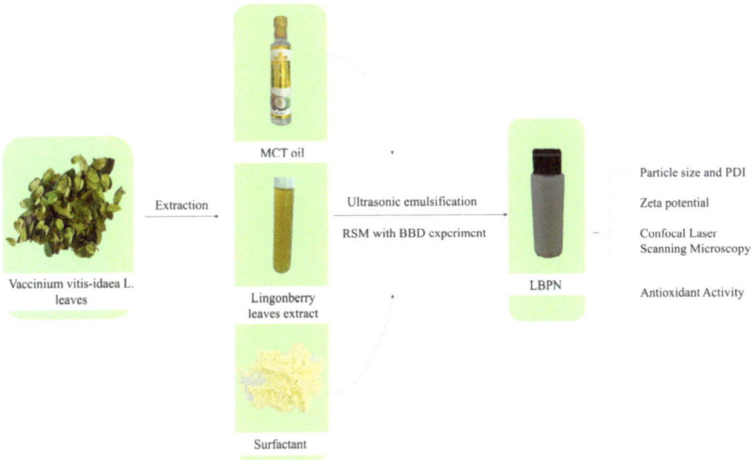

Figure 1. The total experimental contents.

2. Materials and Methods

2.1. Materials

SPI, WPI, and MCT were purchased from Sigma (Shanghai, China). All other chemicals and reagents used were of analytical grade.

The plant material was acquired from Kudu Forestry Bureau, Hulunbeir, Inner Mongolia Autonomous Region. Picking was conducted during the harvest of lingonberry plants, in the middle of September. Then, the plant material was transported to the laboratory, where it was cleaned of residual soil, and the different parts were separated, placed in a container, and stored them in a freezer. Subsequently, the samples were lyophilized using a freeze dryer (SP Scientific, Gardiner, MT, USA).

2.2. Preparation of Lingonberry Leaves Extract

Taking into consideration edible safety and the foundation established by our research team [19], we employed food-grade ethanol for the solvent extraction of polyphenols. The extraction procedure was adapted from the methodology presented in [20], with minor modifications. First, the freeze-dried lingonberry leaves were pulverized into a fine powder using an FW-100 high-speed universal grinder (Braun, Frankfurt, Germany), and then sieved using 100 mesh filter, freeze-stored at −20 °C in a refrigerator, and sealed for storage.

Then, 400 mL of 75% precooled ethanol was added and mixed with an RZ-8012 handheld stirrer at 12,000 r/m for 5 min. The prepared extract was homogenized (BME 100 L, Weiyu,

Shanghai, China) with a high shear for 3 min and stored at −20 °C in a refrigerator prior to preparation of the nanoemulsion.

2.3. Characterization of LBPs by UPLC-TQ-MS

The extract of lingonberry leaves, prepared as described in Section 2.2, was evaporated to dryness using a rotary evaporator at 45 °C under vacuum conditions. Then, the resulting extract was redissolved in 5 mL of methanol and subsequently filtered through a 0.45 μm filter prior to analysis. Identification of phenolic compounds in the extract was performed on an UPLC-TQ-MS system (Waters ACQUITY UPLC^{TM} system, Waters, Milford, CT, USA). The separation of phenolic compounds was conducted in a chromatographic column, Hyperil Glod (100 × 2.1 mm), with a temperature of 25 °C. Ionization Polarity: ESI− and ESI+; spray voltage, 3800/3000 V (+/−); capillary temperature, 320 °C. The positive mobile phase (solvent A) was prepared using 0.1% formic acid (FA) in water and MeOH, while the negative phase (solvent B) was created using 5 mM ammonium acetate in water (ammonia water regulation, pH = 9) and MeOH. The flow rate was set to 0.35 mL/min. Data obtained from UPLC-TQ-MS were analyzed using the Compound Discoverer 3.2 software.

2.4. Preparation of the Nanoemulsion

An ultrasonic homogenizer (ASU-10D, AS one, Matsubara, Japan) was used to produce the nanoemulsion samples. Two natural emulsifiers, soybean isolate and whey isolate, were used as the emulsifiers. A single-factor test was conducted with the mass concentration of emulsifier, oil phase type, oil phase mass fraction, protein ratio, and extract mass fraction as the observed factors. The encapsulation rates of lingonberry nanoemulsion samples prepared under different conditions were determined separately.

Based on the results of a single-factor test, the emulsifier mass concentration, oil phase mass fraction, and extract mass fraction were selected as the three factors to design a response surface test to optimize the formulation of the lingonberry nanoemulsion. Based on the response surface Box–Behnken design principle, the Design-Expert 8.0 software was used for the design and analysis of the response surface test. A three-factor, three-level response surface analysis was used to design the test using the encapsulation rate of lingonberry nanoemulsion as the response value, and the factor levels are shown in Table 1.

Table 1. Box–Behnken experimental factors and levels.

Independent Variable	Actual Levels at Coded Factor Levels		
	−1	0	1
Mass concentration of emulsifier	3%	4%	5%
Oil phase mass fraction	15%	20%	25%
Extract mass fraction	5%	10%	15%

The RSM with BBD experimental design with 3 levels, 3 independent variables, and 17 runs including five center points (Table 2) was used to optimize the operation condition including.

Mass concentration of emulsifier (X1), oil phase mass fraction (X2), and extract mass fraction (X3) of LBP nanoemulsions (LBPNs) were determined using the Design-Expert software (Version 8.0.6, Stat-Ease Inc., Minneapolis, MN, USA):

$$y = \beta_0 + \sum_{i=1}^{3} \beta_i X_i + \sum_{i=1}^{3} \beta_{ii} X_i^2 + \sum_{i=1}^{2} \sum_{j=i+1}^{3} \beta_{ij} X_i X_j + \varepsilon \quad (1)$$

In the present model (Equation (1)), $X1$, $X2$, and $X3$ are defined as independent variables, whereas Y represents the dependent response variable, which is assessed for each unique combination of factorial levels. The model comprises various coefficients: β_0 (intercept), β_i (linear), β_{ii} (quadratic), and β_{ij} (interaction coefficients). These coefficients of the second-order polynomial model are estimated through a multiple regression analysis based on experimental response values. Model fit was evaluated using several metrics as follows: the coefficient of determination (R^2), adjusted coefficient of determination (R^2_{adj}), p-value, and the degree of fit. The model was considered to be a well-fitting model when R^2 approaches 1 and the lack-of-fit p-value exceeds 0.05, signifying statistical insignificance. Additionally, the F-test is employed to ascertain the statistical significance of R^2, while regression coefficients' statistical relevance is determined using the t-test [21].

Table 2. Box–Behnken design.

Run	Protein Concentration (X1,%)	Oil Phase Mass Fraction (X2,%)	Extract Mass Fraction (X3,%)
1	4.00 (0)	25.00 (1)	15.00 (1)
2	3.00 (−1)	15.00 (−1)	10.00 (0)
3	3.00 (−1)	20.00 (0)	5.00 (−1)
4	4.00 (0)	20.00 (0)	10.00 (0)
5	4.00 (0)	20.00 (0)	10.00 (0)
6	3.00 (−1)	25.00 (1)	10.00 (0)
7	5.00 (1)	15.00 (−1)	10.00 (0)
8	4.00 (0)	20.00 (0)	10.00 (0)
9	4.00 (0)	25.00 (1)	5.00 (−1)
10	3.00 (−1)	20.00 (0)	15.00 (1)
11	4.00 (0)	15.00 (−1)	5.00 (−1)
12	4.00 (0)	20.00 (0)	10.00 (0)
13	5.00 (1)	25.00 (1)	10.00 (0)
14	5.00 (1)	20.00 (0)	15.00 (1)
15	5.00 (1)	20.00 (0)	5.00 (−1)
16	4.00 (0)	15.00 (−1)	15.00 (1)
17	4.00 (0)	20.00 (0)	10.00 (0)

2.5. *Characterization of the W/O Nanoemulsion*

2.5.1. Encapsulation Efficiency

The encapsulation efficiency was performed using the Folin–Ciocalteu method described by Mahsa Yazdan-Bakhsh et al. Free surface phenolic compounds were extracted by adding 5 mL of n-hexane to 1 mL of nanoemulsion and centrifuged for 3 min using a centrifuge at 3500× g. Then, 5 mL of ethanol was added to 1 mL of nanoemulsion and the total phenolic compounds were released using ultrasonic shaking for 1 h to disrupt the emulsion encapsulation structure. The encapsulation efficiency was calculated as the difference between the total phenolic content in the nanoparticle suspension and the TPC content in the supernatant [22].

2.5.2. Particle Size and PDI

The mean droplet diameter and polydisperse index (PDI) were measured utilizing a Nano-ZS PALS laser particle size analyzer (Malvern Instruments Ltd., Malvern, UK). The measurements were conducted through the dynamic light scattering method, operational at a light wavelength of 660 nm and a scattering angle fixed at 90°, under a controlled temperature of 25 °C ± 0.1 °C. The PDI, a non-dimensional metric, varies from 0 to 1, with a refractive index set at 1.330 for these measurements.

2.5.3. Zeta Potential

The zeta potential (ZP) values of the samples were ascertained using a Nano-ZS PALS laser particle size analyzer (Malvern Instruments Ltd., Malvern, UK). Prior to analysis, the samples were diluted to a standard concentration of 5 mg/mL with water. This dilution was necessary to minimize the effects of multiple scattering during measurements.

2.5.4. Confocal Laser Scanning Microscopy (CLSM)

The LBPN aqueous and oil phase embedding states were evaluated using a confocal laser scanning microscope (Leica, Heidelberg, Germany) with 40× magnification at 20 °C, and the proteins were treated with Nile Blue (1 mg/mL in anhydrous ethanol, 1:100 v/v) fluorescent dye solution, followed by Nile Red (1 mg/mL in dimethyl sulfoxide, 1:100 v/v). The oil stain treatment was performed according to the method described by [23]: 100 μL of emulsion was mixed with 10 μL of Nile Red and 10 μL of Nile Blue and left for 10 min, and then the proteins were detected using Ar/K and He/Ne dual-channel laser modes with an excitation light of 633 nm, and then the excitation light was changed to 488 nm to detect the oil phase. The software used for the CLSM imaging was ImageJ 2.0.

2.6. Antioxidant Activity of LBPs

2.6.1. DPPH Radical Scavenging Assay

For the DPPH (0.2 mmol/L) assays, the solution was prepared in absolute methanol and stored at 4 °C in darkness. To execute the test, 0.1 mL of ascorbic acid or a sample variant was mixed with 1.5 mL of the DPPH solution. Then, the mixture was incubated in darkness for 30 min. Ascorbic acid's absorbance was monitored at 515 nm, serving as the control ($A_{control}$). The DPPH radical scavenging activity of the sample was calculated as follows:

$$\text{DPPH radical scavenging activity}(\%) = (1 - A_{sample}/A_{control}) \times 100 \tag{2}$$

2.6.2. ABTS•+ Radical Scavenging Assay

The ABTS solution was formulated by combining an ABTS stock solution (7 mmol/L) with $K_2O_8S_2$ (142 mmol/L) at a 1:1 ratio. Then, the solution was stored in a light-deprived environment at ambient temperature for 12 h. Subsequently, 100 μL of the sample was introduced to 2.9 mL of the ABTS+ solution. After mixing, the combined solution was shielded from light and left to react for 6 min at room temperature. The resulting absorbance was measured at 734 nm using a spectrophotometer, with ascorbic acid serving as the control reference ($\mathrm{A_{control}}$). The ABTS radical scavenging activity of the sample was calculated as follows:

$$\text{ABTS radical scavenging activity}(\%) = (1 - \mathrm{A_{sample}}/\mathrm{A_{control}}) \times 100 \tag{3}$$

2.6.3. Reducing Power

The reducing power of the LBPs and LBPN were assessed following the protocol outlined by [24]. A volume of 2.5 mL from the sample, at differing concentrations, was amalgamated with 2.5 mL of potassium hexacyanoferrate buffer (0.2 M, pH 6.6) and 2.5 mL of potassium ferricyanide (1%). The mixture was subjected to an incubation period at 50 °C for 20 min. Post incubation, 2.5 mL of 10% trichloroacetic acid, 2.5 mL of distilled water, and 0.5 mL of 0.1% ferric chloride solution were introduced. The mixture's absorbance was gauged at 700 nm employing a spectrophotometer.

2.7. Statistical Analysis

Each analysis was executed three times, with the resultant data articulated as means ± standard deviation (SD). Notable differences across data sets were scrutinized through the analysis of variance (ANOVA) facilitated by the SPSS Statistics software, version 20 (IBM, New York, NY, USA).

3. Results

3.1. UPLC-TQ-MS Characterization of Extract

The polyphenol extraction rate was calculated based on the method outlined in [25] and was found to be 138.18 mg/g. In this study, in the leaves of lingonberry, 14 phenolic compounds were identified, originating from three phenolic groups: phenolic acids, flavonoids, and terpenoids. The identified 14 compounds are listed in Table 3. The three most abundant active substances in the lingonberry leaves were apigenin 7-rhamnosyl-(1->2)-galacturonide, Bis(4-ethylbenzylidene)sorbitol, and pinocembrin7-O-neohesperidoside3-O-acetate, all belonging to the class of flavonoids. The second most abundant substance in the leaves was a sort of phenolic acid, i.e., caffeic acid. In this study, no anthocyanin substances were identified, possibly due to the anthocyanin substances have been lost due to the long freezing time of the raw material. Nevertheless, in previous reports, a variety of anthocyanins such as cyanidin-3-O-arabinoside, cyanidin-3-O-glucoside, and delphinidie-3-glucoside have been found in lingonberry leaves [26]. According to previous reports, the presence of anthocyanins in lingonberry leaves may be related to the variety, origin, freshness, phenological stage, and ripeness of lingonberries [27]. Anthocyanins were detected only in lingonberry fruits and not in leaves when phenolic substances of ten different varieties of lingonberry such as "Erntekrone", "Koralle", and "Masovia" were analyzed [28]. At the same time, one study reported that anthocyanins were not detected in three types of lingonberry leaves collected from Romania [26]. In contrast, active substances in roots, stems, leaves, and fruits of lingonberries harvested from Inner Mongolia, China were analyzed and two anthocyanins were detected, i.e., empetrin and 3,5-diglucosyldelphinidin [29].

Table 3. The main phenolic compound contents in the leaves of lingonberry using UPLC-TQ-MS.

Chemical Name	Area	RT/min	Molecular Weight	Molecular Formula	Molecular Structure Formula
Apigenin 7-rhamnosyl-(1->2)-galacturonide	1.73×10^9	12.033	592.14407	$C_{27}H_{28}O_{15}$	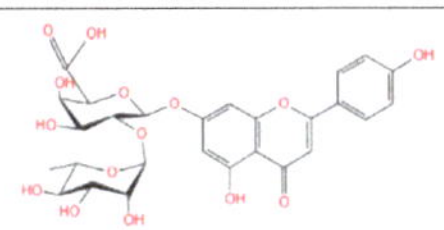
Bis(4-ethylbenzylidene)sorbitol	1.26×10^9	9.157	414.19921	$C_{24}H_{30}O_6$	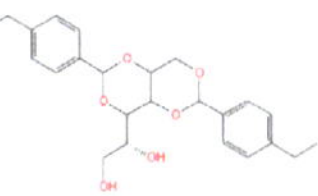
Pinocembrin 7-O-neohesperidoside 3-O-acetate	6.48×10^8	12.596	606.19569	$C_{29}H_{34}O_{14}$	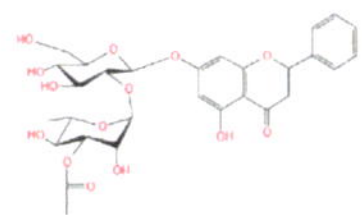
Caffeic acid	4.49×10^8	9.662	162.02992	$C_9H_8O_4$	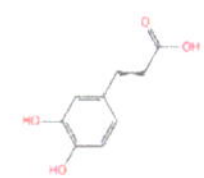
Licochalcone A	2.27×10^8	6.201	360.13567	$C_{21}H_{22}O_4$	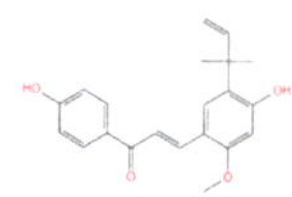
Munduserone	1.76×10^8	0.987	342.11153	$C_{19}H_{18}O_6$	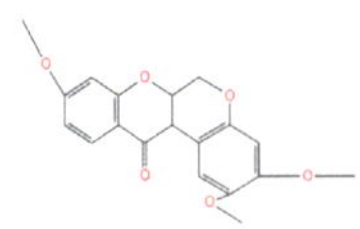
Kaempferol-3-O-α-L-arabidopyranoside	1.29×10^8	7.316	418.0876	$C_{20}H_{18}O_{10}$	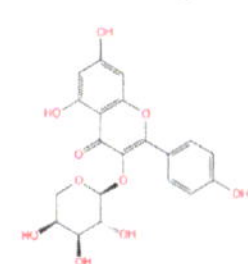
Quercetin-3β-D-glucoside	1.05×10^8	7.135	464.08967	$C_{21}H_{20}O_{12}$	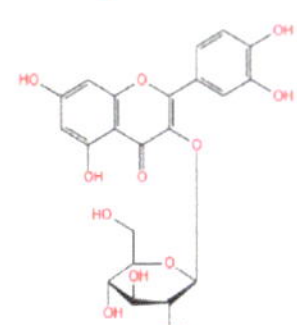

Table 3. *Cont.*

Chemical Name	Area	RT/min	Molecular Weight	Molecular Formula	Molecular Structure Formula
Oleanolic acid	9.03×10^7	10.561	456.35435	$C_{30}H_{48}O_3$	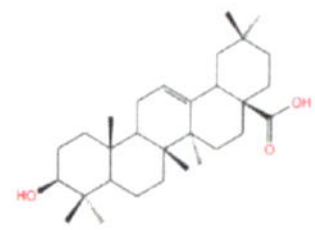
Cycloartomunoxanthone	5.21×10^7	5.695	448.15239	$C_{26}H_{24}O_7$	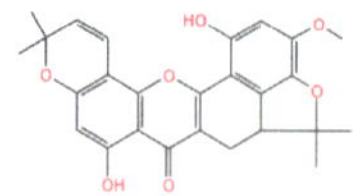
Quercetol B	1.60×10^7	10.416	368.19811	$C_{23}H_{28}O_4$	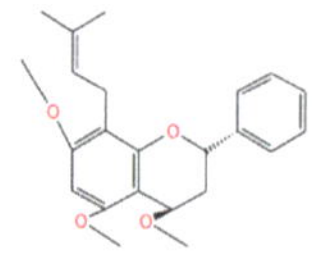
Cycloartocarpin	1.58×10^7	6.83	434.17079	$C_{26}H_{26}O_6$	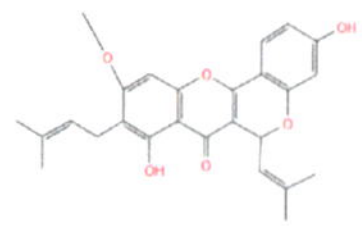
Silybin	1.39×10^7	10.676	482.12114	$C_{25}H_{22}O_{10}$	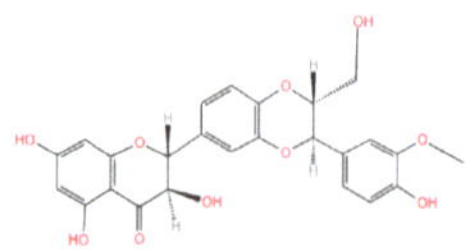
5-Hydroxy-6,6-dimethylpyrano flavone	7.95×10^6	0.994	320.10509	$C_{20}H_{16}O_4$	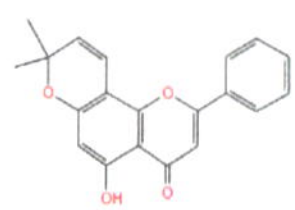

Apigenin 7-rhamnosyl-(1->2)-galacturonide, which is most abundant in the leaf extract, is an apigenin derivative. Apigenin is a secondary plant metabolite, usually found in nature in glycosylated form, and is one of the most abundant and well-studied flavonoids. It can be used as a cancer chemopreventive agent. It has been shown that apigenin-7-O-glucoside is more effective than apigenin in reducing colon cancer cell viability and inducing cell death, and that apigenin-7-O-glucoside is more biologically active than apigenin [30]. With the increasing attention paid to plant-derived diets, the application of natural active substances such as apigenin in food has research value.

The high content of the soluble sugar sorbitol in the extract may be due to the fact that sorbitol is the main form of carbohydrate accumulation in lingonberries. Previous studies have shown that sorbitol is the major sugar component of water-extracted chokeberry leaves, accounting for ca. 80% of the total carbohydrates, with an average content of 145.2 ± 1.8 mg/g in the lyophilized leaves [31].

The higher content of caffeic acid also provides antioxidant activity to the extract. Caffeic acid, as a natural plant-derived antioxidant, has gradually become a hot research topic in recent years [32] with studies conducting in vivo and in vitro antioxidant tests. In in vitro experiments, caffeic acid has shown stronger antioxidant activity than chlorogenic acid, and it has been reported that caffeic acid may play a major role in the protective effect of chlorogenic acid against ischemia reperfusion injury. In addition to the aforementioned active substances, the extract also contains a relatively abundant variety of phenolic compounds, including quercetin, oleanolic acid, and various other phenolic compounds, totaling 14 different types. These findings are in line with previous research on the polyphenolic content of lingonberry leaves [20,33].The total ion current of LBPs is illustrated in Figure 2.

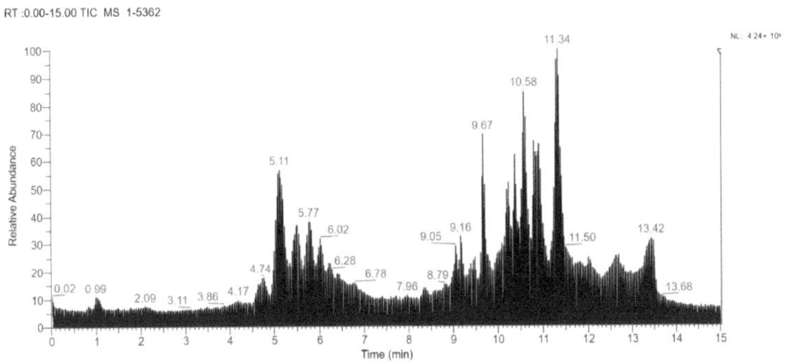

Figure 2. Total ion current (TIC) of LBPs.

3.2. Optimization of the Lingonberry Nanoemulsion Using Response Surface Methodology

A response surface analytical approach was utilized, taking into consideration protein concentration, oil phase mass fraction, and extract mass fraction as independent variables, while the entrapping efficiency was designated as the dependent variable. From this analysis, a regression equation was derived as:

$$\text{Entrapping efficiency} = +73.12 + 0.51 \times A - 1.28 \times B - 3.05 * C + 0.65 \times A \times B - 0.51 \times A \times C + 4.01 \times B \times C - 8.45 \times A^2 - 9.37 \times B^2 - 11.09 \times C^2 \quad (4)$$

where A is the protein concentration, B is the oil phase mass fraction, and C is the extract mass fraction.

Table 4 demonstrates that the model's p-value is below 0.0001, signifying statistical significance. Conversely, the lack-of-fit item has a p-value of 0.9364, denoting an insignificant level, which suggests the validity of the response surface model. The quadratic model obtained was fitted with the data for the responses. Figure 3 illustrates the normal probability plot residuals for the entrapping efficiency. The majority of the results align closely with a straight trajectory, insinuating a normal distribution of the outcomes and their compatibility with the regression model. Consequently, this response surface model adeptly captures the correlation between diverse parameters and the response metric, rendering it suitable for guiding the LBPN preparation phase process. Figure 4, wherein two factors are held constant to assess the interplay of the remaining factors, evidences that maintaining one factor static causes variations in the other two factors, affecting the entrapping efficiency. This negates the possibility of a mere linear correlation between factors and entrapping efficiency. As a result, the response surface methodology was employed to fine tune and discern the paramount conditions for LBPN synthesis. This optimization revealed that at a protein concentration of 4.03%, oil phase mass fraction of 19.50%, and extract mass fraction of 9.22%, the zenith of entrapping efficiency was achieved at 73.4329%. For ensuing studies, given the experiment's feasibility, the conditions were slightly adjusted, i.e., protein concentration to 4%, oil phase mass fraction to 20%, and extract mass fraction to 10%. The entrapping efficiency of the revised LBPN closely mirrored the predictions from the response surface model, cementing the model's precision and dependability. Previous studies have utilized maltodextrin and glucose microcapsules as carriers to prepare lingonberry microcapsules, achieving microencapsulation rates of 79–81% [34].

Table 4. Variance and significant analysis of the Box–Behnken design test.

Source	Sum of Squares	df	Mean Square	F-Value	p-Value
Model	1482.84	9	164.76	555.31	<0.0001
A	2.11	1	2.11	7.12	0.0321
B	13.13	1	13.13	44.26	0.0003
C	74.42	1	74.42	250.83	<0.0001
AB	1.7	1	1.70	5.74	0.0478

Table 4. *Cont.*

Source	Sum of Squares	df	Mean Square	*F*-Value	*p*-Value
AC	1.04	1	1.04	3.51	0.1033
BC	64.16	1	64.16	216.25	<0.0001
A^2	300.80	1	300.8	1013.82	<0.0001
B^2	369.85	1	369.85	1246.54	<0.0001
C^2	518.29	1	518.29	1746.84	<0.0001
Residual	2.08	7	0.30		
Lack of Fit	0.19	3	0.062	0.13	0.9364
Pure Error	1.89	4	0.47		
Cor Total	1484.92	16			

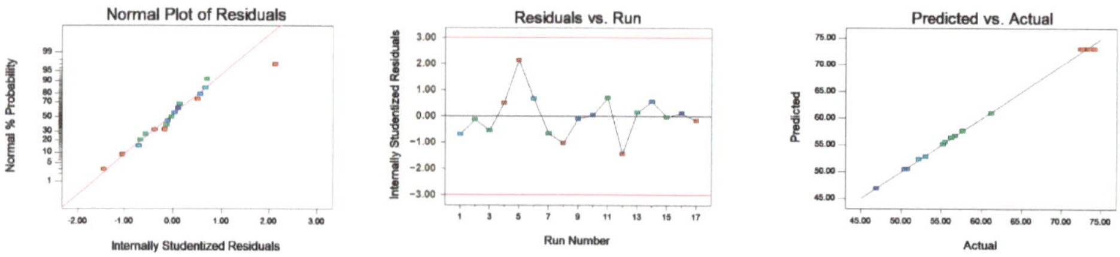

Figure 3. Normal probability plot residual for entrapping efficiency.

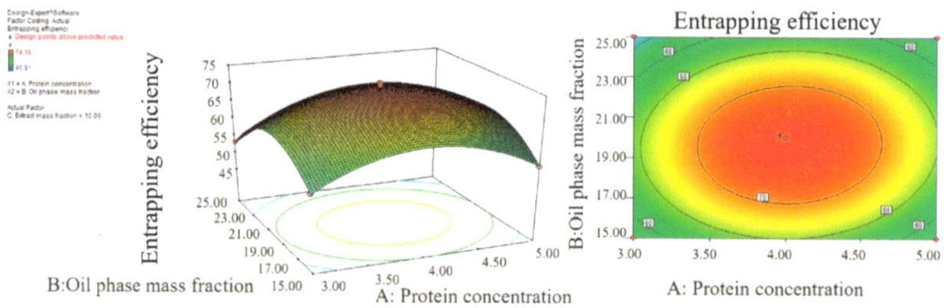

(a) Protein concentration and oil phase mass fraction (extract mass fraction was 10%)

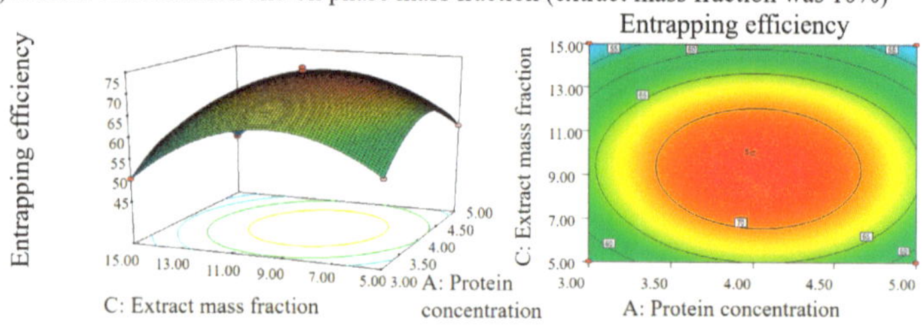

(b) Protein concentration and extract mass fraction (oil phase fraction was 20%)

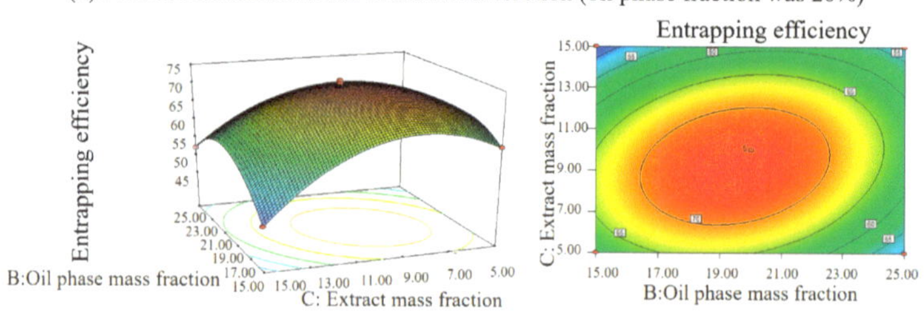

(c) Oil phase mass fraction and extract mass fraction (protein concentration was 4%)

Figure 4. Effect of two factors on the entrapping efficiency of the LBPN.

3.3. Characterization of the Optimal Lingonberry Nanoemulsion

3.3.1. Particle Size and PDI

The LBPN was prepared with the formulation optimized using the response surface methodology, and the entrapping efficiency of the LBPN was close to 73%, with 114.52 ± 0.015 nm and 0.119 ± 0.065 for average particle size and PDI, respectively. As depicted in Figure 5, the average droplet dimension of the LBPN formulation was observed to be within the nanometric domain (<500 nm). In terms of the PDI, a PDI metric below 0.2 signifies that the particles within the nanoemulsion exhibit monodispersity [35].

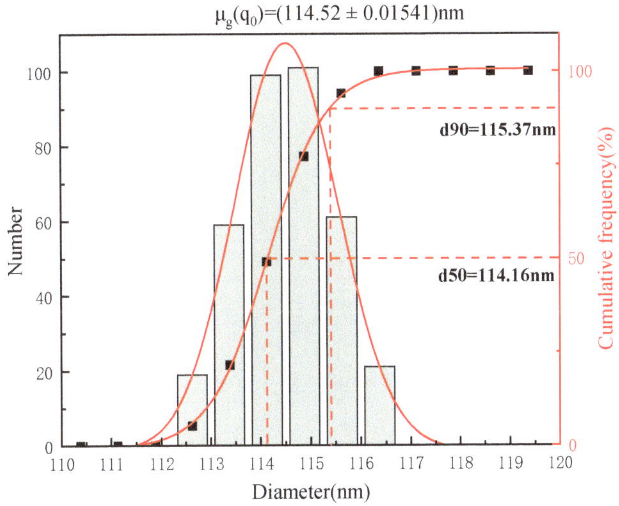

Figure 5. Particle size distribution of the optimally formulated LBPN.

3.3.2. Zeta Potential

Figure 6 reveals that the optimally developed LBPN demonstrated a zeta potential of −23.29 ± 0.4471 mV. The negative zeta potential values for the optimized samples suggest that they are negatively charged, contributing to their physical stability. Zeta potential is integral to colloidal dispersion stability as it reflects the magnitude of repulsion among adjacent particles. Enhanced electrostatic repulsion among the droplets in this context prevents the coalescence of the emulsion particles, thereby stabilizing the system [36].

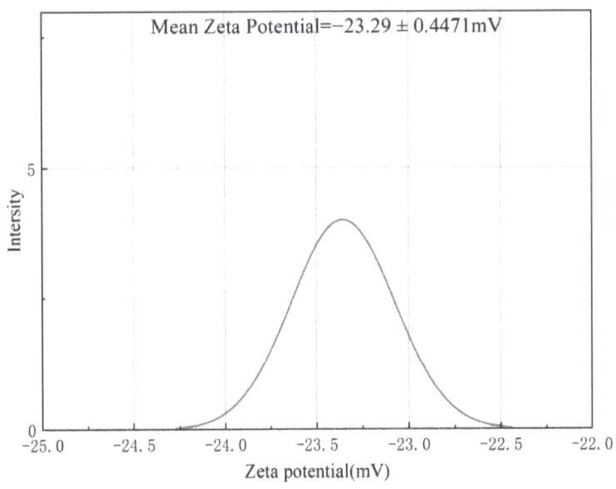

Figure 6. ζ-potential of the optimally formulated LBPN.

3.3.3. Confocal Laser Scanning Microscopy (CLSM)

The microstructure of LBPN is depicted in Figure 7 as shown. The protein and oil phases were labeled with Nile blue and Nile red, respectively, and the SPI/WPI mixed protein phase was green in the image, while the MCT was red. From CLSM imaging, it can be seen that the LBPN emulsion particles are in the shape of regular spheres, with each droplet dispersed in rows, and the mixed proteins wrapped around the oil droplets to form an oil-in-water emulsion structure. It has been shown that fresh emulsions prepared using ultrasonic emulsification have very fine and uniformly dispersed oil droplets, demonstrating that ultrasound is an effective way to prepare emulsions. The high overlap between the green protein phase and the red oil phase proved that the hybrid protein phase of the WPI/SPI was effectively combined with the MCT, indicating good emulsifying ability of the hybrid emulsifier WPI/SPI [37].

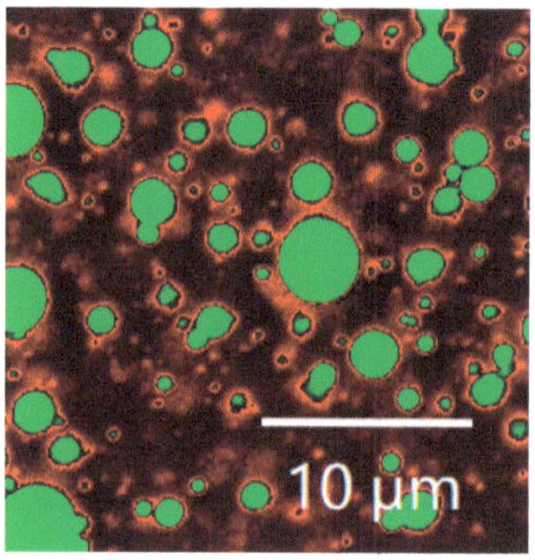

Figure 7. Confocal laser scanning microscopy image of the optimally formulated LBPN.

3.3.4. Antioxidant Activity

The graphic (Figure 8) representation clearly indicates that nanoemulsification substantially augments the DPPH antioxidant activity of LBPs. The ability of the LBPN to quench the DPPH radical was evident through the transition from purple to yellow, a change that was further validated by the observed decrease in absorbance at 515 nm. As depicted in Figure 8, there was a pronounced decrease in the DPPH radical, attributed to the scavenging capacity of the resveratrol solution, the nanoemulsion, and the benchmark antioxidant, ascorbic acid. The percentage inhibition obtained was maximum for ascorbic acid (85.43% $\pm$ 2.85%) compared to the LBPN (79.37% $\pm$ 3.18%) and LBPs (63.58% $\pm$ 2.95%). The utilization of nanoscale emulsions for encapsulating polyphenols greatly enhances their solubility, thus optimizing the antioxidative properties of polyphenols. Research has shown that the preparation of resveratrol as a resveratrol nanoscale emulsion resulted in a DPPH inhibition rate as high as 83.93 $\pm$ 3.81%, an increase of nearly 13% compared to non-encapsulated resveratrol solutions [15]. Kumar et al. [11] employed lecithin and Tween-80 as emulsifiers to prepare a resveratrol-loaded nanoscale emulsion, which remained stable without phase separation even after four months of storage. Under ultraviolet radiation exposure, the nanoscale emulsion significantly slowed down and inhibited the degradation of resveratrol.

Regarding the ABTS assay, the antioxidant activity of the LBPN was markedly superior to that of LBPs. The ABTS assay assesses antioxidant capacity via a single electron transfer mechanism, gauging the reduction of the ABTS radical cation. The incorporation of whey protein isolate (WPI) as a natural emulsifier in the formulation likely enhanced the LBPN's antioxidant properties. The presence of hydrophobic amino acids like alanine, isoleucine, valine, and those with aromatic side chains in the WPI plays a pivotal role in the radical scavenging activities of the protein. Additionally, the homogenization process may have facilitated a more effective dispersion of LBPs within the nanoemulsion, potentially elevating its antioxidant efficacy [38].

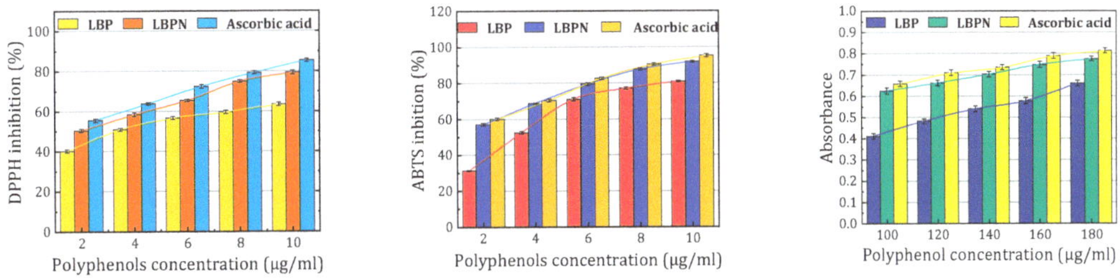

Figure 8. Comparison of antioxidant properties of the LBPN, LBPs, and ascorbic acid.

The reduction capability of both the drug solution and the resveratrol nanoemulsion was benchmarked against ascorbic acid (standard). The reductive ability of the LBPN was evaluated based on the direct reduction of $Fe[(CN)_6]_3$ to $Fe[(CN)_6]_2$. Such reducing properties of a compound typically indicate its antioxidant potential, suggesting that the LBPN may serve as an effective antioxidant. Notably, as depicted in Figure 8, the reducing strength of the LBPN surpassed that of LBPs.

4. Conclusions and Future Research

In conclusion, this study successfully developed a lingonberry polyphenol extract-loaded oil-in-water nanoemulsion using a response surface optimization method. The nanoemulsion exhibited small droplet size, good stability, and enhanced antioxidant activity compared to the extract. These findings contribute to the understanding of the formulation and optimization of nanoemulsions for the delivery of bioactive compounds.

Moving forward, further research can explore the potential applications of this lingonberry nanoemulsion in the food and pharmaceutical industries. The optimized formulation can be utilized to enhance the stability and bioavailability of bioactive compounds, leading to improved functional food products and nutraceuticals. Additionally, this study can be extended to investigate the in vivo efficacy and safety of the nanoemulsion, providing valuable insights for future clinical applications.

Furthermore, it is important to consider the practical implications of this research. The development of a stable and bioactive nanoemulsion opens up opportunities for the utilization of lingonberry polyphenols in various products, such as functional beverages, dietary supplements, and topical formulations. However, it is crucial to address the potential limitations and challenges associated with the scale-up and commercialization of the nanoemulsion, including cost-effectiveness, regulatory considerations, and consumer acceptance. Nanocarriers have been primarily used in drug delivery systems and the development of functional foods. It is, therefore, essential to delve deeper into studies concerning their toxicity and underlying mechanisms of any toxic effects, in order to determine the permissible daily intake of nanoemulsions.The absorption and metabolic processes of nanoscale carriers differ from those of microscale carriers. Currently, the potential toxicity of nanoscale emulsions upon absorption by the human body remains unclear and requires further in-depth investigation, such as a systematic evaluation of the potential risks of the LBPN through toxicological experiments, including clinical studies.

In summary, this study lays the foundation for the application of lingonberry polyphenols in nanoemulsion formulations, highlighting their potential for improving stability, bioavailability, and antioxidant activity. Further research and development efforts are warranted to fully explore the practical applications and benefits of this technology in the food and pharmaceutical industries.

Author Contributions: Conceptualization, S.W. and Z.F.; methodology, S.W., J.W. and Z.F.; software, S.W. and Y.C.; validation, S.W., J.W., M.D. and Z.F.; formal analysis, S.W.; investigation, S.W., Y.C., J.W., M.D. and Z.F.; data curation, S.W.; writing—original draft preparation, S.W., Y.C., M.D. and Z.F.; writing—review and editing, S.W., Y.C., M.D., J.W. and Z.F.; visualization, S.W., Y.C., and J.W.; supervision, S.W., Y.C., M.D., J.W. and Z.F. All authors have read and agreed to the published version of the manuscript.

Funding: This work was supported by the National Natural Science Foundation of China (31170510), the National College Students' innovation and entrepreneurship training program (202210225104), the National Science Foundation of Heilongjiang Province, China (LH 2020C035), the Fundamental Research Funds for the Central Universities (2572019BA09), and the China Postdoctoral Science Foundation Project (2016M600239) for their support. The authors thank these institutions for their contributions to this study.

Data Availability Statement: The datasets generated for this study are available on request to the corresponding author.

Conflicts of Interest: The authors declare no conflict of interest.

References

1. Plunkett, B.J.; Espley, R.V.; Dare, A.P.; Warren, B.W.; Grierson, E.R.P.; Cordiner, S.; Turner, J.L.; Allan, A.C.; Albert, N.W.; Davies, K.M.; et al. MYBA From Blueberry (*Vaccinium Section Cyanococcus*) Is a Subgroup 6 Type R2R3MYB Transcription Factor That Activates Anthocyanin Production. *Front. Plant Sci.* **2018**, *9*, 1300. [CrossRef] [PubMed]
2. Bujor, O.C.; Ginies, C.; Popa, V.I.; Dufour, C. Phenolic compounds and antioxidant activity of lingonberry (*Vaccinium vitis-idaea* L.) leaf, stem and fruit at different harvest periods. *Food Chem.* **2018**, *252*, 356–365. [CrossRef] [PubMed]
3. Bujor, O.C.; Tanase, C.; Popa, M.E. Phenolic Antioxidants in Aerial Parts of Wild Vaccinium Species: Towards Pharmaceutical and Biological Properties. *Antioxidants* **2019**, *8*, 649. [CrossRef] [PubMed]
4. Vilkickyte, G.; Petrikaite, V.; Pukalskas, A.; Sipailiene, A.; Raudone, L. Exploring Vaccinium vitis-idaea L. as a potential source of therapeutic agents: Antimicrobial, antioxidant, and anti-inflammatory activities of extracts and fractions. *J. Ethnopharmacol.* **2022**, *292*, 115207. [CrossRef]
5. Xu, J.; Yang, H.; Nie, C.D.; Wang, T.; Qin, X.Y.; Yang, J.; Chang, Y.H.; Nie, S.M.; Fu, Y.J. Comprehensive phytochemical analysis of lingonberry (*Vaccinium vitis-idaea* L.) from different regions of China and their potential antioxidant and antiproliferative activities. *RSC Adv.* **2023**, *13*, 29438–29449. [CrossRef]
6. Vyas, P.; Curran, N.H.; Igamberdiev, A.U.; Debnath, S.C. Antioxidant properties of lingonberry (*Vaccinium vitis idaea* L.) leaves within a set of wild clones and cultivars. *Can. J. Plant Sci.* **2015**, *95*, 663–669. [CrossRef]
7. Debnath, S.C.; Siow, Y.L.; Petkau, J.; An, D.; Bykova, N.V. Molecular markers and antioxidant activity in berry crops: Genetic diversity analysis. *Can. J. Plant Sci.* **2012**, *92*, 1121–1133. [CrossRef]
8. Foley, S.L.; Debnath, S.C. Influence of in vitro and ex vitro propagation on anthocyanin content and anti-oxidant activity of lingonberries. *J. Hortic. Sci. Biotechnol.* **2007**, *82*, 114–118. [CrossRef]
9. Stefanescu, B.E.; Szabo, K.; Mocan, A.; Crisan, G. Phenolic Compounds from Five Ericaceae Species Leaves and Their Related Bioavailability and Health Benefits. *Molecules* **2019**, *24*, 2046. [CrossRef]
10. Li, J.; Hwang, I.C.; Chen, X.; Park, H.J. Effects of chitosan coating on curcumin loaded nano-emulsion: Study on stability and in vitro digestibility. *Food Hydrocoll.* **2016**, *60*, 138–147. [CrossRef]
11. Kumar, R.; Kaur, K.; Uppal, S.; Mehta, S.K. Ultrasound processed nanoemulsion: A comparative approach between resveratrol and resveratrol cyclodextrin inclusion complex to study its binding interactions, antioxidant activity and UV light stability. *Ultrason. Sonochem.* **2017**, *37*, 478–489. [CrossRef] [PubMed]
12. Zhou, H.L.; Zheng, B.J.; McClements, D.J. Encapsulation of lipophilic polyphenols in plant-based nanoemulsions: Impact of carrier oil on lipid digestion and curcumin, resveratrol and quercetin bioaccessibility. *Food Funct.* **2021**, *12*, 3420–3432. [CrossRef] [PubMed]
13. Pool, H.; Mendoza, S.; Xiao, H.; McClements, D.J. Encapsulation and release of hydrophobic bioactive components in nanoemulsion-based delivery systems: Impact of physical form on quercetin bioaccessibility. *Food Funct.* **2013**, *4*, 162–174. [CrossRef]
14. Huang, H.; Belwal, T.; Liu, S.B.; Duan, Z.H.; Luo, Z.S. Novel multi-phase nano-emulsion preparation for co-loading hydrophilic arbutin and hydrophobic coumaric acid using hydrocolloids. *Food Hydrocoll.* **2019**, *93*, 92–101. [CrossRef]
15. Sharma, B.; Iqbal, B.; Kumar, S.; Ali, J.; Baboota, S. Resveratrol-loaded nanoemulsion gel system to ameliorate UV-induced oxidative skin damage: From in vitro to in vivo investigation of antioxidant activity enhancement. *Arch. Dermatol. Res.* **2019**, *311*, 773–793. [CrossRef]
16. Ruengdech, A.; Siripatrawan, U. Application of catechin nanoencapsulation with enhanced antioxidant activity in high pressure processed catechin-fortified coconut milk. *LWT-Food Sci. Technol.* **2021**, *140*, 110594. [CrossRef]
17. Pongsumpun, P.; Iwamoto, S.; Siripatrawan, U. Response surface methodology for optimization of cinnamon essential oil nanoemulsion with improved stability and antifungal activity. *Ultrason. Sonochem.* **2020**, *60*, 104604. [CrossRef]

18. Niknam, S.M.; Kashaninejad, M.; Escudero, I.; Sanz, M.T.; Beltran, S.; Benito, J.M. Preparation of Water-in-Oil Nanoemulsions Loaded with Phenolic-Rich Olive Cake Extract Using Response Surface Methodology Approach. *Foods* **2022**, *11*, 279. [CrossRef]
19. Fan, Z.; Chen, K.; Liu, Y.; Yang, L.; Fu, Y. Correlation between Antioxidant Activity in Vitro and Active Components of Different Solvent Extracts from Lingonberr. *Food Sci.* **2017**, *38*, 138–144.
20. Tian, Y.; Puganen, A.; Alakomi, H.L.; Uusitupa, A.; Saarela, M.; Yang, B.R. Antioxidative and antibacterial activities of aqueous ethanol extracts of berries, leaves, and branches of berry plants. *Food Res. Int.* **2018**, *106*, 291–303. [CrossRef]
21. Raviadaran, R.; Chandran, D.; Shin, L.H.; Manickam, S. Optimization of palm oil in water nano-emulsion with curcumin using microfluidizer and response surface methodology. *LWT-Food Sci. Technol.* **2018**, *96*, 58–65. [CrossRef]
22. Yazdan-Bakhsh, M.; Nasr-Esfahani, M.; Esmaeilzadeh-Kenari, R.; Fazel-Najafabadi, M. Evaluation of antioxidant properties ofHeracleum Lasiopetalumextract in multilayer nanoemulsion with biopolymer coating to control oxidative stability of sunflower oil. *J. Food Meas. Charact.* **2021**, *15*, 1014–1023. [CrossRef]
23. Wang, D.Q.; Qi, B.K.; Xu, Q.Q.; Zhang, S.; Xie, F.Y.; Li, Y. Effect of salt ions on an ultrasonically modified soybean lipophilic protein nanoemulsion. *Int. J. Food Sci. Technol.* **2021**, *56*, 6719–6731. [CrossRef]
24. Nikkhah, E.; Khayami, M.; Heidari, R. In vitro antioxidant activity of berry (Morus alba var.nigra). *Int. J. Plant Prod.* **2009**, *3*, 15–18.
25. Drozdz, P.; Seziene, V.; Pyrzynska, K. Phytochemical Properties and Antioxidant Activities of Extracts from Wild Blueberries and Lingonberries. *Plant Foods Hum. Nutr.* **2017**, *72*, 360–364. [CrossRef] [PubMed]
26. Ştefanescu, B.-E.; Calinoiu, L.F.; Ranga, F.; Fetea, F.; Mocan, A.; Vodnar, D.C.; Crișan, G. Chemical Composition and Biological Activities of the Nord-West Romanian Wild Bilberry (*Vaccinium myrtillus* L.) and Lingonberry (*Vaccinium vitis-idaea* L.) Leaves. *Antioxidants* **2020**, *9*, 495. [CrossRef] [PubMed]
27. Hajazimi, E.; Landberg, R.; Zamaratskaia, G. Simultaneous determination of flavonols and phenolic acids by HPLC-CoulArray in berries common in the Nordic diet. *LWT-Food Sci. Technol.* **2016**, *74*, 128–134. [CrossRef]
28. Vilkickyte, G.; Motiekaityte, V.; Vainoriene, R.; Raudone, L. Promising cultivars and intraspecific taxa of lingonberries (*Vaccinium vitis-idaea* L.): Profiling of phenolics and triterpenoids. *J. Food Compos. Anal.* **2022**, *114*, 104796. [CrossRef]
29. Zhang, D.; Adelina, N.M.; Fan, Z.L.; Liu, J.R. Phytochemical profile and biological activities from different parts of Vaccinium vitis-idaea. *J. Berry Res.* **2022**, *12*, 445–462. [CrossRef]
30. Smiljkovic, M.; Stanisavljevic, D.; Stojkovic, D.; Petrovic, I.; Vicentic, J.M.; Popovic, J.; Grdadolnik, S.G.; Markovic, D.; Sankovic-Babic, S.; Glamoclija, J.; et al. Apigenin-7-o-glucoside versus apigenin: Insight into the modes of anticandidal and cytotoxic actions. *Excli J.* **2017**, *16*, 795–807. [CrossRef]
31. Zielinska, A.; Bryk, D.; Paradowska, K.; Wawer, I. Aronia melanocarpa Leaves as a Source of Chlorogenic Acids, Anthocyanins, and Sorbitol, and Their Anti-Inflamtory Activity. *Pol. J. Food Nutr. Sci.* **2020**, *70*, 409–418. [CrossRef]
32. Sato, Y.; Itagaki, S.; Kurokawa, T.; Ogura, J.; Kobayashi, M.; Hirano, T.; Sugawara, M.; Iseki, K. In vitro and in vivo antioxidant properties of chlorogenic acid and caffeic acid. *Int. J. Pharm.* **2011**, *403*, 136–138. [CrossRef]
33. Kostka, T.; Ostberg-Potthoff, J.J.; Stärke, J.; Guigas, C.; Matsugo, S.; Stojanov, L.; Velickovska, S.K.; Winterhalter, P.; Esatbeyoglu, T.; Mireeski, V. Bioactive Phenolic Compounds from Lingonberry (*Vaccinium vitis-idaea* L.): Extraction, Chemical Characterization, Fractionation and Cellular Antioxidant Activity. *Antioxidants* **2022**, *11*, 467. [CrossRef] [PubMed]
34. Stefanescu, B.E.; Nemes, S.A.; Teleky, B.E.; Calinoiu, L.F.; Mitrea, L.; Martau, G.A.; Szabo, K.; Mihai, M.; Vodnar, D.C.; Crisan, G. Microencapsulation and Bioaccessibility of Phenolic Compounds of *Vaccinium Leaf* Extracts. *Antioxidants* **2022**, *11*, 674. [CrossRef] [PubMed]
35. Izquierdo, P.; Feng, J.; Esquena, J.; Tadros, T.F.; Dederen, J.C.; Garcia, M.J.; Azemar, N.; Solans, C. The influence of surfactant mixing ratio on nano-emulsion formation by the pit method. *J. Colloid Interface Sci.* **2005**, *285*, 388–394. [CrossRef]
36. Chen, L.; Lin, X.J.; Xu, X.W.; Chen, Y.; Li, K.; Fan, X.Y.; Pang, J.; Teng, H. Self-nano-emulsifying formulation of Sonchus oleraceus Linn for improved stability: Implications for phenolics degradation under in vitro gastro-intestinal digestion Food grade drug delivery system for crude extract but not single compound. *J. Funct. Foods* **2019**, *53*, 28–35. [CrossRef]
37. Ye, H.X.; Chen, T.S.; Huang, M.; Ren, G.R.; Lei, Q.F.; Fang, W.J.; Xie, H.J. Exploration of the Microstructure and Rheological Properties of Sodium Alginate-Pectin-Whey Protein Isolate Stabilized Β-Carotene Emulsions: To Improve Stability and Achieve Gastrointestinal Sustained Release. *Foods* **2021**, *10*, 1991. [CrossRef] [PubMed]
38. Feng, Y.Y.; Yuan, D.X.; Cao, C.A.; Kong, B.H.; Sun, F.D.; Xia, X.F.; Liu, Q. Changes of in vitro digestion rate and antioxidant activity of digestion products of ethanol-modified whey protein isolates. *Food Hydrocoll.* **2022**, *131*, 107756. [CrossRef]

foods

MDPI

Article

Physicochemical Quantitative Analysis of the Oil–Water Interface as Affected by the Mutual Interactions between Pea Protein Isolate and Mono- and Diglycerides

Ziyan Wang [1,2], Jingwen Li [3], Chao Peng [3], Bin Li [4], Qian Shen [1,*] and Yijie Chen [4,*]

1 Hubei Key Laboratory for Processing and Transformation of Agricultural Products and College of Food Science and Engineering, Wuhan Polytechnic University, Wuhan 430023, China; wzymelody@163.com
2 School of Food Science and Nutrition, University of Leeds, Leeds LS2 9JT, UK
3 National Facility for Protein Science in Shanghai, Shanghai Advanced Research Institute, Chinese Academy of Sciences, Shanghai 201210, China; lijw@sari.ac.cn (J.L.); pengchao@sari.ac.cn (C.P.)
4 College of Food Science and Technology, Huazhong Agricultural University, Wuhan 430070, China; libinfood@mail.hzau.edu.cn
* Correspondence: shenqian@whpu.edu.cn (Q.S.); yijiechen@mail.hzau.edu.cn (Y.C.)

Abstract: As a commercially available ingredient, the mono- and diglycerides (MDG) were widely used in a plant protein-based emulsion to provide effective, functional, emulsifying properties. The simultaneous addition of the MDG and pea protein isolate (PPI) was investigated by the methods of interfacial rheology and quantitative protein proteomics. The physicochemical quantitative analysis of the oil–water interface revealed an interfacial stability mechanism for the protein adsorption layer. For a low MDG concentration, the interfacial quantities of vicilin and albumin were increased, which could be attributed to the adsorption rate. For a high MDG concentration, both vicilin and albumin were displaced by MDG and desorbed from the interface, while legumin was more difficult to displace due to its slow adsorption and the complex structure of protein molecules. The protein molecules with the structural rearrangement interacted with MDG, exhibiting potential effects on the interfacial film structure. Combined with some nanotechnologies, the new comprehension of protein-emulsifier interactions may promote food delivery systems. The research aims to develop an in-depth analysis of interfacial proteins, and provide more innovative and tailored functionalities for the application of the plant protein emulsion.

Keywords: pea protein isolate; mono- and diglycerides; competition adsorption; quantitative protein proteomics

Citation: Wang, Z.; Li, J.; Peng, C.; Li, B.; Shen, Q.; Chen, Y. Physicochemical Quantitative Analysis of the Oil–Water Interface as Affected by the Mutual Interactions between Pea Protein Isolate and Mono- and Diglycerides. *Foods* **2024**, *13*, 176. https://doi.org/10.3390/foods13010176

Academic Editors: Shuai Chen and Qing Guo

Received: 30 November 2023
Revised: 26 December 2023
Accepted: 28 December 2023
Published: 4 January 2024

1. Introduction

In the food processing industry, many emulsifiers of drink products consisted of proteins and low molecular weight surfactants. This kind of application aims for a long-term stability and satisfactory shelf-life. As one of the typical plant proteins, the pea protein is specifically characterized by its hypoallergenic and nutritional properties. In our previous studies, the pea protein isolate (PPI) showed a prominent emulsifying and foaming ability [1,2]. Considering the interface stability in the long term, synthesized emulsifiers were adopted in the formulas, such as protein and low molecular weight (LMW) surfactants. The mono- and diglycerides of fatty acids commonly named E471 (MDG, in European Union regulation) as food additives were widely applied in foods, such as bread, cream, and drinks [3]. In recent years, the mono- and diglycerides (MDG) were re-evaluated to add in foods for infants and all population groups [4,5]. MDG was one of the nonionic surfactants and dissolved in the oil phase. In the industry, as a commercially available ingredient, MDG was used in petrolatum-based topical emulsions to provide effective, bifunctional, and emulsifying properties [6]. Since the multiple components in food systems, proteins appear with a variety of low molecular weight (LMW) surfactants to stabilize food emulsions or foams. Among these, proteins reacted by forming a viscoelastic adsorbent layer on the surface of the oil droplets, thereby forming a physical barrier to avoid droplet coalescence. As for the physicochemical properties, MDG was applied as an emulsifier due to its polar duality within the molecular structure. The glycerin corresponded to the hydrophilic part, and the aliphatic chains formed the lipophilic counterpart. These properties imparted multiple, effective MDG features for emulsion and foam stabilization [7]. In practical food applications, there are usually mixtures of proteins and LMW surfactants competing for the interfacial region. The LMW surfactants may disrupt the viscoelastic protein adsorbent layer, leading to a reduction in the interfacial stability. In order to avoid the dynamic instability, MDG was added in the formulation step due to its ability to adsorb certain plant-based products at the interface with good efficiency. Therefore, different interface stabilization mechanisms between proteins and LMW surfactants occurred based on complex interface behaviors.

Generally, most low molecular weight (LMW) surfactants could reduce interfacial tension effectively with their rapid adsorption rate [8]. The protein component may form a highly viscoelastic interface network around oil droplets. In theory, the simultaneous addition of PPI and MDG could provide steric hindrance or electrostatic repulsions, preventing coalescence and flocculation between oil droplets, and cooperatively improving the long-term stability of the emulsion [9]. During the process, the protein displacement caused by the LMW surfactants might influence the interfacial stability. Some studies have shown that viscoelastic properties of interfacial films formed by sodium caseinate were affected by the MDG addition, suggesting that the ratio between sodium caseinate and MDG played an essential and critical role in the emulsion [10,11]. Therefore, the displacement of interfacial protein caused by the protein level may consequently change the steric and electrostatic repulsive forces or the conformational properties at the protein-adsorbed layer.

In general, the interfacial properties of protein-surfactant mixed films were investigated by the drop volume tensiometer and multiple microscopy technology [12,13]. It was reported that MDG was added with the phospholipids to promote the stability of the oil–water interface [14]. Generally, these two types of emulsifiers have been widely combined in the food formulas, which could significantly improve the quality and shelf-life of butter [15]. The critical value of the surfactant volume fraction and its associated relaxation time was determined for the emulsion rheological characterization [16]. The formation of the interfacial film was associated with the interactions between film-forming components and the protein displacement by the surfactant. Moreover, this study aimed to investigate the emulsifying stability of PPI-MDG, and the concentration effect of MDG on the interfacial protein by the semi-quantitative method (sodium dodecyl sulfate-polyacrylamide gel electrophoresis, SDS-PAGE). The effect of the MDG addition on the interfacial properties was further analyzed by the quantitative protein proteomics, involving the protein displacement and consequent conformational change.

In this study, we investigated the adsorption behaviors of PPI and MDG to explore the protein quantitative relationships at the oil–water interface. The main point focused on the competitive adsorption of the interface film in order to reveal the stabilization mechanism of food surfactants in applications. Moreover, the nonlinear interfacial rheology was determined to reveal the film interactions. With the stabilizing effects of the PPI-MDG in the interfacial system, the study further provided the theoretical supports for the formulation optimization of plant protein-based dairy products. Moreover, the results were of great significance in the manufacture and consumption development.

2. Materials and Methods

2.1. Materials

PPI and its major component proteins (vicilin, legumin, and albumin) were prepared from the pea protein powder (Yantai, Shandong, China) according to the previous operations [17,18]. MDG was provided by Aladdin Reagent Company and dissolved in the oil phase. The reagents used for the mass spectrometry experiment were from Sigma–Aldrich Co (Shanghai, China). All other reagents were of an analytical reagent grade (AR grade).

2.2. Emulsion Preparation

The pea protein isolate (PPI) was dissolved in deionized water at a concentration range of 0.1–1% (*w*/*v*). Then, MDG was added once in the medium chain triglycerides (MCT) before the stirring process at 50 °C for 10 min. The protein solution and oil phase mixed at 9:1 was pre-emulsified with a high-speed disperser at 8000 r/min for 2 min. Finally, the emulsion was prepared by the high-pressure homogenizer at 40 MPa for 3 cycles (almost 2 min).

2.3. Particle Size

The particle size of emulsion droplets was measured by the Malvern Mastersizer 2000 (Worcestershire, UK) with the laser diffraction. Due to the wet dispersion, the refractive indices were 1.45 for the droplets and 1.33 for the dispersant. The volume-weighted mean diameter (D_{43}) was obtained through a particle size distribution with three repeated measurements.

2.4. The Percentage of Adsorbed Proteins (AP)

The percentage of adsorbed proteins (AP) was determined according to the relevant methods [19]. The emulsion (8 mL) was centrifuged in 15 mL centrifuge tubes (10,000× *g*, 30 min, 4 °C). After centrifugation, the supernatant was collected by the needle filter at 0.45 μm. Then, the protein concentration in the supernatant (C_S) and the initial protein solution (C_P) was measured with the bicinchoninic acid (BCA) protein assay. The percentage of adsorbed proteins was calculated:

$$\text{Percentage of adsorbed proteins}(\%) = \frac{C_p - C_s}{C_p} \times 100 \quad (1)$$

2.5. SDS-PAGE

The SDS-PAGE profile of interfacial proteins was determined to show the protein subunit changes at different concentrations. After the centrifugation at 10,000× *g* for 30 min, the upper layer of the cream was collected and then dissolved in a phosphate buffer. Protein samples were mixed with the β-mercaptoethanol (β-ME) at 5:1 (*v*/*v*) for an equal concentration, and then subjected to a boiling water bath for 5 min. In addition, the stacking gel and the resolving gel contained 5% and 12% acrylamide, respectively. The constant voltage during the gel electrophoresis was set to 100 V. Finally, the gel was scanned and imaged using a gel imaging analysis system (Universal Hood II, Bio-Rad, Hercules, CA, USA).

2.6. Interface Rheological Properties

2.6.1. Interfacial Tension

The interfacial tension at the oil–water interface stabilized by proteins and MDG was determined using an interfacial rheometer (TECLIS, Civrieux-d'Azergues, France). Firstly, the tip of the syringe was immersed into the cuvette containing the protein solution, which were placed on an optical platform between the light source and the high-speed charge-coupled camera [20]. An automated motor device was used to form a droplet at the syringe tip. Meanwhile, the camera was connected to the video screen for a real-time image. The interfacial tension was measured by fitting a model to the droplet shape. Thus, the reduction in the oil–water interfacial tension, caused by adding proteins and MDG, could be analyzed. The whole period was 2–3 h from the adsorption of proteins and MDG to the interfacial stabilization. During the process, the measuring temperature was 25 °C.

2.6.2. Amplitude Sweep

The dilatational viscoelastic modulus was measured using the above method at a 0.1 Hz frequency with a 10% amplitude [21]. After the stable adsorption for 3 h, the linear or nonlinear viscoelastic region of the protein and MDG at the oil–water interface was determined through the amplitude sweep mode. This mode was measured with an amplitude range of 10–30% at a 0.1 Hz frequency. The experiment was performed in triplicate.

2.6.3. Lissajous Plots

According to the related methods, the analysis of Lissajous curves was to reflect the nonlinear rheological response at the interface [22,23]. The oscillation process contained two different changes (expansion and compression) for three active periods and two balanced periods. Therefore, the relationship of interfacial pressure (π) versus deformation (A) was considered to reveal the properties of interfacial film.

$$\Delta\pi = \pi - \pi_0, \Delta A = (A - A_0)/A_0 \tag{2}$$

where the π and A was the droplet's interfacial pressure and interface area at a certain time point, respectively. Accordingly, the π_0 and A_0 was the initial value.

During the measurement, the interface was compressed and extended in a sinusoidal process. In the extension part of the cycle, the dilatational moduli at the largest extension (E_{LE}) and the minimum extension (E_{ME}) were determined, respectively. The S-factor (S_E) during the extension was calculated according to the following formula:

$$S_E = (E_{LE} - E_{ME})/E_{LE} \tag{3}$$

In addition, in the compression part of the cycle, the dilatational moduli at the largest compression (E_{LC}) and the minimum compression (E_{MC}) were used for the calculation of the S-factor (S_C).

$$S_C = (E_{LC} - E_{MC})/E_{LC} \tag{4}$$

2.7. Interfacial Quantitation of Adsorbed Proteins

Regarding the quantitative analysis of the interfacial protein, the most commonly adopted method was the stable isotope dimethyl labeling [1]. The protein samples were firstly digested by trypsin, then labeled by dimethyl (0, +4, +8) quantitation and desalted by a BAKERBOND C18 column. Through these treatments, interfacial protein samples from a 0.5% PPI-prepared emulsion were set as light labeling (0, Light); interfacial protein samples from a 0.5% PPI-0.1% MDG-prepared emulsion were set as intermediate labeling (+4, Intermediate); interfacial protein samples from a 0.5% PPI-0.5% MDG-prepared emulsion were set as heavy labeling (+8, Heavy). Then, samples (L, M, and H) were detected to compare the quantitative difference. The peptide abundance was performed by a mixed quadrupole-TOF LC-MS/MS mass spectrometer (TripleTOF 5600+, AB Sciex, Shimadzu, Kyoto, Japan) equipped with a sodium stream spray ion source, as described in a previous work [1,24]. Thus, the relative abundance of protein was collected from the ratio of the apex signal intensity of survey scan.

2.8. Statistical Analysis

The results by independent trials were presented as the means ± standard deviations (SD). All treatments were replicated 3 times unless there was a special statement. An analysis of variance (ANOVA) of experiment data was conducted with SPSS 26.0 Statistics. The statistical differences were set at a significant standard ($p < 0.05$).

3. Results and Discussion

3.1. The Effect of MDG Addition on the Interfacial Adsorption

Generally, MDG was separated from diglycerides and triglycerides through membrane distillation, and used in bread or margarine for long-term storage. PPI used with MDG was further investigated by an emulsion stability analysis to explore the application of the protein and low molecule weight (LMW) surfactants. As shown in Figure 1, the protein concentration increased from 0.1% to 1%, the droplet size of the emulsion decreased firstly and then increased with the increasing concentration of PPI from 0.1% to 1%. In comparison with the percentage of adsorbed proteins (AP, %), the AP value was the highest at 54.41% when the PPI concentration was 0.5%. After adding MDG, the droplet size was increased. When the PPI concentration was greater than 0.5%, the AP value gradually decreased with the MDG addition. When the protein concentration was 1%, the interface showed the saturation adsorption. This phenomenon was explained by the compact layer after protein adsorption to the interface. The inaccessibility of more protein molecules to the interface prevented the effective adsorption of protein and the intermolecular interactions [25]. According to the principle of LMW emulsifiers, the result was mainly caused by interfacial displacement.

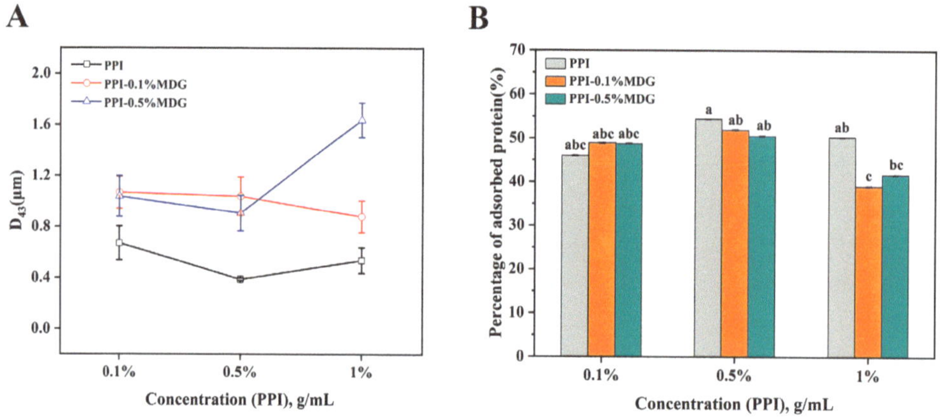

Figure 1. Changes of particle size (D43, **A**) and percentage of adsorbed proteins (AP, **B**) at different concentrations of PPI and MDG. The same lowercase letters marked in the figure represent no significant difference; different lowercase letters indicate significant differences.

The interfacial proteins from the PPI-MDG-stabilized emulsion were collected for the SDS-PAGE measurement, the semi-quantitative method. Through this detection, the differences in the subunit composition of PPI were determined to indicate the protein displacement at the oil–water interface. In Figure 2, the color of protein bands was too light at 0.1% PPI compared with other samples. With the MDG addition, the interfacial protein bands almost disappeared. When the PPI concentration was increased to 0.5%, the SDS-PAGE profiles showed that the interfacial adsorbed proteins (vicilin and legumin B) were decreased according to the missed corresponding bands. The semi-quantitative results suggested that vicilin (25–35 kDa) and legumin (13–20 kDa) were displaced by MDG at the interface. This result was observed by the SDS-PAGE profiles; meanwhile, an accurate measurement confirmed its necessity.

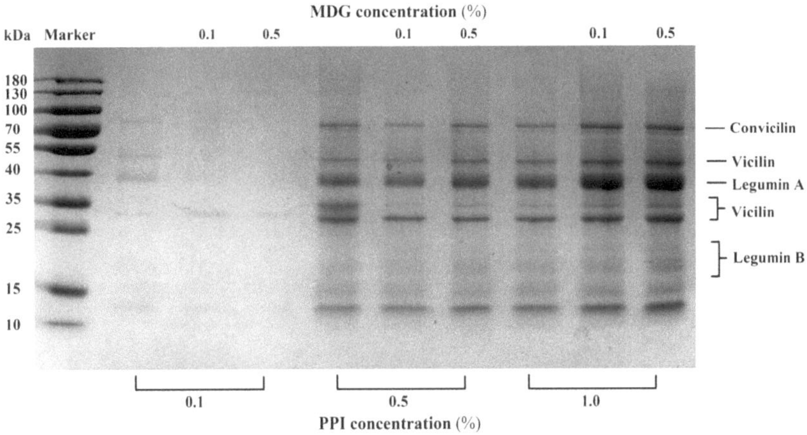

Figure 2. SDS-PAGE profile of adsorbed proteins (Ads.) at the oil–water interface stabilized with PPI and MDG at different concentrations.

3.2. Dilatational Rheological Properties

According to the interfacial rheology results in Figure 3, the interfacial tension of samples was decreased at 0.5% MDG. Meanwhile, the interfacial modulus was lower compared with the 0.1% MDG addition. The effect of MDG on the interface tension was more apparent than PPI, which was the typical property of LMW surfactants. The complex modulus decreased significantly with the increase in MDG concentration. In Figure 3B, the complex modulus of 0.5%PPI-0.1%MDG was 43.84 mN/m. Generally, the larger the interfacial complex modulus, the higher the mechanical strength, and this phenomenon was necessary to be further explored [26]. The diffusion rate of 0.1% PPI–0.5% MDG indicated that the protein diffused quickly after the MDG addition. Based on the interfacial kinetic analysis, the interfacial tension decreased at the initial period was attributed to the MDG addition. Therefore, this indicated that proteins with MDG reached interfacial equilibrium more rapidly compared with only protein samples. The faster adsorption and slower rearrangement of the 0.5% PPI–0.1% MDG suggested more adsorbed protein at the oil–water interface and less displacement by the MDG. The result caused less desorption and unfolding of proteins at the interface [27]. At 0.5% PPI, the addition of MDG with less protein rearrangement suggested that the frequency of protein–protein interactions decreased. At the final stage, the interfacial pressure of the 0.5% MDG sample was increased, and the maximum value of the interfacial pressure was 18.78 mN/m for the 0.1% PPI–0.5% MDG. The gradual increase in E with the adsorption time was explained by an intermolecular interaction and protein adsorptions at the interface [28,29]. The result showed that MDG addition played a dominant part in the interfacial tension, and its potential mechanism on the interfacial film was further analyzed from the nonlinear rheology.

According to the relevant research, the properties of an interfacial protein film could define the competitive adsorption behaviors at the oil–water interface [30]. As shown in Figure 4, the complex modulus of the PPI-MDG stabilized interface was determined at an amplitude of 10–30%. The dependence of the interfacial viscoelastic modulus on the interfacial pressure was further estimated. The interfacial complex modulus was decreased with the increasing amplitude, suggesting the nonlinear viscoelastic region with the amplitude range.

Based on the above analysis, it was determined that the interfacial elasticity of the adsorbed protein film decreased after the protein displacement by MDG. The result indicated that the structural change of the interfacial film was caused by the interface deformation and followed by the interfacial protein rearrangement during the process. It has been reported that the proteins changed their conformation upon adsorption at the oil-water interfaces. This result supported the conclusion.

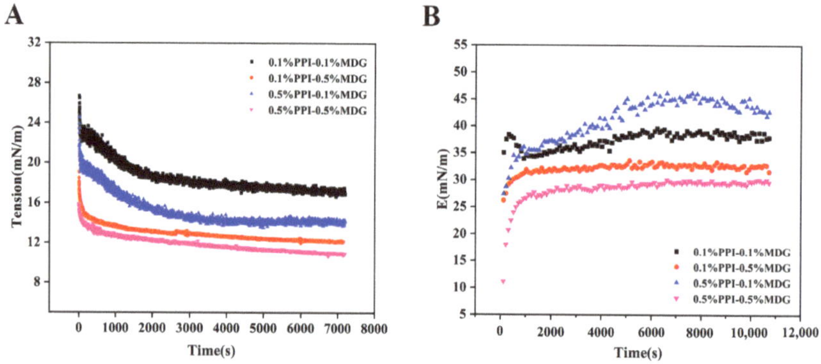

Figure 3. Interface tension (**A**) and interface dilatational modulus (**B**) as a function of time with different concentrations of PPI and MDG at the oil–water interface.

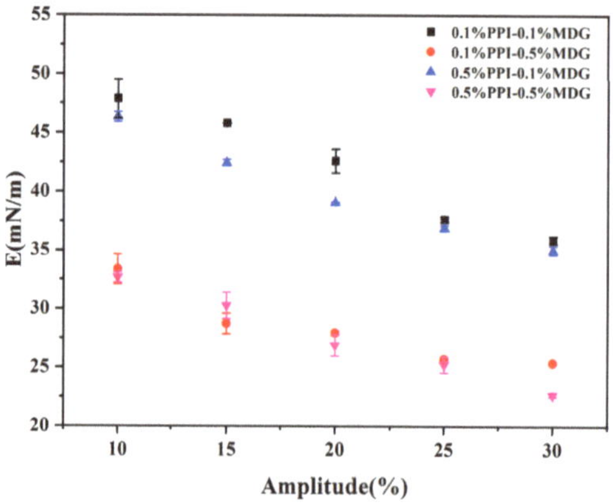

Figure 4. Interface dilatational modulus as a function of amplitude for the oil–water interface stabilized by MDG and PPI at different concentrations. Frequency: 0.1 Hz; deformation amplitude: 10–30%.

3.3. Lissajous Curves

In order to further understand the interfacial behavior of PPI-MDG, it was essential to investigate the relationship between the interfacial pressure and interfacial deformation during the amplitude sweep. Therefore, the Lissajous curves were described for the nonlinear interfacial rheological response, and the surface pressure was described as a function of the deformation. According to the previous result, a completely linear shape indicated a purely elastic behavior, while a completely spherical shape indicated a purely viscous behavior of the interface [31,32]. In Figure 5, at lower PPI concentrations, the interfacial proteins were gradually displaced with the increasing MDG. Thus, the interfacial protein film showed a looser distribution, and the compression-hardening response was weakened. At high concentrations of PPI, the interface indicated the dense film structure. In the case of protein displacement, the interfacial film exhibited more significant compression hardening and extension softening. When the film was deformed or disturbed, the protein interactions were weakened to maintain the interfacial stability, or even lead to the interface rupture [33,34].

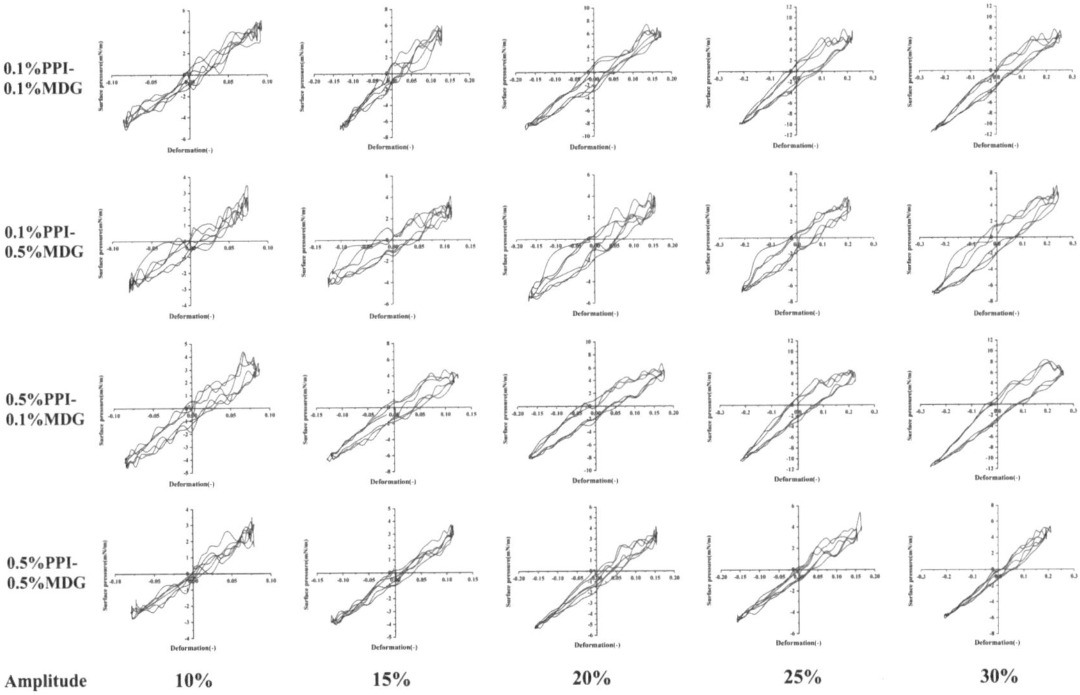

Figure 5. Lissajous plots obtained during amplitude sweeps (10–30%) at the oil–water interface stabilized by PPI and MDG at different concentrations.

The lower MDG concentration was used for the faster relaxation of the interfacial tension. The surfactant interfacial properties were related to the interfacial film structure. The interfacial behaviors were attributed to the protein exchanges at the interface, as well as the rearrangement of adsorbed protein at the interfacial film. As shown in Figure 6, the positive value of S factor represented the interface-hardening response, while a negative value of S factor represented the interface-softening response. During the process, all samples showed strain-softening behavior, and the degree of strain softening increased with the increasing amplitude. Similarly, strain-hardening behavior was presented during compression, and the degree of strain hardening increased with the increasing amplitude [35]. Some significant differences in the shape of Lissajous curves were observed at 0.5% PPI. The Lissajous curves narrowed as the MDG concentration gradually increased, suggesting that the interfacial responses were more direct and rapid. The interactions between proteins and LMW surfactants resulted from the hydrophobic or electrostatic interactions, thus affecting the initial structure and interfacial conformation of protein molecules [36,37]. The protein interfacial rearrangement at the oil–water interface was caused by the protein-inherit flexibility and protein hydrophobicity.

Since LMW surfactants could gradually displace the protein molecules, the addition of MDG might influence the complexation process and the protein composition of the adsorbed layer. At sufficiently high LMW surfactant concentrations (depending on the natural properties of the protein and the small-molecule surfactant), protein molecules could be completely displaced by the LMW surfactants, considering the competitive adsorption mechanism [23]. Combined with the nonlinear rheological behavior, the changes of the interfacial film structure were related with the decrease in interfacial proteins and the weakening of interfacial film strength [38].

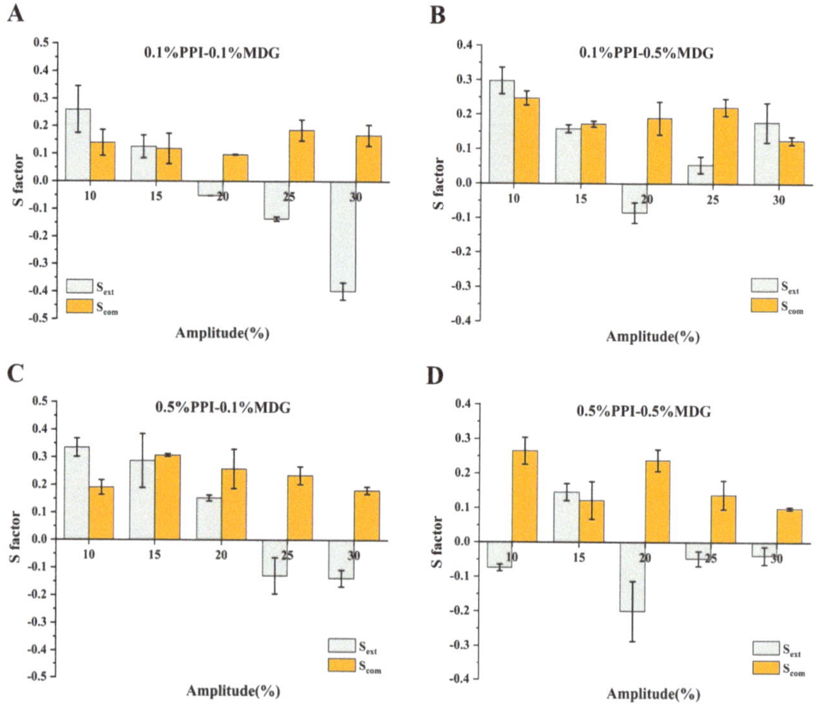

Figure 6. The (**A**–**D**) indicated the different samples. The S factor during extension (S_E) and compression (S_C) determined from amplitude sweeps (10–30%) at the oil–water interface stabilized by PPI and MDG.

3.4. *Quantitative Analysis of Interfacial Proteins*

The interfacial protein composition was determined by mass spectrometry technology (LC-MS/MS) with the method of stable isotope dimethyl labeling. Based on the above experimental results, three sample groups were selected for comparison. In the measurement, the sample of 0.5%PPI was set as the control group, the light standard (L). The sample of 0.5% PPI0.1% MDG was the medium standard (M), and the sample of 0.5% PPI0.5% MDG was the heavy standard (H). The adsorbed protein types and relative contents of three groups at the oil–water interface were detected to further illustrate the relative level of protein displacement by MDG.

Vicilin (7S) is the trimeric molecule in which each monomer is composed by some subunits. The physicochemical and conformational properties of vicilin depended on its amino acid composition (especially the relative ratio of acidic to basic charged amino acids). Moreover, this ratio influenced its tertiary conformation, quaternary conformation, protein flexibility and thermal stability. In Figure 7, at the quantitative analysis of 0.5% PPI0.5% MDG samples, the vicilin contents were significantly decreased. Combined with the interfacial adsorption results, the increased MDG concentration at 0.5% PPI resulted in the decrease in the interfacial adsorption rate and adsorption capacity. Simultaneously, when the MDG concentration was 0.1%, the vicilin increased. In Table S1, when the MDG concentration was 0.1%, the relative content of vicilin (N4-N5) increased the N17 and N19 as up-regulated proteins. In addition, legumin (11S) is one type of protein with a hexameric structure, in which each monomer contains subunits linked by disulfide bonds. In the 0.5% PPI0.1% MDG samples, the legumin content averagely decreased by 23.27%, indicating that legumins were gradually displaced from the interface. It was speculated that more binding sites appeared at the interface, and other kinds of proteins adsorbed to the interface slowly. There remained a dynamic equilibrium during the process.

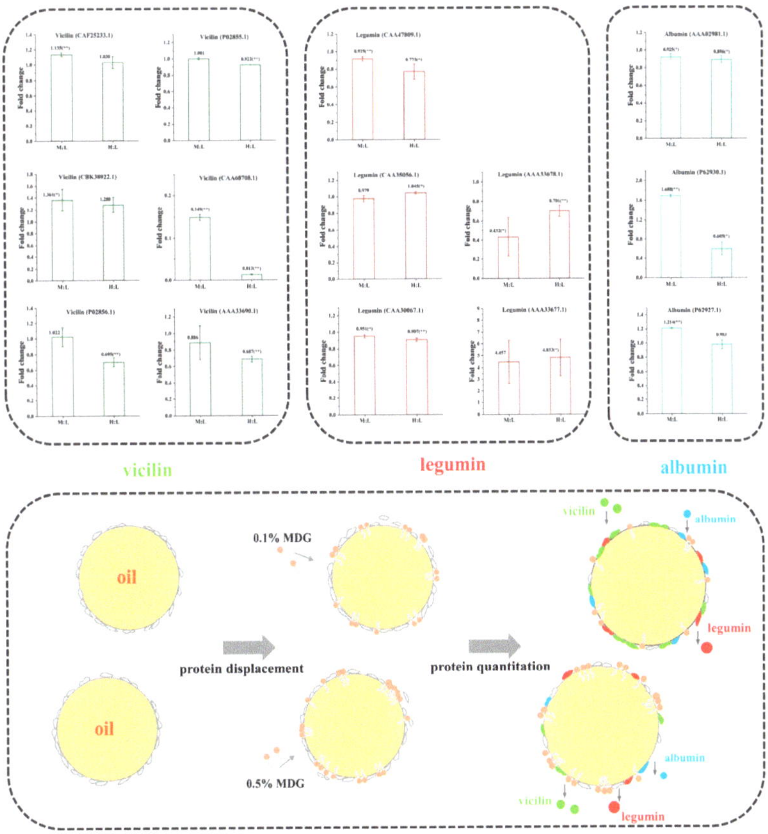

Figure 7. The fold changes of M:L (0.5% PPI–0.1% MDG:0.5% PPI) and H:L (0.5% PPI–0.5% MDG:0.5% PPI) (*, $p < 0.05$; **, $p < 0.01$).

In the comparation of M:L, the content of albumin (2S) increased. When the MDG concentration increased to 0.5%, the 2S decreased by 13.6%, on average. In summary, at a low concentration of MDG, the vicilin and albumin at the interface showed higher content and were preferentially adsorbed at the interface. Moreover, the legumin with the larger molecular weight suggested a slow adsorption rate. At 0.5% MDG, both vicilin and albumin were displaced by MDG and desorbed from the interface, while legumin was more difficult to be displaced due to its slow adsorption and the complex structure of protein molecules [39].

According to the quantitative results, the interfacial rheological behaviors between the component proteins of PPI and MDG were further investigated. The results showed that the interfacial complex modulus increased firstly and then decreased at low concentration of MDG [40]. In Figure 8, the greatest decrease in interfacial tension was observed at 11S when the MDG concentration was 0.5%. The complex moduli of 7S0.5% MDG and 11S0.5% MDG were stable in the range of 16.5–17.5 mN/m, which was higher than that of 2S0.5% MDG. The complex modulus at the interface of the 2S–0.5% MDG samples first increased and then decreased rapidly, suggesting the significant decrease in the content of albumin. In combination with the high concentration of MDG in the quantitative results, the significant desorption of albumin was the explanation for the decrease in the complex modulus at the oil–water interface [41,42].

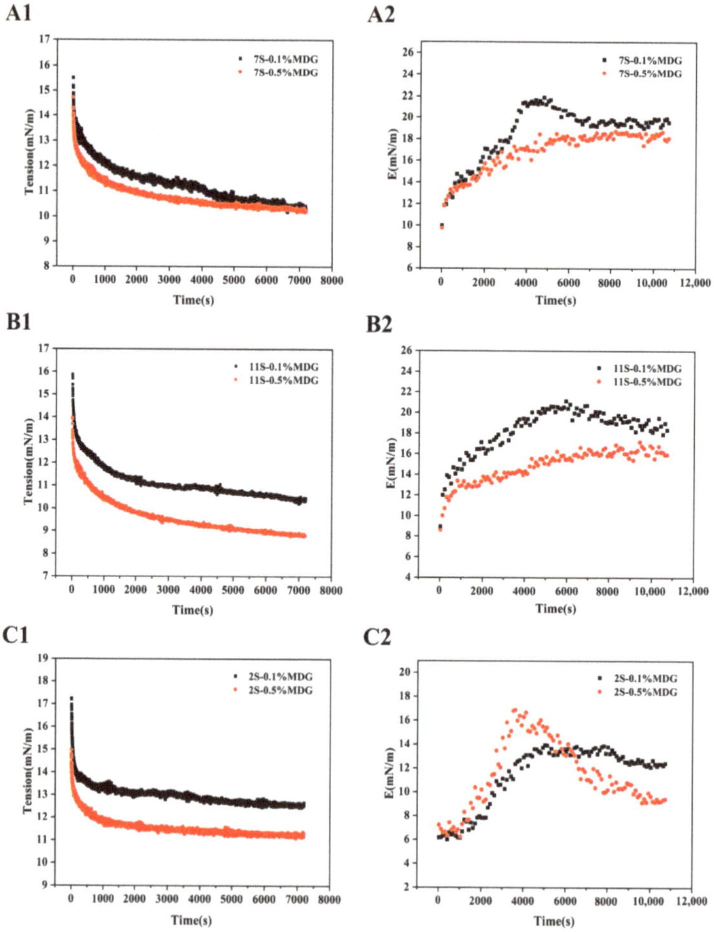

Figure 8. Interfacial tension and dilatational viscoelastic modulus (E) versus time (s) of 7S (**A1**,**A2**), 11S (**B1**,**B2**) and 2S (**C1**,**C2**) at the oil–water interface.

4. Conclusions

During the co-adsorption of PPI and MDG, an interface rheological analysis revealed a significant competitive adsorption. With the increase in MDG concentration, the interfacial proteins were gradually displaced and more loosely distributed, suggesting that the compression-hardening response was weakened. The protein molecules with the structural rearrangement interacted with MDG and showed the potential effects on the interfacial film structure. At low concentrations of MDG, vicilin and albumin were found to be higher at the interface and adsorbed preferentially at the interface. At 0.5% MDG, both vicilin and albumin were displaced by MDG and desorbed from the interface. Meanwhile, legumin with a complex molecular structure was too complicated to displace. Therefore, when LMW surfactants and proteins were used as complex emulsifiers, their synergistic effects at the oil–water interface were the essential factors in selecting the appropriate concentration range.

Supplementary Materials: The following supporting information can be downloaded at: https://www.mdpi.com/article/10.3390/foods13010176/s1, Table S1: Adsorbed proteins identified by LC-MS/MS in the emulsion that was prepared by PPI and MDG at different concentrations and their relative abundance.

Author Contributions: Conceptualization, Q.S. and Y.C.; methodology, Z.W. and Q.S.; software, J.L.; formal analysis, Z.W. and C.P.; resources, Y.C.; writing—original draft preparation, Z.W.; writing—review and editing, Q.S. and Y.C.; supervision, B.L.; project administration, B.L. and Y.C.; funding acquisition, B.L. and Y.C. All authors have read and agreed to the published version of the manuscript.

Funding: This work was supported by the National Key R&D Program of China (No. 2022YFD2101302), National Natural Science Foundation of China (Grant No. 31972019), Hubei Key R&D Program (2023BBB130), Fundamental Research Funds for the Central Universities (Grant No. 2662023PPY003) and Open Fund of Key Laboratory of Fermentation Engineering (Ministry of Education) (Grant No. 202209FE01).

Institutional Review Board Statement: Not applicable.

Informed Consent Statement: Not applicable.

Data Availability Statement: The data presented in this study are available on request from the corresponding author. The data are not publicly available due to privacy.

Conflicts of Interest: The authors declare no conflict of interest.

References

1. Xiong, T.; Ye, X.D.; Su, Y.T.; Chen, X.; Sun, H.; Li, B.; Chen, Y.J. Identification and quantification of proteins at adsorption layer of emulsion stabilized by pea protein isolates. *Colloid Surf. B-Biointerfaces* **2018**, *171*, 1–9. [CrossRef]
2. Shen, Q.; Zheng, W.; Han, F.; Zuo, J.; Dai, J.; Tang, C.; Song, R.; Li, B.; Chen, Y. Air-water interfacial properties and quantitative description of pea protein isolate-Tween 20. *Food Hydrocoll.* **2023**, *139*, 108568. [CrossRef]
3. Stampfli, L.; Nersten, B. Emulsifiers in bread making. *Food Chem.* **1995**, *52*, 353–360. [CrossRef]
4. Younes, M.; Aquilina, G.; Castle, L.; Engel, K.H.; Fowler, P.; Fernandez, M.J.F.; Furst, P.; Gurtler, R.; Husoy, T.; Manco, M.; et al. Opinion on the re-evaluation of mono- and diglycerides of fatty acids (E 471) as food additive in foods for infants below 16 weeks of age and follow-up of their re-evaluation as food additives for uses in foods for all population groups. *EFSA J.* **2021**, *19*, e06885. [CrossRef]
5. Vurma, M.; DeMichele, S.; Lee, D.; Wang, F.; Tso, P. In vivo assessment of lipophilic nutrients absorption mechanisms using human milk and infant formula with novel mono-and diglycerides based emulsification technologies. *J. Pediatr. Gastroenterol. Nutr.* **2016**, *62*, 205.
6. Ali, S.; Tiwari, A.; Yeoh, T.; Doshi, P.; Kelkar, N.; Shah, J.C.; Seth, J.R. Crystallization and Rheology of Mono-and Diglycerides and Their Role in Stabilization of Emulsion Droplets in Model Topical Ointments. *Langmuir* **2022**, *38*, 8502–8512. [CrossRef]
7. Lee, L.Y.; Chin, N.L.; Lim, C.H.; Yusof, Y.A.; Talib, R.A. Saturated distilled monoglycerides variants in gel-form cake emulsifiers. *Agric. Agric. Sci. Procedia* **2014**, *2*, 191–198. [CrossRef]
8. Hasenhuettl, G.L.; Hartel, R.W. *Food Emulsifiers and Their Applications*; Springer: Berlin/Heidelberg, Germany, 2008; Volume 19.
9. McSweeney, S.L.; Healy, R.; Mulvihill, D.M. Effect of lecithin and monoglycerides on the heat stability of a model infant formula emulsion. *Food Hydrocoll.* **2008**, *22*, 888–898. [CrossRef]
10. Marriage, B.J.; Williams, J.A.; Choe, Y.S.; Maki, K.C.; Vurma, M.; DeMichele, S.J. Mono-and diglycerides improve lutein absorption in healthy adults: A randomised, double-blind, cross-over, single-dose study. *Br. J. Nutr.* **2017**, *118*, 813–821. [CrossRef] [PubMed]
11. Munk, M.; Larsen, F.; Van Den Berg, F.; Knudsen, J.; Andersen, M. Competitive displacement of sodium caseinate by low-molecular-weight emulsifiers and the effects on emulsion texture and rheology. *Langmuir* **2014**, *30*, 8687–8696. [CrossRef] [PubMed]
12. Kang, W.; Yin, X.; Yang, H.; Zhao, Y.; Huang, Z.; Hou, X.; Sarsenbekuly, B.; Zhu, Z.; Wang, P.; Zhang, X. Demulsification performance, behavior and mechanism of different demulsifiers on the light crude oil emulsions. *Colloids Surf. A Physicochem. Eng. Asp.* **2018**, *545*, 197–204. [CrossRef]
13. Baret, J.C.; Kleinschmidt, F.; El Harrak, A.; Griffiths, A.D. Kinetic aspects of emulsion stabilization by surfactants: A microfluidic analysis. *Langmuir* **2009**, *25*, 6088–6093. [CrossRef] [PubMed]
14. Monié, A.; David, A.; Clemens, K.; Malet-Martino, M.; Balayssac, S.; Perez, E.; Franceschi, S.; Crepin, M.; Delample, M. Enzymatic hydrolysis of rapeseed oil with a non-GMO lipase: A strategy to substitute mono-and diglycerides of fatty acids and improve the softness of sponge cakes. *LWT* **2021**, *137*, 110405. [CrossRef]
15. Pourfarzad, A.; Shokouhi Kisomi, R. Effect of lecithin and mono-and di-glyceride on quality and shelf life of hazelnut butter: Chemometric approach. *Pol. J. Food Nutr. Sci.* **2020**, *70*, 399–408. [CrossRef]
16. Sinzato, Y.Z.; Dias, N.J.S.; Cunha, F.R. An experimental investigation of the interfacial tension between liquid-liquid mixtures in the presence of surfactants. *Exp. Therm. Fluid Sci.* **2017**, *85*, 370–378. [CrossRef]
17. Lam, A.; Can Karaca, A.; Tyler, R.; Nickerson, M. Pea protein isolates: Structure, extraction, and functionality. *Food Rev. Int.* **2018**, *34*, 126–147. [CrossRef]
18. Burger, T.G.; Zhang, Y. Recent progress in the utilization of pea protein as an emulsifier for food applications. *Trends Food Sci. Technol.* **2019**, *86*, 25–33. [CrossRef]

19. Hebishy, E.; Buffa, M.; Guamis, B.; Blasco-Moreno, A.; Trujillo, A.-J. Physical and oxidative stability of whey protein oil-in-water emulsions produced by conventional and ultra high-pressure homogenization: Effects of pressure and protein concentration on emulsion characteristics. *Innov. Food Sci. Emerg. Technol.* **2015**, *32*, 79–90. [CrossRef]
20. Berry, J.D.; Neeson, M.J.; Dagastine, R.R.; Chan, D.Y.; Tabor, R.F. Measurement of surface and interfacial tension using pendant drop tensiometry. *J. Colloid Interface Sci.* **2015**, *454*, 226–237. [CrossRef]
21. Wang, J.-M.; Xia, N.; Yang, X.-Q.; Yin, S.-W.; Qi, J.-R.; He, X.-T.; Yuan, D.-B.; Wang, L.-J. Adsorption and dilatational rheology of heat-treated soy protein at the oil–water interface: Relationship to structural properties. *J. Agric. Food Chem.* **2012**, *60*, 3302–3310. [CrossRef]
22. Sagis, L.M.; Fischer, P. Nonlinear rheology of complex fluid–fluid interfaces. *Curr. Opin. Colloid Interface Sci.* **2014**, *19*, 520–529. [CrossRef]
23. Bykov, A.; Liggieri, L.; Noskov, B.; Pandolfini, P.; Ravera, F.; Loglio, G. Surface dilational rheological properties in the nonlinear domain. *Adv. Colloid Interface Sci.* **2015**, *222*, 110–118. [CrossRef] [PubMed]
24. Boersema, P.J.; Raijmakers, R.; Lemeer, S.; Mohammed, S.; Heck, A.J. Multiplex peptide stable isotope dimethyl labeling for quantitative proteomics. *Nat. Protoc.* **2009**, *4*, 484–494. [CrossRef] [PubMed]
25. Ambrosone, L.; Mosca, M.; Ceglie, A. Impact of edible surfactants on the oxidation of olive oil in water-in-oil emulsions. *Food Hydrocoll.* **2007**, *21*, 1163–1171. [CrossRef]
26. Ravera, F.; Loglio, G.; Kovalchuk, V.I. Interfacial dilational rheology by oscillating bubble/drop methods. *Curr. Opin. Colloid Interface Sci.* **2010**, *15*, 217–228. [CrossRef]
27. Miller, R.; Ferri, J.K.; Javadi, A.; Krägel, J.; Mucic, N.; Wüstneck, R. Rheology of interfacial layers. *Colloid Polym. Sci.* **2010**, *288*, 937–950. [CrossRef]
28. Jiang, F.; Chen, C.; Wang, X.; Huang, W.; Jin, W.; Huang, Q. Effect of Fibril Entanglement on Pickering Emulsions Stabilized by Whey Protein Fibrils for Nobiletin Delivery. *Foods* **2022**, *11*, 1626. [CrossRef] [PubMed]
29. Van Den Berg, M.E.; Kuster, S.; Windhab, E.J.; Sagis, L.; Fischer, P. Nonlinear shear and dilatational rheology of viscoelastic interfacial layers of cellulose nanocrystals. *Phys. Fluids* **2018**, *30*, 072103. [CrossRef]
30. Porras, M.; Solans, C.; González, C.; Martínez, A.; Guinart, A.; Gutiérrez, J.M. Studies of formation of W/O nano-emulsions. *Colloids Surf. A Physicochem. Eng. Asp.* **2004**, *249*, 115–118. [CrossRef]
31. Joyner, H.S. Nonlinear (large-amplitude oscillatory shear) rheological properties and their impact on food processing and quality. *Annu. Rev. Food Sci. Technol.* **2021**, *12*, 591–609. [CrossRef]
32. Sharif, H.R.; Williams, P.A.; Sharif, M.K.; Abbas, S.; Majeed, H.; Masamba, K.G.; Safdar, W.; Zhong, F. Current progress in the utilization of native and modified legume proteins as emulsifiers and encapsulants–A review. *Food Hydrocoll.* **2018**, *76*, 2–16. [CrossRef]
33. Kotsmar, C.; Grigoriev, D.; Xu, F.; Aksenenko, E.; Fainerman, V.; Leser, M.; Miller, R. Equilibrium of adsorption of mixed milk protein/surfactant solutions at the water/air interface. *Langmuir* **2008**, *24*, 13977–13984. [CrossRef] [PubMed]
34. Yang, J.; Thielen, I.; Berton-Carabin, C.C.; van der Linden, E.; Sagis, L.M. Nonlinear interfacial rheology and atomic force microscopy of air-water interfaces stabilized by whey protein beads and their constituents. *Food Hydrocoll.* **2020**, *101*, 105466. [CrossRef]
35. Rühs, P.A.; Affolter, C.; Windhab, E.J.; Fischer, P. Shear and dilatational linear and nonlinear subphase controlled interfacial rheology of β-lactoglobulin fibrils and their derivatives. *J. Rheol.* **2013**, *57*, 1003–1022. [CrossRef]
36. Fainerman, V.; Lucassen-Reynders, E.; Miller, R. Description of the adsorption behaviour of proteins at water/fluid interfaces in the framework of a two-dimensional solution model. *Adv. Colloid Interface Sci.* **2003**, *106*, 237–259. [CrossRef] [PubMed]
37. Lad, M.D.; Ledger, V.M.; Briggs, B.; Green, R.J.; Frazier, R.A. Analysis of the SDS− lysozyme binding isotherm. *Langmuir* **2003**, *19*, 5098–5103. [CrossRef]
38. Bos, M.A.; Van Vliet, T. Interfacial rheological properties of adsorbed protein layers and surfactants: A review. *Adv. Colloid Interface Sci.* **2001**, *91*, 437–471. [CrossRef] [PubMed]
39. Li, J.; Zhang, B.; Ye, J.; Sun, F.; Liu, Y.; Yang, N.; Nishinari, K. Nonlinear dilatational rheology of different protein aggregates at the oil–water interface. *Soft Matter* **2022**, *18*, 2383–2393. [CrossRef]
40. Liu, X.Y.; Bennema, P. The relation between macroscopic quantities and the solid–fluid interfacial structure. *J. Chem. Phys.* **1993**, *98*, 5863–5872. [CrossRef]
41. Zhang, B.; Lei, M.; Huang, W.; Liu, G.; Jiang, F.; Peng, D.; Huang, Q.; Jin, W. Improved Storage Properties and Cellular Uptake of Casticin-Loaded Nanoemulsions Stabilized by Whey Protein-Lactose Conjugate. *Foods* **2021**, *10*, 1640. [CrossRef]
42. li Zhai, J.; Day, L.; Aguilar, M.-I.; Wooster, T.J. Protein folding at emulsion oil/water interfaces. *Curr. Opin. Colloid Interface Sci.* **2013**, *18*, 257–271. [CrossRef]

foods

MDPI

Article

Differences in the Chromogenic Effect of Corn Starch and Potato Starch on Paprika Red Pigment and Structural Characterisation

Fan Su, Yongqiang Wu, Yanping Cao and Shaojia Wang *

Beijing Advanced Innovation Center for Food Nutrition and Human Health (BTBU), School of Food and Health, Beijing Higher Institution Engineering Research Center of Food Additives and Ingredients, Beijing Technology and Business University, Beijing 100048, China; sf19980817@163.com (F.S.); wuyongqiang0095@163.com (Y.W.); caoyp@th.btbu.edu.cn (Y.C.)

* Correspondence: wangshaojia@btbu.edu.cn

Abstract: The present study aims to investigate the chromogenic effect and the interaction between starch-pigment complexes of corn starch (CS) and potato starch (PS) complexed with paprika red pigment. Compared to PS, CS showed 12.5 times higher adsorption capacity for paprika red pigment. Additionally, the a* value of CS-P (26.90 ± 0.23) was significantly higher than that of PS-P (22.45 ± 1.84), resulting in a corn starch-paprika red pigment complex (CS-P) with a more intense red colour. The addition of paprika red pigment significantly decreased the particle size and porosity of CS by 48.14 ± 5.29% and 17.01 ± 3.80%, respectively. Conversely, no significant impact on PS was observed. Additionally, the Fourier transform infrared (FT-IR) spectroscopy results revealed that the starch molecules and paprika red pigment were bound to each other through strong hydrogen bonds. X-diffraction (XRD) results indicated that the starch-paprika red pigment complexes have a V-shaped structure. Furthermore, the relative crystallinity of the complexes between starch and red pepper pigment showed an increasing trend, however, the relative crystallinity of CS increased significantly by 11.77 ± 0.99–49.21 ± 3.67%. Consequently, the CS-P colouring was good.

Keywords: corn starch; potato starch; paprika red pigment; chromogenic effect; structural characterisation

Citation: Su, F.; Wu, Y.; Cao, Y.; Wang, S. Differences in the Chromogenic Effect of Corn Starch and Potato Starch on Paprika Red Pigment and Structural Characterisation. *Foods* **2024**, *13*, 191. https://doi.org/10.3390/foods13020191

Academic Editor: Moo-Yeol Baik

Received: 29 November 2023
Revised: 26 December 2023
Accepted: 30 December 2023
Published: 6 January 2024

1. Introduction

Paprika red pigment, consisting of capsorubin, capsanthin, β-carotene, zeaxanthin, and β-cryptoxanthin, is a high-quality natural dye mainly extracted from dried chili peppers [1]. As a precursor of vitamin A, paprika red pigment possesses various physiological functions. Studies have shown that paprika red pigment has the potential to regulate lipid metabolism, thereby reducing the risk of cardiovascular diseases, cancer, and other chronic diseases [2]. Paprika red pigment has been commercialized as colourants, feed supplements, and nutritional products due to its wide distribution, bright colour, high safety, and anti-oxidation, and is widely used in food, medical, feed, cosmetics, and other fields [3]. For instance, paprika red pigment imparts higher initial redness to fresh red sausage and fresh chorizo (red-line meat products), which remains unchanged until the end of the product's shelf life [4]. Paprika red pigment could ameliorate the detrimental effects of diet-induced obesity by improving impaired lipid metabolism [5].

With the emerging growth of the social economy, people's consumption level has increased considerably, accelerating the demand for food with high standards for appearance and sensory quality and meeting healthier dietary requirements. Studies have indicated that red environments are often more stimulating for consumers, leading to an augmented interest in purchasing products. Hence, developing products with appealing colours and appearance is crucial within the food sector. The food industry has been working on the development of highly stable, attractive colours, and low-cost synthetic dyes to improve the appearance of food products. Nevertheless, synthetic food additives, such as carmine and quinoline yellow, have the potential to cause health issues such as allergies, irritability, and even more severe conditions like cancer. Therefore, in recent years, research on the development and application of natural dye has received considerable attention [6]. Although paprika red pigment has a wide range of applications in the food industry, its high lipophilicity, low water solubility, and light sensitivity significantly hinder its application as a natural food additive [7]. To address these issues, efforts were made to encapsulate paprika red pigment in lipid nanoparticles, liposomes, and emulsions [8]. Paprika red pigment may also interact with proteins [9], polysaccharides [10], and other inorganic nanoparticles [11] to form complexes with high stability and activities. Particularly, polysaccharides are promising delivery systems due to their good biocompatibility and biodegradability [12]. For example, in a previous study, the stability of the colour index in yoghurt stained with the complex formed by paprika red pigment and β-cyclodextrin was higher than that of yoghurt stained with paprika red pigment alone [13]. Additionally, the capsaicin-chitosan colloidal complex had high colloidal stability and high apparent solubility [14]. Ethyl cellulose-capsaicin composite membrane has the advantages of ease in production, eco-friendliness, and high antimicrobial activity [10]. Therefore, polysaccharides can be used as potential carriers of natural pigments such as paprika red pigment with wide research prospects.

Starch, derived from the polymerisation of glucose molecules, is a polysaccharide widely used in food as a good wall material. It is reported that the interaction of β-carotene and sweet potato starch could improve the bioaccessibility and bioavailability of β-carotene and starch molecules [15]. In a previous study, a curcumin emulsion delivery system was developed using debranched starch as a carrier, which offered improved stability and solubility of curcumin compared to the system developed using Tween 80 and lectin [16]. CS can be used as a carrier for grape seed proanthocyanidins [17]. Seaweed polyphenols bind to CS and promote its gelation [18]. Starches and pigments are mainly connected by non-covalent bonds [19]. A previous study on the interaction between sorghum proanthocyanidin and amylose and demonstrated that sorghum proanthocyanidin interacts effectively with amylose through hydrophobic and hydrogen bonding [20]. The study discovered that lutein binds to soybean starch through hydrophobic forces, resulting in a complex that significantly enhances lutein stability [21]. The tremella fuciformis polysaccharide could interact with PS mainly through hydrogen bonds [22]. The barley β-glucan interacts with PS through hydrogen bonding, thereby improving the gelling properties of PS [23]. To date, some interactions of capsaicin with indica starch and high-amylose corn starch have been reported internationally [24,25]. However, no studies have reported the complex mechanisms between paprika red pigment and corn starch (CS), the most dominant in the starch industry, and between paprika red pigment and potato starch (PS), the second most important starch in the industry. Therefore, in this study, corn starch-paprika red pigment (CS-P) and potato starch-pigment (PS-P) complexes were prepared from paprika red pigment, CS, and PS, and their physicochemical properties were characterised using scanning electron microscopy (SEM), pore space, high-performance liquid chromatography (HPLC), particle size, Fourier transform infrared spectroscopy (FT-IR), and x-ray diffraction (XRD) technique. Furthermore, the adsorption capacity and colour rendering effect of CS and PS on paprika red pigment were determined. The experimental results provide a theoretical basis for developing a natural colourant that can be used as a substitute for nitrite.

2. Materials and Methods

2.1. Materials

Paprika red pigment E150 (>98%) was provided by M&G Biotechnology Group Co., Ltd. (Handan, China). Corn starch (>99%) was purchased from Yishui Dadi Corn Development Co., Ltd. (Yishui, China). Potato starch (>99%) was purchased from Yantai Shuangta Food Co., Ltd. (Yantai, China). Amylopectin was obtained from Shanghai Aladdin Biochemical Technology Co., Ltd. (Shanghai, China). Amylose was purchased from Sigma-Aldrich Co., Ltd. (St. Louis, MO, USA). Methanol (MeOH) (Chromatographic grade) and methyl tert-butyl ether (MTBE) (Chromatographic grade) were purchased from Beijing Mairuida Technology Co., Ltd. (Beijing, China). Capsanthin standard sample (>97%), capsorubin standard sample (>97%), zeaxanthin standard sample (>97%), and β-cryptoxanthin standard sample (>97%) were obtained from Shanghai Huicheng Biotechnology Co., Ltd. (Shanghai, China). β-Carotene standard sample (>97%) was purchased from Shanghai Aladdin Biochemical Technology Co., Ltd. (Shanghai, China). The reagents acetone, n-hexane, and ethyl ether were of analytical grade and were purchased from Sinopharm Group Chemical Reagent Co., Ltd. (Shanghai, China). Sodium chloride, anhydrous sodium sulfate, sodium hydroxide, potassium iodide, and iodine were purchased from Fuchen (Tianjin) Chemical Reagent Co., Ltd. (Tianjin, China).

2.2. Preparation of Starch-Paprika Red Pigment Complexes

Preparation of starch-paprika red pigment complexes was performed as follows: First, 2 g of different types of starch (corn starch, potato starch, amylose, and amylopectin) and 50 mg paprika red pigment were mixed with 28 mL of MeOH to make a starch-paprika red pigment suspension. Then, the starch-paprika red pigment suspension was heated at 70 °C (DF-101S, Gongyi Yuhua instrument Co., Ltd., Gongyi, China) for 30 min with continuous stirring and then cooled to indoor temperature. Subsequently, the starch-paprika red pigment suspension was centrifuged at 4 °C and 8000 r/min (JXN-30, Beckman Coulter Co., Ltd., Brea, CA, America) for 20 min, and the precipitate was collected. The precipitate was washed twice with MeOH to remove the free material that was not bound to the starch. Finally, the starch-paprika red pigment complexes were recovered by drying in a vacuum freeze dryer (LGJ-1C-56, Beijing Yatai Colon Instrument Technology Co., Ltd., Beijing, China).

2.3. Colour Evaluation of Starch-Paprika Red Pigment Complexes

Colour values (L*, a*, b*) of PS-P and CS-P were determined via a spectrophotometer (CR-800, Beijing Kemerunda Instruments Co., Ltd., Beijing, China). The hue (h) of samples were calculated using the following equation [26],

$$h = \tan^{-1}\frac{b^*}{a^*}$$

where L, a, and b represent the sample brightness, red/green value, and yellow/blue value, and L*, a*, and b* represent the control brightness, red/green value, and yellow/blue value.

2.4. Determination of Reflectance and Scattering Rate of Starch-Paprika Red Pigment Complexes

The starch-paprika red complexes were placed in a quartz cuvette measuring 1 cm, and their reflectance was measured using the reflectance mode of the spectrophotometer (CR-800, Beijing Kemerunda Instruments Co., Ltd., Beijing, China) within a range of 400–700 nm. The transmission mode of the spectrophotometer was utilized to determine the transmittance and haze of the samples, and based on the following equation, the scattering rate of the starch-paprika red pigment complexes was calculated.

$$\text{Scattering rate}(\%) = \text{transmittance} \times \text{haze}$$

2.5. Analysis of Pigments Adsorbed by Starch in Paprika Red Pigment

2.5.1. Pretreatment of Samples

Using the method by Kim et al. [27], appropriate adjustments were made to perform the following analysis. For the analysis of carotenoids, extraction was carried out using acetone. For the extraction, 1 g of dried sample was mixed with 20 mL of acetone and shaken for 20 min, and was repeated until the solution was colourless. The supernatants were merged and the solvents were evaporated to dryness using a rotary evaporator (RE-

52AA, Shanghai Yarong Company Biochemical Instrument Factory, Shanghai, China) at 35 °C, dissolved with 15 mL ether/n-hexane [1:1(w/w)], then incubated with 15 mL MeOH and 5 mL 30% KOH/MeOH at room temperature for 2 h 30 min in the dark, and this was repeated until the solution was colourless. The supernatants were then washed several times with distilled water until the pH was neutral, and 5 mL of 10% NaCl and 5 mL of 2% Na_2SO_4 were added and discarded as the separated hydrophilic phase. After evaporating the collected extracts, the residue was dissolved in MeOH/MTBE (1 mL) [1:1(w/w)] and passed through a 0.22 µm microporous filter membrane until ready for use. Standard curves were plotted using standards of capsorubin, capsanthin, zeaxanthin, β-cryptoxanthin, and β-carotene that were >97% pure. The calculation of the adsorption ratio of starch to paprika red pigment was performed utilizing the subsequent formula,

$$Adsorption\ \text{ratio}(\%) = \frac{\text{S}}{\text{S}_{\text{Total}}} \times 100\%$$

where S represents the content of individual pigments in the starch-paprika red complexes and Stotal represents the total pigment content in the starch-paprika red complexes.

2.5.2. Analysis of Pigments Content in CS-P and PS-P

The content of five pigments in CS-P and PS-P complexes was determined via HPLC (Waters 2695, Waters Corporation, Milford, MA, USA). Using the method by Murillo et al. [28], with some modifications, separation was performed using a Venusil XBP C30 column (4.6 × 250 mm, 5 µm) (Tianjin Bona Ijar Technology Co., Ltd., Tianjin, China) at 25 °C, and the mobile phase was a binary solvent consisting of phase A (MeOH/MTBE/W, 81:15:4, $v/v/v$) and phase B (MeOH/MTBE/W, 6:90:4, $v/v/v$). The UV wavelength was set to 450 nm. The injection volume was set to 1 µL and the flow rate was 1 mL/min. The gradients were programmed as follows (%B): first, 0–45 min, 0–50% of B; second, 45–50 min, 50–0% of B; and finally, 50–60 min, 0% of B.

2.6. Determination of Complexing Index (CI) of Samples

The values of *CI* of CS-P, PS-P, amylose-P, and amylopectin-P were measured using the method described by Wang et al. [29] with a slight modification. Briefly, 1 mL of the samples mentioned above were mixed with 200 µL I2/KI [1.3% (w/w) I2 and 2.0% (w/w) KI in deionized water] solution, and the absorbance was measured at 620 nm via a Ultraviolet-visible Spectrophotometer (SHIMADZU-1280, Shimadzu Corporation, Kyoto, Japan). The complex index was calculated using the following equation [30].

$$CI(\%) = \frac{Abs_{starch} - Abs_{starch-paprika\ red\ pigment}}{Abs_{starch}} \times 100\%$$

2.7. Scanning Electron Microscopy (SEM) Analysis

The microscopic morphology of the pretreated samples was observed using a S4800 SEM (Hitachi Production Co., Ltd., Tokyo, Japan). The dried samples were mounted on a copper stake with a double-sided carbon tag and coated with platinum for 80 s. SEM images were taken at an accelerating voltage of 15 kV and captured using the accompanying software (Digital Micrograph 3.4).

2.8. Particle Size Analysis

The suspended particle sizes of PS, CS, PS-P, and CS-P were determined via particle size analyzer (Mastersizer 2000, Malvern UK Go., Ltd., London, UK). The optical path of the cuvettes used was 1 cm, and the refractive index and viscosity of the samples were calculated using standard calculation tools provided by the software. During the process, the sample was balanced for at least 1 min, and at least 13 consecutive readings were taken.

2.9. Mercury Porosimetry Analysis

Using the method by Włodarczyk-Stasiak et al. [31], the measurements were carried out using Mercury piezometers (AutoPore V 9600, Micromeritics Corporation, Norcross, GA, USA). Briefly, 1 g of starches and starch-capsaicin complexes was dried at 105 °C, placed in a dilatometer, outgassed underhigh vacuum, and filled with mercury under a pressure range of 0.1 to 61,000 psi.

2.10. Fourier Transform Infra Red (FT-IR) Spectroscopy

The FT-IR spectrometer (IS 5, Thermo Fisher Scientific, Waltham, MA, USA) was used to scan the FT-IR spectra of PS, CS, PS-P, and CS-P at full wavelength (4000–400 cm^{-1}) at room temperature. The dried samples mentioned above and 100 mg KBr were mixed in the mixing mortar to make the powder more uniform. The spectrum was recorded using 32 scans with a resolution of 4 cm^{-1}. Each sample was measured 3 times in parallel.

2.11. X-ray Diffraction

X-ray diffraction analyses of PS, CS, PS-P, and CS-P were performed using a SE diffractometer (Nihon Riken Electric Co., Ltd., Niigata, Japan) operated at 40 kV and 40 mA with graphite-filtered Cu Ka radiation and a q compensating slit. The relative intensity was recorded in a scattering angle range (2θ) of 5–60° with a scintillation counter at a scanning speed of 0.02°/min, 17.7 s/step.

2.12. Statistical Analysis

Results are shown as mean ± standard deviation (n = 3). Statistical analysis was carried out using SPSS System Software 22.0. Significant differences between the individual means were tested using t-test and Duncan's multiple range ($p < 0.05$). All graphs involved in this article were drawn using OriginLab 2021 software, and all tables were drawn using Office Word 2019 software.

3. Results

3.1. CS-P and PS-P Colour Comparison

The interactions between CS and PS with paprika red pigment were assessed using a binary mixed system consisting of starch-paprika red pigment complexes. Figure 1A displays the adsorption capacity of CS and PS for paprika red pigment. The results showed that there was a significant difference in the adsorption capacity of CS and PS for paprika red pigment, with CS-P being more reddish. Colour space is a geometric representation of colour in three dimensions, which can be calculated and derived from the initial stimulus values of red, green, and blue. Chromatic aberration is a viable measure to assess the extent of colour variation between different substances [32]. Therefore, in this study, the L*, a*, and b* values were used to verify the colour differences among the samples. The CS-P L*, a*, and b* values exceeded those of PS-P (Table 1), and the total colour difference ΔE was considerably higher than that of PS-P. This outcome is consistent with the findings in Figure 1A. The hue (h) is a qualitative characteristic of colour, traditionally defined as the property of red and green. It provides a clearer indication of the degree to which starch is bound to pigments, proteins, polyphenols, and other substances [33]. As shown in Figure 1B, the h-value for PS was higher than that for CS, and both were greater than 90. Compared to CS and PS, the h-value for both CS-P and PS-P were significantly lower at 47.88 ± 0.50 and 60.22 ± 0.60, respectively, indicating that starches formed complexes with paprika red pigment and the binding capacity of CS for paprika red pigment was stronger than that of PS. Starch is formed by numerous polymer chains containing many hydroxyl groups and arrangements. The paprika red pigment also possesses many hydrophilic hydroxyl groups [34]. Therefore, it can be inferred from the above results that the interaction between starch macromolecules and paprika red pigment might be attributed to the hydrogen bonding or van deer Waals forces, facilitating the encapsulation of paprika red pigment into the starch particles [32].

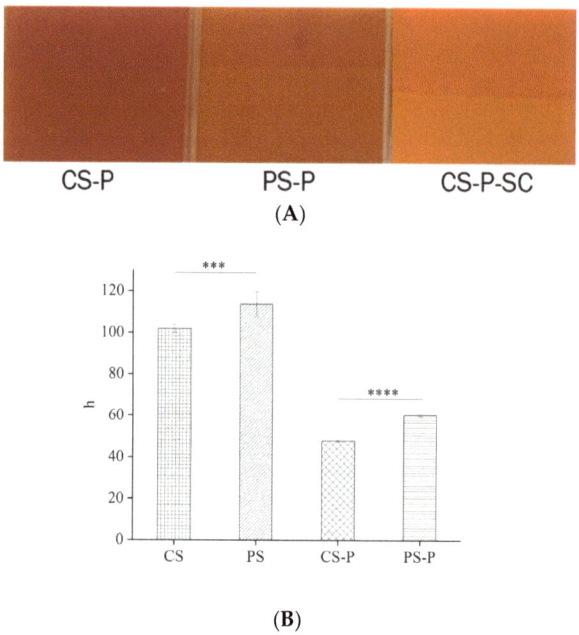

Figure 1. Comparison of the colour rendering effect of starch-paprika red pigment complexes (**A**) and h-values of starches and starch-paprika red complexes (**B**) are shown. CS: Corn starch, PS: Potato starch, CS-P: Corn starch-paprika red pigment, PS-P: Potato starch-paprika red pigment, and CS-P-SC: Corn starch-paprika red pigment-sodium caseinate. Significant difference was defined as $p < 0.05$ (*** $p < 0.001$, **** $p < 0.0001$).

Table 1. Values of colour indices L*, a*, b*, h and ΔE of CS-P, PS-P and CS-P-SC complexes.

	L*	a*	b*	ΔE
PS-P	42.63 ± 0.41 [b]	22.45 ± 1.84 [b]	20.83 ± 1.40 [b]	6.75 ± 0.58
CS-P	46.01 ± 0.41 [a]	26.90 ± 0.23 [a]	25.39 ± 0.50 [a]	12.57 ± 0.63 ***
CS-P-SC	38.80 ± 0.11 [c]	18.44 ± 0.67 [c]	15.69 ± 0.47 [c]	-

ΔE denotes the total colour difference calculated using the CS-P-SC dispersion system as a control. CS-P: Corn starch-paprika red pigment, PS-P: Potato starch-paprika red pigment, and CS-P-SC: Corn starch-paprika red pigment-sodium caseinate. Values with different letters in the same column are significantly different ($p \leq 0.05$). Significant difference was defined as $p < 0.05$ (*** $p < 0.001$). Values are expressed as mean ± standard deviation ($n = 3$).

3.2. Analysis of the Optical Properties of Starches and Starch-Paprika Red Pigment Complexes

Reflectance and scattering could describe the different phenomena occurring on the surface of an object and are used to describe the interaction between light and the object. For most substances, objects with high reflectance reflect light brightly and appear lighter, while objects with low reflectance absorb more light and appear darker [35]. Based on the microstructure analysis of CS and PS (Figure 2), the CS particles were found to be smaller and rougher, whereas the PS particles were larger and smoother. Subsequently, the reflectance of CS was higher than that of PS. Figure 3A indicates clear absorption peaks around 700 nm for both CS-P and PS-P, suggesting a complex formation between starches and paprika red pigment. Moreover, CS-P exhibited higher reflectance compared to PS-P, resulting in a brighter and redder appearance for CS-P. The reflectivity of the complexes increased with increased CS concentration (Figure 3C), implying that a higher concentration of CS yields a redder colour for the complexes. Literature indicates that substances with high reflectivity demonstrate low scattering, and the results of this experiment confirm this hypothesis (Figure 3B,D).

3.3. HPLC Analysis of Adsorption of Paprika Red Pigment from CS and PS

HPLC uses column packing to selectively retain the components in the material to achieve the effect of separation of material components [36]. Moreover, HPLC is often used to detect the colourants, such as synthetic and natural colourants in green-coloured foodstuffs, curcuminoids, and synthetic edible pigments in beverages, etc. [37–39]. Therefore, in this study, the adsorption of CS and PS to the five pigments with the highest paprika red pigment was determined using HPLC. As shown in Table S1, the standard curve is linear ($R^2 > 0.9900$). The adsorption capacity and adsorption ratio show the distribution of the substance on the adsorbent during the adsorption process. A greater adsorption capacity suggests that the adsorbent has the ability to adsorb a higher quantity of substances. A high adsorption ratio may indicate that the adsorbed substance is adsorbed effectively by the adsorbent [40]. As shown in Table 2, both CS and PS adsorbed all five pigments. However, CS adsorbed 85.44 ± 1.10 μg/g, representing 12.5 times higher adsorption capacity than PS. This result suggests that CS has superior adsorption capabilities for paprika red pigments. The adsorption ratios showed that both CS-P and PS-P had high capsanthin percentages, indicating that CS and PS could effectively adsorb capsanthin, and the red pigment is the primary colour-presenting substance in the starch-paprika red pigment complexes. Furthermore, CS had a significantly higher adsorption ratio of capsanthin than PS, implying that CS-P is redder than PS-P. Research suggests that the adsorption capacity of starch for pigment molecules is related to the amount of amylose present [41]. Therefore, it can be inferred that the variances in amylose content between CS and PS might lead to differences in paprika red pigment adsorption by these starches.

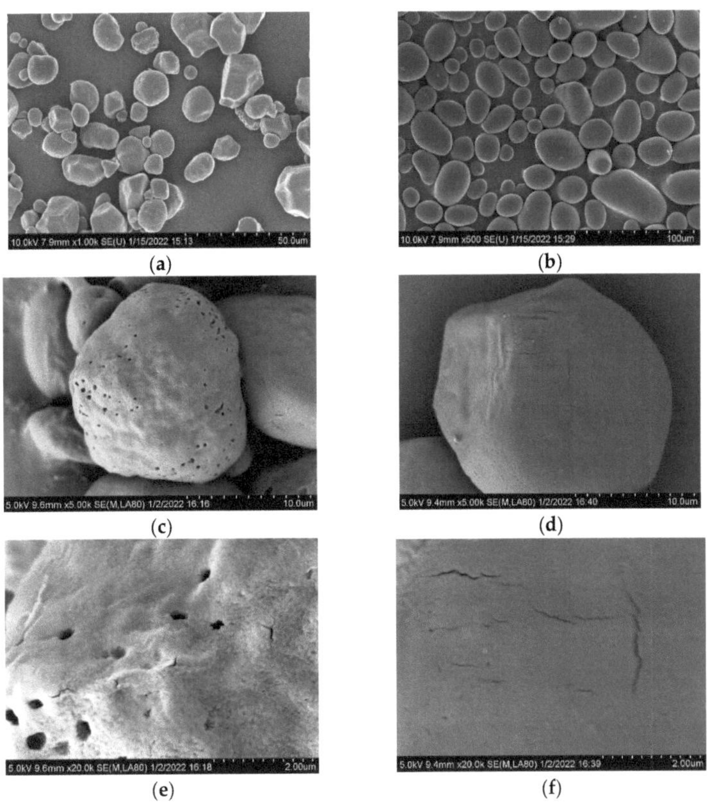

Figure 2. Microscopic morphology of CS (**a**,**c**,**e**) and PS (**b**,**d**,**f**), with ×1 k (**a**,**b**), ×5 k (**c**,**d**), and ×20 k (**e**,**f**) magnification. CS: corn starch, PS: potato starch.

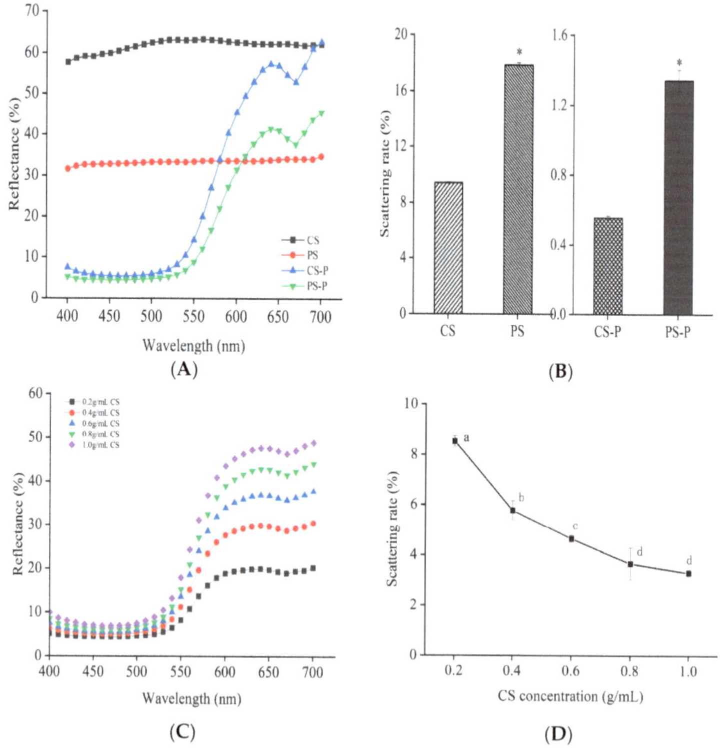

Figure 3. Reflectance spectra of starches and starch-paprika red pigment complexes (**A**), scattering rates of starches and starch-paprika red pigment complexes (**B**), effects of CS concentration on the reflection spectrum of CS-P (**C**), and effects of CS concentration on the scattering rate of CS-P (**D**) are shwon. CS: Corn starch, PS: Potato starch, CS-P: Corn starch-paprika red pigment, and PS-P: Potato starch-paprika red pigment. Significant difference was defined as * $p < 0.05$. Different letters indicate significant differences ($p < 0.05$).

Table 2. Adsorption analysis of carotenoids in paprika red pigment by CS and PS.

		Adsorption Capacity (μg/g)	Adsorption Ratio/%
CS	capsorubin	3.56 ± 0.03 [e]	4.57 ± 0.50 [d]
	capsanthin	40.63 ± 0.65 [a]	48.24 ± 1.79 [a]
	zeaxanthin	6.50 ± 0.08 [d]	8.78 ± 1.37 [cd]
	β-cryptoxanthin	10.00 ± 0.15 [c]	11.85 ± 1.87 [c]
	β-carotene	24.75 ± 0.19 [b]	29.81 ± 4.84 [b]
	Total	85.44 ± 1.10	100
PS	capsorubin	0.40 ± 0.01 [d]	7.37 ± 1.20 [c]
	capsanthin	2.75 ± 0.25 [a]	42.61 ± 2.94 [a]
	zeaxanthin	0.48 ± 0.08 [cd]	9.09 ± 1.90 [c]
	β-cryptoxanthin	0.75 ± 0.06 [c]	11.60 ± 2.24 [c]
	β-carotene	2.27 ± 0.30 [b]	34.14 ± 5.36 [b]
	Total	6.65 ± 1.53	100

CS: Corn starch, PS: Potato starch. Different letters indicate significant differences ($p < 0.05$). Values (mean ± SD) were calculated using the results from three independent experiments.

3.4. CI values of the Starch-Paprika Red Pigment Complexes

The adsorption capacity of amylose and amylopectin for paprika red pigment was determined through the *CI* method. *CI* is based on starch-iodine complex formation in starch representing the degree of starch complexed with binding agents. The higher *CI* values indicate the enhanced degree of formation of V-type inclusion complexes between paprika red pigment and amylose molecules [42]. Herein, the adsorption capacity of CS and PS for paprika red pigment was compared, and the results showed that the capacity of CS to adsorb paprika red pigment was significantly higher than that of PS with the *CI* values of 77.23 ± 5.72% and 53.62 ± 5.90%, respectively. Secondly, the ability of amylose and amylopectin to adsorb paprika red pigment was examined, demonstrating that amylose was more readily bound to the paprika red pigment with a *CI* value of 79.97 ± 3.13%, which was 4 times more than that of amylopectin (Figure 4). This phenomenon might be attributed to the formation of starch molecules. Amylose is a primary linear polysaccharide with α-(1-4)-linked D-glucose units. The complexing agents (lipids, emulsifiers, flavor compounds, etc.) can induce amylose to form a single helix structure. The single helix structure features a notably spacious hydrophobic cavity, wherein the inner surface holds a lipophilic core comprised of C-H groups, while the outer surface consists of polar hydroxyl groups. Additionally, the lipophilic core of the single helix allows for the entrance of the aliphatic group of the complexing agent [43]. Therefore, the single helical structure of the cavity formed by the amylose could adsorb more paprika red pigment. Amylopectin consists of many short chains of glucose linked by α-1,4-glycosidic bonds, and these short chains are linked together by α-1,6-glycosidic bonds at the reducing end, making it a highly branched macromolecule. The weak complexing ability of amylopectin is attributed to the short chain lengths of amylopectin branches and their steric hindrance [44]. CS contained a considerable amount of straight-chain starch, in contrast to PS, which contained up to 79% branched-chain starch. Consequently, the *CI* of CS was significantly higher than that of PS, consistent with previous hypotheses.

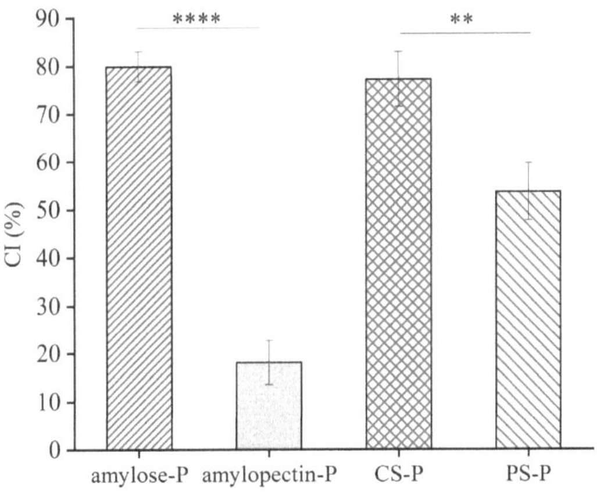

Figure 4. Complexing index of different starch-paprika red complexes. CS: Corn starch, PS: Potato starch, CS-P: Corn starch-paprika red pigment, and PS-P: Potato starch-paprika red pigment. Significant difference was defined as $p < 0.05$ (** $p < 0.01$, **** $p < 0.0001$). Values are expressed as mean ± standard deviation ($n = 3$).

3.5. Particle Size Analysis of Starches and Starch-Paprika Red Pigment Complexes

Particle size plays a vital role in determining the interaction between pigments and macromolecules. The average particle size of CS and PS was significantly higher than the mean particle size of the starch-paprika red pigment complexes (Table 3), which might be because the starches became gelatinous after gelatinisation during the complex formation, resulting in the adhesion between the particles and the loss of the original morphology [13]. The reduced starch particle size could be attributed to the formation of smaller composite particles during amylose release from the starch granules. Particle size can be an indicator of complex stability. Generally, the smaller the particle size, the greater the stability of the complex system [45]. The particle size of the complexes formed by starch and pigment decreased, assuming that the starch-paprika red pigment complexes were more stable. Additionally, the particle size of CS-P was smaller compared to PS-P, indicating that CS-P was more stable than PS-P. The *CI* experiment result indicated that corn starch had a strong binding ability for paprika red pigment, resulting in a higher adsorption capacity compared to PS-P. As a result, the particle size of CS-P was smaller. Similarly, it was reported that the addition of lycopene reduced the particle size of amylose, and with increasing ferulic acid concentration, the particle size of CS decreased [46]. Liang et al. fabricated nanocomplexes of chitosan hydrochloride (CHC), carboxymethyl chitosan (CMC), and anthocyanins (ACNs) through electrostatic interaction to improve the stability of ACNs. At the optimal ratio of 1.2 g CHC to 1.0 g CMC (*w*/*w*, 8 mg of ACNs), the ACNs-loaded CHC/CMC nanocomplexes showed high encapsulation efficiency with a smaller particle size [47]. This phenomenon could be attributed to the fact that after the starches interact with the small molecule, water is expelled from the complex in the presence of the hydrophobic cavity.

Table 3. Particle size, pore area, and porosity of starches and starch-paprika red pigment complexes.

	Particle Size (μm)	Pore Area (m^2)	Porosity (%)
CS	10.16 ± 0.12 ^{b}	0.48 ± 0.06 ^{a}	55.44 ± 0.65 ^{a}
PS	27.54 ± 0.28 ^{a}	0.19 ± 0.05 ^{b}	40.44 ± 0.75 ^{c}
CS-P	5.27 ± 0.60 ^{c}	0.48 ± 0.02 ^{a}	46.05 ± 2.12 ^{b}
PS-P	26.34 ± 0.22 ^{a}	0.25 ± 0.06 ^{b}	41.63 ± 1.81 ^{c}

CS: Corn starch, PS: Potato starch, CS-P: Corn starch-paprika red pigment, and PS-P: Potato starch-paprika red pigment. Values with different letters in the same column are significantly different ($p \leq 0.05$). Values (mean ± SD) were calculated using the results from three independent experiments.

3.6. *Pore Analysis of Starches and Starch-Paprika Red Pigment Complexes*

The pore structure affects the adsorption properties, solubility, and storage of substances. The pore area represents a substance's internal pore structure and helps evaluate its adsorption properties. Porosity is a useful tool for determining the porosity and permeability of a substance [48]. Generally, substances with large pore areas and high porosity exhibit more adsorption sites, resulting in an increased ability to adsorb the targeted substances effectively. Starch has a certain pore structure, so studying and evaluating the pore area and porosity of starch are imperative to expand the starch application and modification. According to Table 3, CS displayed greater pore area and porosity than PS. This result can be explained by the microstructure of starch. As shown in Figure 2, the PS granules had a smooth surface with no signs of pores and cracks. In contrast, the CS particles had more surface pores and cracks, endowing them with better adsorption properties. Therefore, CS-P was redder than PS-P. Lower porosity improves the substance's stability due to the reduced number of voids and pores within the substance. This reduction limits the pathways for air and moisture to enter and contributes to a reduction in some reactions, including oxidation, decomposition, and deterioration [49]. The porosity of CS-P was significantly lower than that of CS after the complexation of starch pigment. However, there was no significant difference between the porosity of PS-P and PS. These results suggest that the inclusion of paprika red pigment improved the stability of CS but had minimal impact on the stability of PS. This experimental result was consistent with the results of *CI* and particle size analyses. Su et al. found that the porosity of tapioca starch decreased following the complexation of pigments from sugarcane juice with tapioca starch [50]. These results offer theoretical guidance for developing an effective adsorbent.

3.7. *Structural Characterisation of Starchs and Starch-Paprika Red Pigment Complexes*

3.7.1. FT-IR Spectroscopy Analysis

FT-IR spectra are primarily used to observe the differences in the crystalline and amorphous regions between the complexes and the natural starch and the bonding forces [51]. Normally, starch shows typical bands and peaks in FT-IR spectra. Therefore, in this study, the effect of paprika red pigment on the structure of starch was explored using FTIR spectroscopy. The infrared spectra of CS, PS, CS-P, and PS-P complexes are illustrated in Figure 5. The typical common absorption peaks are as follows: The absorption peak near 572 cm^{-1} refers to the skeletal mode vibration of starch, the absorption peak at 986 cm^{-1} refers to the backbone vibration of the asymmetric ring mode (α-1,4-glycosidic bonds (C-O-C) of starch, the absorption peaks at 1161 cm^{-1} refers to the C-O bond and the stretching vibration of the C-C bond, the band at 1462 cm^{-1} refers to the bending vibration of CH2 [52], and the band around 1654 cm^{-1} refers to the absorption peak in the amorphous region of adsorbed water in starch. The band at 2156 cm^{-1} was probably due to the presence of free water in starch. An intermediate intensity peak at 2930 cm^{-1} refers to the antisymmetric stretching oscillation of CH2, and the band around 3000 cm^{-1} to 3600 cm^{-1} was dominated by hydrogen bond stretching vibrations and absorption, which was a sensitive indicator to characterize the strength of hydrogen bonds [53]. As shown in Figure 5A, no new absorption peaks appeared, and no absorption peaks disappeared for the starch-paprika red pigment complex compared to CS and PS, indicating no chemical bond modification or covalent bond formation between starch and paprika red pigment. Overall, the addition of paprika red pigment had little effect on the chemical bonding of CS and PS molecules. As shown in Figure 5B,C, after the addition of paprika red pigment, the absorption peaks of CS-P and PS-P complexes were enhanced at 3421.82 cm^{-1}, indicating the formation of a strong hydrogen bond between starches and paprika red pigment [54].

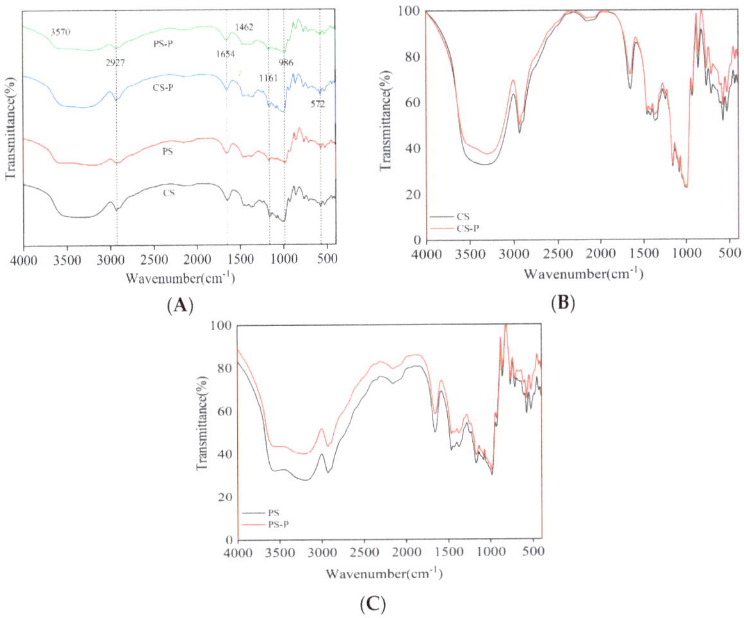

Figure 5. FT−IR spectra of starches and starch−paprika red complexes (**A**), Green line: PS−P, Blue line: CS−P, Red line: PS, Black line: CS. FT−IR spectra of CS and CS−P complex (**B**), and FT−IR spectra of PS and PS−P complex (**C**) are shown. CS: corn starch, PS: potato starch, CS−P: corn starch−paprika red complex, and PS−P: potato starch−paprika red complex.

3.7.2. X-Diffraction Analysis

XRD plays a crucial role in identifying the formation of starch complexes with polyphenols, lipids, and anthocyanins [55]. Therefore, in this study, the binding mechanism between starches and paprika red pigment was determined via XRD. The XRD analysis revealed that type A starch exhibited notable peaks at 2θ = 15.05°, 17.09°, 17.92°, and 23.00° while type B starch exhibited discernible absorption peaks at 2θ = 15.26°, 17.21°, 19.75°, 22.32°, and 24.08° [56]. As shown in Figure 6A,C, CS exhibited strong characteristic peaks at 2θ = 15.50°, 17.09°, and 23.00° and PS exhibited characteristic peaks at 2θ = 17.21°, 19.75°, and 22.32°, indicating that CS exhibits a typical A-type diffraction peak while PS displays a typical b-type diffraction peak. Additionally, the starch V-shaped structure showed characteristic broad peaks at 13.00° and 20.00°. The CS-P and PS-P complexes showed characteristic broad peaks at 20.00° (Figure 6B,D). It can be seen that the starch formed a V-shaped structure after binding to the paprika red pigment, which was consistent with the findings of Liu et al. The presence of paprika red pigment significantly altered the crystal structure of starch. A higher proportion of ligand material increases the probability of forming a V-shaped structure [57]. As shown in Figure 6E, CS-P had a relative crystallinity of 49.21 ± 3.67%, which was significantly higher than that of CS with 11.77 ± 0.99%. Additionally, PS-P had a slightly higher relative crystallinity of 52.74 ± 1.75% than that of PS with 50.93 ± 2.74%. These results suggested that the resultant structure was more stable and compact, with increased relative crystallinity due to the complexation of starch with paprika red pigment.

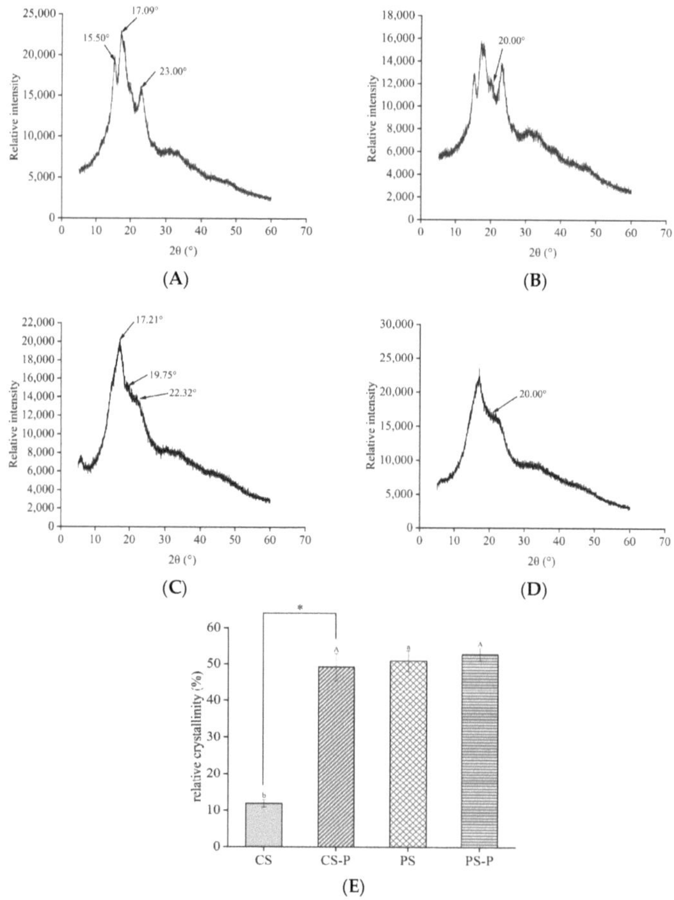

Figure 6. X-ray diffraction of starchs and starch–paprika red complexes. (**A**–**D**) represent CS, CS-P, PS, and PS-P, respectively. The relative crystallinity of starches and starch–paprika red pigment complexes (**E**) is shown. CS: corn starch, PS: potato starch, CS-P: corn starch–paprika red complex, and PS-P: potato starch–paprika red complex. * indicates a significant difference between CS and CS-P ($p < 0.05$). Different small letters indicate significant differences between CS and PS ($p < 0.05$). Different capital letters represent significant differences between CS-P and PS-P ($p < 0.05$).

4. Conclusions

In summary, the CS-P and PS-P complexes were synthesized using CS, PS, and paprika red pigment, and their structures were analyzed. The results showed that both CS and PS had the ability to adsorb paprika red pigment, with CS exhibiting the highest ability. CS-P exhibited a brighter and redder appearance than PS-P. The particle size and pore structure analyses demonstrated that the stability of the starch-paprika red pigment complexes was surpassed compared to the starch molecules. Moreover, CS-P displayed greater stability than PS-P. The FT-IR and XRD analyses demonstrated that strong hydrogen bond interactions were formed between the starch molecules and paprika red pigment. The starch-paprika red pigment complexes displayed a V-type structure with increased relative crystallinity. However, the relative crystallinity of CS-P was greater than that of PS-P, suggesting that the stability of CS-P was superior to that of PS-P. Overall, CS-P has a better colour appearance than PS-P. This study provides a theoretical basis for further exploitation of paprika red pigment and starch. In the future, the starch-pigment complexes can be used in the delivery systems to develop more optimal controlled-release and targeted drug delivery systems. Despite the positive outcomes, the study has many limitations that need to be addressed. Firstly, the colour difference in capsanthin in different starches is influenced by various factors, such as the properties of starch particles, the optical properties of the dispersion system, and the selective adsorption of red and yellow pigments in capsaicin. However, quantifying the contribution of each factor remains a challenge. In addition, further investigation into the colouring effect of different starches on capsanthin is imperative.

Supplementary Materials: The following supporting information can be downloaded at: https://www.mdpi.com/article/10.3390/foods13020191/s1, Table S1: Standard curves for five pigments.

Author Contributions: F.S.: writing—review and editing, methodology, and writing—original draft; Y.W.: data curation, methodology, and investigation; Y.C.: validation and project administration; S.W.: conceptualization, project administration, and supervision. All authors have read and agreed to the published version of the manuscript.

Funding: This research was supported by the National Natural Science Foundation of China (31871808), School Level Cultivation Fund of Beijing Technology and Business University for Distinguished and Excellent Young Scholars (BTBUYP2020), and Cultivation and Development of Innovation Base (Z171100002217019).

Institutional Review Board Statement: Not applicable.

Informed Consent Statement: Not applicable.

Data Availability Statement: All original data included in this paper are available from the authors upon reasonable request. The data are not publicly available due to privacy.

Acknowledgments: We are very grateful to Lijun Qi and Wei Gao for their support and help with the experiment.

Conflicts of Interest: The authors declare no conflicts of interest.

References

1. Arimboor, R.; Natarajan, R.B.; Menon, K.R.; Chandrasekhar, L.P.; Moorkoth, V. Red pepper (*Capsicum annuum*) carotenoids as a source of natural food colors: Analysis and stability—A review. *J. Food Sci. Technol.* **2015**, *52*, 1258–1271. [CrossRef]
2. Hawley, W.R.; Witty, C.F.; Daniel, J.M.; Dohanich, G.P. Choline acetyltransferase in the hippocampus is associated with learning strategy preference in adult male rats. *Behav. Brain Res.* **2015**, *289*, 118–124. [CrossRef] [PubMed]
3. Kim, J.S.; Lee, W.M.; Rhee, H.C.; Kim, S. Red paprika (*Capsicum annuum* L.) and its main carotenoids, capsanthin and beta-carotene, prevent hydrogen peroxide-induced inhibition of gap-junction intercellular communication. *Chem. Biol. Interact.* **2016**, *254*, 146–155. [CrossRef] [PubMed]
4. Gómez, R.; Alvarez-Orti, M.; Pardo, J.E. Influence of the paprika type on redness loss in red line meat products. *Meat Sci.* **2008**, *80*, 823–828. [CrossRef] [PubMed]

5. Kim, J.S.; Ha, T.Y.; Kim, S.; Lee, S.J.; Ahn, J.Y. Red paprika (*Capsicum annuum* L.) and its main carotenoid capsanthin ameliorate impaired lipid metabolism in the liver and adipose tissue of high-fat diet-induced obese mice. *J. Funct. Foods* **2017**, *31*, 131–140. [CrossRef]
6. Echegaray, N.; Guzel, N.; Kumar, M.; Guzel, M.; Hassoun, A.; Lorenzo, J.M. Recent advancements in natural colorants and their application as coloring in food and in intelligent food packaging. *Food Chem.* **2023**, *404*, 134453. [CrossRef] [PubMed]
7. Vertzoni, M.; Kartezini, T.; Reppas, C.; Archontaki, H.; Valsami, G. Solubilization and quantification of lycopene in aqueous media in the form of cyclodextrin binary systems. *Int. J. Pharm.* **2006**, *309*, 115–122. [CrossRef]
8. Wu, X.; Xu, N.; Cheng, C.; McClements, D.J.; Chen, X.; Zou, L.; Liu, W. Encapsulation of hydrophobic capsaicin within the aqueous phase of water-in-oil high internal phase emulsions: Controlled release, reduced irritation, and enhanced bioaccessibility. *Food Hydrocoll.* **2022**, *123*, 107184. [CrossRef]
9. Xiao, J.X.; Huang, G.Q.; Wang, S.Q.; Sun, Y.T. Microencapsulation of capsanthin by soybean protein isolate-chitosan coacervation and microcapsule stability evaluation. *J. Appl. Polym. Sci.* **2013**, *131*, 39671. [CrossRef]
10. Su, X.; Yang, Z.; Tan, K.B.; Chen, J.; Huang, J.; Li, Q. Preparation and characterization of ethyl cellulose film modified with capsaicin. *Carbohydr. Polym.* **2020**, *241*, 116259. [CrossRef]
11. Si, W.; Gao, Y.; Mei, X.; Wu, C.; Li, J.; Zhang, J. Mesoporous silica nanoparticles loaded with capsaicin and their oxidation resistance in meat preservation. *Food Chem.* **2021**, *344*, 128737. [CrossRef]
12. Balde, A.; Kim, S.K.; Benjakul, S.; Nazeer, R.A. Pulmonary drug delivery applications of natural polysaccharide polymer derived nano/micro-carrier systems: A review. *Int. J. Biol. Macromol.* **2022**, *220*, 1464–1479. [CrossRef] [PubMed]
13. Gomes, L.M.M.; Petito, N.; Costa, V.G.; Falcão, D.Q.; de Lima Araújo, K.G. Inclusion complexes of red bell pepper pigments with β-cyclodextrin: Preparation, characterisation and application as natural colorant in yogurt. *Food Chem.* **2014**, *148*, 428–436. [CrossRef] [PubMed]
14. Tran, T.T.; Hadinoto, K. A new solubility enhancement strategy of capsaicin in the form of high-payload submicron capsaicin-chitosan colloidal complex. *Colloid Surf. A* **2017**, *520*, 62–71. [CrossRef]
15. Zhang, K.; Wu, Z.; Tang, D.; Lv, C.; Luo, K.; Zhao, Y.; Liu, X.; Huang, Y.; Wang, J. Development and Identification of SSR Markers Associated with Starch Properties and β-Carotene Content in the Storage Root of Sweet Potato (*Ipomoea batatas* L.). *Front. Plant Sci.* **2016**, *7*, 223. [CrossRef]
16. Feng, T.; Hu, Z.; Wang, K.; Zhu, X.; Chen, D.; Zhuang, H.; Yao, L.; Song, S.; Wang, H.; Sun, M. Emulsion-based delivery systems for curcumin: Encapsulation and interaction mechanism between debranched starch and curcumin. *Int. J. Biol. Macromol.* **2020**, *161*, 746–754. [CrossRef]
17. Wang, H.L.; Lv, J.; Jiang, S.W.; Niu, B.C.; Pang, M.; Jiang, S.T. Preparation and characterization of porous corn starch and its adsorption toward grape seed proanthocyanidins. *Starch-Starke* **2016**, *68*, 1254–1263. [CrossRef]
18. Gisbert, M.; Aleixandre, A.; Sineiro, J.; Rosell, C.M.; Moreira, R. Interactions between Ascophyllum nodosum Seaweeds Polyphenols and Native and Gelled Corn Starches. *Foods* **2022**, *11*, 1165. [CrossRef]
19. Yi, J.; Fan, Y.; Yokoyama, W.; Zhang, Y.; Zhao, L. Characterization of milk proteins-lutein complexes and the impact on lutein chemical stability. *Food Chem.* **2016**, *200*, 91–97. [CrossRef]
20. Barros, F.; Awika, J.M.; Rooney, L.W. Interaction of tannins and other sorghum phenolic compounds with starch and effects on in vitro starch digestibility. *J. Agric. Food Chem.* **2012**, *60*, 11609–11617. [CrossRef]
21. Yifan, L.; Bo, Z.; Huishan, S.; Xiangzhen, G.; Xiangxiang, S.; Qian, Z.; Xiuyun, Z.; Zhuangzhuang, S.; Wenhao, L. Sodium caseinate and acetylated mung bean starch for the encapsulation of lutein: Enhanced solubility and stability of lutein. *Foods* **2021**, *11*, 65. [CrossRef]
22. Yang, F.; Du, Q.L.; Miao, T.; Zhang, X.; Xu, W.; Jia, D.Y. Interaction between potato starch and polysaccharide. *Food Hydrocoll.* **2022**, *127*, 107509. [CrossRef]
23. Zhang, S.; Yue, M.; Yu, X.; Wang, S.; Zhang, J.; Wang, C.; Ma, C. Interaction between potato starch and barley β-glucan and its influence on starch pasting and gelling properties. *Int. J. Biol. Macromol.* **2023**, *253*, 126840. [CrossRef]
24. Tao, X.; Zhan, L.; Huang, Y.; Li, P.; Liu, B.; Chen, P. Preparation, characterization and evaluation of capsaicin-loaded indica rice starch nanoparticles. *Food Chem.* **2022**, *386*, 132692. [CrossRef] [PubMed]
25. Isaschar-Ovdat, S.; Shani-Levi, C.; Lesmes, U. Capsaicin stability and bio-accessibility affected by complexation with high-amylose corn starch (HACS). *Food Funct.* **2021**, *12*, 6992–7000. [CrossRef]
26. Romano, G.; Baranyai, L.; Gottschalk, K.; Zude, M. An Approach for Monitoring the Moisture Content Changes of Drying Banana Slices with Laser Light Backscattering Imaging. *Food Bioprocess Technol.* **2008**, *1*, 410–414. [CrossRef]
27. Kim, J.S.; An, C.G.; Park, J.S.; Lim, Y.P.; Kim, S. Carotenoid profiling from 27 types of paprika (*Capsicum annuum* L.) with different colors, shapes, and cultivation methods. *Food Chem.* **2016**, *201*, 64–71. [CrossRef] [PubMed]
28. Murillo, E.; Turcsi, E.; Szabo, I.; Mosquera, Y.; Agocs, A.; Nagy, V.; Gulyas-Fekete, G.; Deli, J. Carotenoid Composition of the Fruit of Red Mamey (*Pouteria sapota*). *J. Agric. Food Chem.* **2016**, *64*, 7148–7155. [CrossRef]
29. Wang, S.; Chao, C.; Cai, J.; Niu, B.; Copeland, L.; Wang, S. Starch-lipid and starch-lipid-protein complexes: A comprehensive review. *Compr. Rev. Food Sci. Food Saf.* **2020**, *19*, 1056–1079. [CrossRef]
30. Oyeyinka, S.A.; Singh, S.; Ma, Y.; Amonsou, E.O. Effect of high-pressure homogenization on structural, thermal and rheological properties of bambara starch complexed with different fatty acids. *RSC Adv.* **2016**, *6*, 80174–80180. [CrossRef]
31. Włodarczyk-Stasiak, M.; Mazurek, A.; Jamroz, J.; Hajnos, M.; Sokołowska, Z. Influence of physico-chemical modification of waxy corn starch on changes in its structure. *Food Hydrocoll.* **2017**, *70*, 201–210. [CrossRef]

32. Fan, M.; Lian, W.; Li, Y.; Qian, H.; Zhang, H.; Rao, Z.; Wang, L. Evaluation of the physicochemical properties and in vitro digestibility of the complex formed between rice starch and a novel pigment from *Vaccinium bracteatum* Thunb. leaf. *Food Chem.* **2021**, *374*, 131627. [CrossRef] [PubMed]
33. Grebenyuk, P.E.; Chmutin, A.M.; Chuiko, V.A. Algorithm for Hue contrast transformation in the Lab color space. *Izv. Vysših Učebnyh Zaved. Priborostr.* **2018**, *61*, 71–77. [CrossRef]
34. Wu, Y.; Lin, Q.; Chen, Z.; Xiao, H. The interaction between tea polyphenols and rice starch during gelatinization. *Food Sci. Technol. Int.* **2011**, *17*, 569–577. [CrossRef] [PubMed]
35. Gueli, A.M.; Bonfiglio, G.; Pasquale, S.; Troja, S.O. Effect of Particle Size on Pigments Colour. *Color Res. Appl.* **2017**, *42*, 236–243. [CrossRef]
36. Ntrallou, K.; Gika, H.; Tsochatzis, E. Analytical and Sample Preparation Techniques for the Determination of Food Colorants in Food Matrices. *Foods* **2020**, *9*, 58. [CrossRef] [PubMed]
37. Mathiyalagan, S.; Mandal, B.K.; Ling, Y.C. Determination of synthetic and natural colorants in selected green colored foodstuffs through reverse phase-high performance liquid chromatography. *Food Chem.* **2019**, *278*, 381–387. [CrossRef]
38. Lee, J.H.; Choung, M.G. Determination of curcuminoid colouring principles in commercial foods by HPLC. *Food Chem.* **2011**, *124*, 1217–1222. [CrossRef]
39. Li, R.; Wang, Y.; Tan, J.; Tang, S.H.; Jiang, Z.T.; Di, S.D.; Geng, Y. Simultaneous determination of synthetic edible pigments in beverages by titania-based RP-HPLC. *Arab. J. Chem.* **2020**, *13*, 3875–3881. [CrossRef]
40. Zhang, B.; Cui, D.P.; Liu, M.Z.; Gong, H.H.; Huang, Y.J.; Han, F. Corn porous starch: Preparation, characterization and adsorption property. *Int. J. Biol. Macromol.* **2012**, *50*, 250–256. [CrossRef]
41. Zhu, F.; Cui, R. Comparison of physicochemical properties of oca (*Oxalis tuberosa*), potato, and maize starches. *Int. J. Biol. Macromol.* **2020**, *148*, 601–607. [CrossRef] [PubMed]
42. Raza, H.; Liang, Q.; Ameer, K.; Ma, H.; Ren, X. Dual-frequency power ultrasound effects on the complexing index, physicochemical properties, and digestion mechanism of arrowhead starch-lipid complexes. *Ultrason. Sonochem.* **2022**, *84*, 105978. [CrossRef] [PubMed]
43. Pérez, S.; Bertoft, E. The molecular structures of starch components and their contribution to the architecture of starch granules: A comprehensive review. *Starch-Starke* **2010**, *62*, 389–420. [CrossRef]
44. Wang, S.; Wang, J.; Yu, J.; Wang, S. Effect of fatty acids on functional properties of normal wheat and waxy wheat starches: A structural basis. *Food Chem.* **2015**, *190*, 285–292. [CrossRef]
45. Wang, K.; Hong, Y.; Gu, Z.B.; Cheng, L.; Li, Z.F.; Li, C.M. Stabilization of Pickering emulsions using starch nanocrystals treated with alkaline solution. *Int. J. Biol. Macromol.* **2020**, *155*, 273–285. [CrossRef] [PubMed]
46. Karunaratne, R.; Zhu, F. Physicochemical interactions of maize starch with ferulic acid. *Food Chem.* **2016**, *199*, 372–379. [CrossRef] [PubMed]
47. Ge, J.; Yue, P.X.; Chi, J.P.; Liang, J.; Gao, X.L. Formation and stability of anthocyanins-loaded nanocomplexes prepared with chitosan hydrochloride and carboxymethyl chitosan. *Food Hydrocoll.* **2018**, *74*, 23–31. [CrossRef]
48. Guo, L.; Li, J.; Li, H.; Zhu, Y.; Cui, B. The structure property and adsorption capacity of new enzyme-treated potato and sweet potato starches. *Int. J. Biol. Macromol.* **2019**, *144*, 863–873. [CrossRef]
49. Sujka, M.; Jamroz, J. Characteristics of pores in native and hydrolyzed starch granules. *Starch-Starke* **2010**, *62*, 229–235. [CrossRef]
50. Yin, J.Y.; Fang, K.; Li, J.B.; Du, N.; Hu, D.B.; Cao, D.Y.; Tian, R.; Deng, L.G.; Li, K. Competitive adsorption mechanisms of pigments in sugarcane juice on starch-based magnetic nanocomposites. *Int. J. Biol. Macromol.* **2023**, *231*, 123134. [CrossRef]
51. Chen, L.; Tian, Y.; Sun, B.; Cai, C.; Ma, R.; Jin, Z. Measurement and characterization of external oil in the fried waxy maize starch granules using ATR-FTIR and XRD. *Food Chem.* **2017**, *242*, 131–138. [CrossRef] [PubMed]
52. Jivan, M.J.; Madadlou, A.; Yarmand, M. An attempt to cast light into starch nanocrystals preparation and cross-linking. *Food Chem.* **2013**, *141*, 1661–1666. [CrossRef] [PubMed]
53. Fang, J.M.; Fowler, P.A.; Sayers, C.; Williams, P.A. The chemical modification of a range of starches under aqueous reaction conditions. *Carbohyd. Polym.* **2003**, *55*, 283–289. [CrossRef]
54. Warren, F.J.; Gidley, M.J.; Flanagan, B.M. Infrared spectroscopy as a tool to characterise starch ordered structure—A joint FTIR–ATR, NMR, XRD and DSC study. *Carbohydr. Polym.* **2016**, *139*, 35–42. [CrossRef]
55. Deng, N.; Deng, Z.; Tang, C.; Liu, C.M.; Luo, S.J.; Chen, T.T.; Hu, X.T. Formation, structure and properties of the starch-polyphenol inclusion complex: A review. *Trends Food Sci. Technol.* **2021**, *112*, 667–675. [CrossRef]
56. Govindaraju, I.; Zhuo, G.-Y.; Chakraborty, I.; Melanthota, S.K.; Mal, S.S.; Sarmah, B.; Baruah, V.J.; Mahato, K.K.; Mazumder, N. Investigation of structural and physico-chemical properties of rice starch with varied amylose content: A combined microscopy, spectroscopy, and thermal study. *Food Hydrocoll.* **2022**, *122*, 107093. [CrossRef]
57. Cohen, R.; Orlova, Y.; Kovalev, M.; Ungar, Y.; Shimoni, E. Structural and functional properties of amylose complexes with genistein. *J. Agric. Food Chem.* **2008**, *56*, 4212–4218. [CrossRef]

foods

MDPI

Article

Enhancing Encapsulation Efficiency of Chavir Essential Oil via Enzymatic Hydrolysis and Ultrasonication of Whey Protein Concentrate–Maltodextrin

Nasrin Beigmohammadi [1], Seyed Hadi Peighambardoust [1,*], Asad Mohammad Amini [2] and Kazem Alirezalu [1]

1 Department of Food Science, College of Agriculture, University of Tabriz, Tabriz 5166614766, Iran; nasrinbeigmohammadi@yahoo.com (N.B.); kazem.alirezalu@tabrizu.ac.ir (K.A.)

2 Department of Food Science and Engineering, Faculty of Agriculture, University of Kurdistan, Sanandaj 6617715175, Iran; a.mohammadamini@uok.ac.ir

* Correspondence: peighambardoust@tabrizu.ac.ir; Tel.: +98-413-339-2052; Fax: +98-413-334-5332

Abstract: This study focused on the characterization of emulsions and microparticles encapsulating Chavir essential oil (EO) by application of modified whey protein concentrate–maltodextrin (WPC-MD). Different physical, chemical, morphological, thermal, and antioxidant properties and release behavior of spray-dried microparticles were assessed. Antioxidant, solubility, emulsifying, and foaming activities of modified WPC were increased compared to those of primary material. The results indicated that the particle size distribution varied depending on the type of carriers used, with the smallest particles formed by hydrolyzed WPC (HWPC). Binary blends of modified WPC-MD led to improved particle sizes. The spray-drying yield ranged from 64.1% to 85.0%, with higher yields observed for blends of MD with sonicated WPC (UWPC). Microparticles prepared from primary WPC showed irregular and wrinkled surfaces with indentations and pores, indicating a less uniform morphology. The UWPC as a wall material led to microparticles with increased small cracks and holes on their surface. However, HWPC negatively affected the integrity of the microparticles, resulting in broken particles with irregular shapes and surface cracks, indicating poor microcapsule formation. Encapsulating EO using WPC-MD increased the thermal stability of EO significantly, enhancing the degradation temperature of EO by 2 to 2.5-fold. The application of primary WPC (alone or in combination with MD) as wall materials produced particles with the lowest antioxidant properties because the EO cannot migrate to the surface of the particles. Enzymatic hydrolysis of WPC negatively impacted microparticle integrity, potentially increasing EO release. These findings underscore the crucial role of wall materials in shaping the physical, morphological, thermal, antioxidant, and release properties of spray-dried microparticles, offering valuable insights for microencapsulation techniques.

Keywords: core material; chavir essential oil; encapsulation; hydrolysis; ultrasonication; stability

Citation: Beigmohammadi, N.; Peighambardoust, S.H.; Mohammad Amini, A.; Alirezalu, K. Enhancing Encapsulation Efficiency of Chavir Essential Oil via Enzymatic Hydrolysis and Ultrasonication of Whey Protein Concentrate–Maltodextrin. *Foods* **2024**, *13*, 1407. https://doi.org/10.3390/foods13091407

Academic Editors: Shuai Chen and Qing Guo

Received: 13 April 2024
Revised: 24 April 2024
Accepted: 29 April 2024
Published: 3 May 2024

1. Introduction

Ferulago angulata is a perennial plant with small yellow flowers locally known as 'Chavir' or 'Chavil', which grows in the western provinces of Iran (Kermanshah and Kurdistan) [1]. Traditionally, the aerial parts of the plant are added to edible fats, butter, and meat products by locals to increase their shelf life while also enhancing their flavor [2]. The essential oil (EO) extracted from the flowers of Chavir is a rich source of bioactive compounds with potent antioxidant and antibacterial properties, in addition to other medicinal benefits [3,4]. Based on reports in the literature, these components mostly belong to monoterpenes and sesquiterpenes categories. More specifically, compounds such as α-pinene, α-phellandrene, β-ocimene, p-cymene, methyl carvacrol, myrcene, and terpinolene have been reported in the EO from *Ferulago angulata* [5]. In general, EOs are highly concentrated extracts that contain volatile compounds responsible for the characteristic fragrance and therapeutic properties of the plant. Some biological activities such as antimicrobial, antioxidant, and therapeutic properties such as anthelmintic, antifungal, and antiviral have been reported for EO from Chavir [3,6,7]. The concentration of EO in aromatic herbal plants can vary depending on factors such as the plant species, growing conditions, and extraction method. According to previous reports, the EO yield of the aerial parts and seeds of *Ferulago angulata* collected from western Iran (Kermanshah province) was 0.63 and 3.2% (*v*/*w*) based on the dry weight, respectively [3].

Despite the beneficial biological properties of essential oils, their direct application in food can pose challenges. High concentrations of EOs used to leverage their antimicrobial effects can lead to undesirable organoleptic effects in food [8]. To address this issue, researchers have explored the encapsulation of EOs in stable micro- or nano-particles as a viable approach to enhance their utilization, stability, and efficacy [9]. Moreover, due to the volatile nature and the instability of EO components, it is necessary to preserve them against possible adverse environmental or processing conditions while maintaining their bioavailability via microencapsulation. There are different microencapsulation techniques proposed with their benefits and influential parameters [10]. Spray-drying is the prevalent method used for solidifying emulsions using suitable carriers based on proteins and polysaccharides in their intact or modified forms [11]. A decade overview and prospect of spray drying encapsulation of bioactive compounds from fruit products showed that the type and the ratio of encapsulating material had a significant impact on the physiochemical properties of the microcapsules, and spray drying encapsulation could open up new possibilities for the controlled delivery of beneficial compounds to the host [12].

Whey proteins are suitable material for the microencapsulation of bioactive compounds via a spray-drying process. However, low solubility, aggregation, agglomeration, sedimentation, and denaturation of these proteins under acidic conditions and/or under mechanical and thermal stress may cause loss of their functional properties and limit their use as an emulsifier and carrier [13]. Therefore, modification of proteins' structure by non-thermal techniques, such as high-pressure processing (HPP), pulsed electric field (PEF), cold plasma, irradiation, and ultrasonication, have been proposed [14,15]. The application of high-power ultrasound would enhance the surface activity of most proteins, leading to improved foaming and emulsifying properties [16]. Partial hydrolysis of proteins produced hydrolysates or peptides with enhanced functional characteristics and biological activities [17]. Proteases break peptide bonds and liberate low molecular weight protein fractions with improved solubility and biological and surface activities, as compared to their primary proteins [18]. This structural modification and improved functional properties of protein hydrolysates positively affect their amphiphilic characteristics by their tendency to interact at the water/oil interfaces, leading to stabilizing emulsions [19]. Moreover, as a result of enzymatic, structural changes in proteins, they effectively show film-forming properties that can be utilized as wall materials for the microencapsulation of many foods' bioactive components. Maltodextrin (MD), a commonly used food additive, is among the wall materials that are used for spray-drying microencapsulation of active bioactive substances due to its availability and lower price, biocompatibility, biodegradability, and non-toxicity. The combination of MD with other hydrocolloids, gums, and protein hydrolysates can positively affect the physicochemical characteristics of obtained microparticles. In this respect, spray-drying encapsulation efficiency and release properties of particles can be improved [20].

Nevertheless, there is limited research on the application of enzymatically or power ultrasound-modified WPC as wall materials for spray-drying EO-loaded microparticles. One of the primary goals of this study was to use binary blends of MD with WPC that are modified by enzymatic hydrolysis and ultrasonication as carriers in spray-drying microencapsulation. The combination of MD with protein hydrolysates can positively affect the physicochemical characteristics of obtained microparticles. Therefore, the main objectives of this study were the following: (1) to evaluate the effectiveness of physically or enzymatically modified WPC and its combination with MD as carriers for microencapsulation of herbal essential oil during spray-drying; (2) to examine the properties of emulsions produced from the binary blends of modified WPC with MD; and (3) to explore the diverse technological and functional attributes of spray-dried powders containing the essential oil.

2. Materials and Methods

2.1. Materials

Aerial parts (leaf and stem) of Chavir (*Ferulago angulate*) were collected from the Dallahou mountains (Kermanshah, Iran) by three local farmers over two days. The plant's authenticity was confirmed by expert botanists from the University of Kurdistan (Sanandaj, Iran). The freshly collected Chavir was then dried in the shade in preparation for essential oil extraction. MD with DE of 15–20 (18.8 DE, test result by provider) and WPC (Hilmar 9010, 92.0% protein, dry basis) were purchased from Dalian Future International Co. (Dalian, China) and Hilmar Ingredients (Hilmar, CA, USA), respectively. The Alcalase enzyme from *Bacillus licheniformis* (specific activity $\geq$ 0.75 Anson units/mL) was acquired from Sigma-Aldrich (St. Louis, MI, USA), and other chemicals were provided from Merck (Darmstadt, Germany).

2.2. Extraction and Characterization of EO

A Clevenger-type apparatus applying the steam-distillation method was used to extract the EO from the aerial parts of the plant according to the detailed method reported previously [21]. The collected EO (1.2 $\pm$ 0.1% v/w, wet basis) was purified following freezing at -20 °C and stored in airtight dark containers for subsequent analysis. A GC-MS system (Agilent 7890A gas chromatography with 5975C mass spectrometer, Agilent Technologies, Santa Clara, CA, USA) equipped with a DB-35MS column (30 m long, ID 0.32 mm) was employed to identify the chemical constituents of the EO according to the method reported previously [6].

2.3. Whey Protein Modification

The enzymatic hydrolysis of 10% (w/v) whey protein aqueous solution in potassium phosphate buffer (pH = 7.6) was performed using Alcalase enzyme (100:2 ratio) at a constant temperature of 38 °C for 60 min under continuous stirring (300 rpm). Following hydrolysis, the enzyme activity was halted by placing the reaction chamber in a hot water bath at 96 °C for 20 min. This sample is encoded as HWPC (hydrolyzed WPC). The obtained mixture was centrifugated at 8000$\times$ g for 15 min, and the supernatant comprising hydrolysates was collected and subjected to freeze-drying until subsequent use. For power ultrasonic treatments, 100 mL of WPC (5% w/v) in phosphate buffer (pH adjusted to 7.4) was sonicated using a probe-type device (UHP-400, Hielscher, Teltow, Germany) with a titanium probe (diameter 13 mm) with operating power of 400 W and frequency of 20 kHz with a pulsed mode of 1.0 s on/1.0 s off for 30 min. The solution temperature was kept below 15 °C using an ice bath during the sonication. The sonicated WPC solution (called UWPC) was freeze-dried and stored frozen until the next use.

2.4. Emulsion Preparation and Characterization

Different formulations of wall materials and their binary systems were prepared according to Table 1.

The total solids content of emulsions was fixed at 30% (for use as feed for spray-drying), in which the binary systems were mixed at a 1:1 ratio. The emulsion preparation procedure consisted of adding wall material including MD, WPC, hydrolyzed WPC (H-WPC), ultrasound-treated WPC (U-WPC), or their blends to distilled water. The solution was brought to continuous stirring at ambient temperature for 30 min, shaking for 24 h using an orbital shaker at room temperature. Then, an appropriate amount of EO (based on EO to WM ratios suggested in Table 1) was added to the latter solution, and it was homogenized at 7500 rpm for 4 min using Ultra-Turrax T18 digital (IKA, Staufen Im Breisgau, Germany). The sample codes for emulsions are mentioned in Table 1. The stability of emulsions, as a critical measure for spray-drying, was measured by the creaming index (CI) according to the detailed description by Frascareli et al. [22]. The droplet size and Zeta potential of emulsions were characterized based on the dynamic light scattering (DLS) method using Zetasizer Nano ZS (Malvern Panalytical, Malvern, UK) at 24 °C. The average droplet size was reported as the Z-average diameter of particles.

Table 1. The formulations of emulsions prepared using different wall materials (WM) and EO ratios.

Sample Code	Wall Materials	EO:WM Ratio
MD2	Maltodextrin	1:2
MD3	Maltodextrin	1:3
WPC2	Whey protein concentrate	1:2
WPC3	Whey protein concentrate	1:3
HWPC2	Hydrolyzed WPC	1:2
HWPC3	Hydrolyzed WPC	1:3
UWPC2	Ultrasound-treated WPC	1:2
UWPC3	Ultrasound-treated WPC	1:3
MD-WPC2	Maltodextrin + WPC	1:2
MD-WPC3	Maltodextrin + WPC	1:3
MD-HWPC2	Maltodextrin + hydrolyzed WPC	1:2
MD-HWPC3	Maltodextrin + hydrolyzed WPC	1:3
MD-UWPC2	Maltodextrin + ultrasound-treated WPC	1:2
MD-UWPC3	Maltodextrin + ultrasound-treated WPC	1:3

2.5. Spray-Drying Microencapsulation

A pilot-scale spray dryer (Maham Sanaat Co., Neyshabur, Iran) was used to solidify prepared emulsions. The spray-drier chamber (100 cm diameter × 200 cm height) was used with a rotary atomizer (5 cm diameter) in a co-current flow layout. A volumetric feed pump (fluid capacity of 3 L/h and a pressure of 1 bar) was used to feed the atomizer. In preliminary experiments, drying parameters were examined, and the best drying parameters were selected as follows. The inlet and outlet air temperatures were set to 160 ± 5 °C and 80 ± 5 °C, respectively. Atomizer rotational speed of 15,000 rpm, feed flow rate of 10 mL/min, feed temperature of 30 ± 2 °C, and atomizer pressure of 4.5 ± 0.2 bar were kept constant during all spray-drying trials. Based on the preliminary results, emulsions prepared with higher ratios of essential oil/wall materials (EO:WM) showed larger droplets with all wall materials used compared to their corresponding samples with lower EO concentrations. For this reason, spray-drying treatments were carried only with feed emulsions of lower EO content (EO:WM ratio of 1:3). This feed composition was kept constant for all spray-drying treatments. Dried emulsion powders were collected from the bottom of the drying chamber and kept in air-tight containers inside the refrigerator (4 ± 0.1 °C) until used. The weight ratio of spray-dried powders to the initial solid contents in the feed solution of the emulsions was stated as production yield.

2.6. Droplet and Particle Size and Zeta Potential

The droplet/particle sizes and the Zeta potentials of emulsions and spray-dried microparticles were assessed using the dynamic light scattering (DLS) technique with a Zetasizer (Nanosizer 3000, Malvern Instruments, Malvern, UK) at ambient temperature with an angle of 90°. Prior to measurements, the samples were diluted 100-fold with distilled water. The experiments were conducted in triplicate, and the volume mean diameter was utilized to indicate the average emulsion/particle size.

2.7. Encapsulation Efficiency of Emulsions and Spray-Dried Microparticles

The essential oil impregnated inside emulsion droplets or spray-dried microspheres is regarded as encapsulation efficiency (EE). To measure EE of emulsions, aliquots of 5 mL emulsion and n-hexane were mixed under magnetic stirring for 30 min to remove the EO content on the surface of droplets. The mixture was then centrifugated for 3 min at

3000 rpm, and the solvent phase was discarded. This procedure was repeated twice. In the next step, the obtained emulsion was mixed with 10 mL deionized water and mixed for 5 min, blended with 5 mL n-hexane while stirring for 5 min, followed by centrifugating at 5000 rpm for 10 min. The solvent phase containing EO was collected, and its absorbance was determined at 270 nm using a UV-2100 spectrophotometer (UNICO, Franksville, WI, USA). The concentration of EO was calculated based on a calibration curve, and the percentage ratio of the loaded EO to its initial concentration was determined as the encapsulation efficiency (EE) of the emulsions [23]. For the spray-dried particles, the same procedure was followed, except that 500 mg of powder was used instead of 5 mL of emulsion [24].

2.8. Density, Flowability, and Cohesiveness of Obtained Powders

The bulk density (ρ_b) and tapped density (ρ_t) of the powders were determined by the ratio of their mass to the occupied volume, measured using a graduated cylinder before and after tapping the cylinder, respectively. The Carr index (*CaI*) and Hausner ratio (*HR*) were computed to assess the flowability and cohesiveness of the microspheres, as per Equations (1) and (2), respectively.

$$CaI(\%) = \frac{(\rho_t - \rho_b)}{\rho_t} \times 100 \quad (1)$$

$$HR = \frac{\rho_t}{\rho_b} \quad (2)$$

2.9. Solubility, Hygroscopicity, and Wettability

For the measurement of water solubility, 1 g of powder was dispersed in 15 mL of distilled water, vortexed for 3 min, and then centrifuged at 4000 rpm for 15 min. The supernatant was collected and dried at 110 ± 5 °C until a constant weight was achieved. The percentage ratio of the dry weight of solubilized materials to the initial sample weight was considered as the solubility [25]. To measure hygroscopicity, a powder sample (predetermined weight) was placed in a desiccator (75% relative humidity obtained by saturated NaCl solution) for 7 days. The hygroscopicity was defined as weight gain (g) of water per 100 g sample. To assess wettability, an amount of 50 mg powder was sprinkled onto the water surface, and the time taken (in seconds) for complete submergence of the particles was recorded as the wettability parameter.

2.10. Field Emission Scanning Electron Microscopy (FE-SEM)

The surface morphology of spray-dried powders was characterized using SEM (MIRA3 TESCAN, Brno, Czech Republic). The powder samples were mounted on aluminum plates and coated with a thin layer of gold using a vacuum ion sputter (SC7620 sputter-coaters, Quorum Technologies, Laughton, UK) for SEM observation.

2.11. Color Analysis

The color characteristics of the spray-dried powders were evaluated using a digital image processing technique as outlined in a previously described method [26]. Color parameters were calculated using Equations (3)–(5):

$$C^* = \sqrt{(a^*)^2 + (b^*)^2} \quad (3)$$

$$WI = 100 - \sqrt{(100 - L^*)^2 + (a^*)^2 + (b^*)^2} \quad (4)$$

$$YI = \frac{142.86 \times b^*}{L^*} \quad (5)$$

where L^* = lightness, a^* = redness/greenness, b^* = yellowness/blueness, and C^*, WI, and YI are chroma, whiteness, and yellowness indices of the samples, respectively.

2.12. Fourier Transform-Infrared Spectroscopy (FT-IR)

The powder samples were blended with potassium bromide (KBr) at a ratio of 1:100 and compressed into disk shapes. These sample discs were analyzed using an FT-IR spectrophotometer (Shimadzu 8400, Tokyo, Japan) in transmission mode across the wavenumber range of 400–4000 cm^{-1}.

2.13. Thermal Properties

The thermal properties of free EO and spray-dried powders encapsulating EO were analyzed by a simultaneous thermal analyzer (SDT-Q600, TA instruments, Lindon, UT, USA). An amount of approximately 15 mg samples were transferred to aluminum pans, hermetically sealed and heated from 30 to 600 °C at 10 °C/min under a stream of Argon atmosphere with 50 mL/min flow rate.

2.14. Antioxidant Properties

The antioxidant capacity of spray-dried encapsulated microparticles, along with that of free EO, was measured using DPPH and $ABTS^+$ radicals' scavenging activity assays. A method by Shahi et al. [27] was used to measure the inhibition percent for both DPPH and $ABTS^+$ radicals (Equation (6)), assuming that the absorptions for DPPH and $ABTS^+$ radicals were read at 517 nm and 734 nm, respectively:

$$Inhibition(\%) = 1 - \frac{A_{sample}}{A_{blank}} \times 100 \quad (6)$$

2.15. In-Vitro Release Behavior

The amount of EO released from spray-dried particles was assessed using dialysis bags in a buffer phosphate medium [28]. In this measurement, 40 mg of spray-dried powder was suspended in 4 mL of phosphate buffer at a pH of 7.3 (as release media), and the suspension was transferred to a dialysis bag (D0530, Sigma Aldrich, Hamburg, Germany) previously conditioned in the same media. The bag was immersed in the release media (20 mL) and kept at a constant temperature of 35 °C with gentle shaking for a while. To measure released EO, specific volumes of release media were taken (substituted with an equivalent volume of fresh buffer) at defined time intervals of up to 300 min. Then, UV–vis spectrometry (Varian Cary 500, Agilent Technologies, Santa Clara, CA, USA) at 270 nm was used to measure the amount of released EO from dried particles. The cumulative concentration of released EO was calculated during the whole release time.

2.16. Statistical Analyses

The impact of applying partial enzymatic hydrolysis or ultrasound treatment of WPC and using their binary combination with MD as wall material treatments on different properties of spray-dried microparticles was investigated through factorial experiments using a completely randomized design (CRD). Mean comparisons were conducted using Duncan's test at a significance level of 5%. Statistical analyses were performed using SAS software version 9.2. All preparations and measurements were independently conducted in triplicate to ensure reliability.

3. Results and Discussion

3.1. Characterization of Essential Oil from Chavir

The yield of extracted EO from Chavir was 1.2 ± 0.1% (v/w, wet basis), similar to the values obtained in previous reports [4]. GC-MS analysis of EO identified 21 chemical compounds comprising 89.6% of the total chemical composition. Among other constituents of EO, cis-β-ocimene with 23.25%, α-pinene with 14.51%, bornyl acetate with 11.25%, and cis- and trans-verbenol with 7.17% were the most abundant compounds in the EO (Table 2). This finding aligns with the outcomes documented in previous studies [3,21,29].

Table 2. Chemical composition of EO from Chavir (*Ferulago angulate*).

Component	Retention Time	Percentage
α-Pinene	3.95	14.51
Camphene	4.43	3.66
β-Pinene	5.02	3.27
β-Myrcene	5.41	3.96
δ-3-Carene	5.75	1.31
D-Limonene	6.38	3.41
cis-β-Ocimene	6.98	23.25
γ-Terpinene	7.30	4.17
Linalool	8.53	2.27
allo-Ocimene	8.97	2.13
cis-Verbenol	9.95	1.85
trans-Verbenol	10.29	5.32
Borneol	10.79	1.10
α-Terpineol	10.80	0.79
α-Phellandren-8-ol	11.10	1.19
Bornyl acetate	13.20	11.25
α-Muurolene	17.15	2.47
Methyl eugenol	17.47	0.89
Cadinol	18.10	0.45
β-Caryophyllene	22.35	0.58
Spathulenol	20.33	1.76
Total		89.59

3.2. Physical Characteristics of Emulsions Containing EO

The droplet size of emulsions containing *Ferulago angulata* EO prepared using MD and WPC (modified or unmodified) varied within a range of 203.2–1185.0 nm (Table 3). The enzyme hydrolysis of whey protein resulted in emulsions with slightly smaller droplet sizes (243–289 nm) than those prepared using untreated WPC with an emulsion droplet size of 253–338 nm. On the other hand, power ultrasonication of WPC led to an increase in droplet size of emulsions with higher amounts of EO, which was opposite to samples with lower EO content. Based on the results, it was evident that the droplet size of emulsions containing higher EO content was larger (289 nm in HWPC2 sample to 1185 nm in MD-HWPC2 sample, Table 3) compared with their corresponding samples with lower EO concentration, which fairly varied in the range of 203 nm in UWPC3 sample to 253 nm in MD-WPC3 sample. The largest size was achieved for binary systems of MD-HWPC2 (with 1185 nm) and MD-WPC2 (with 849 nm). Generally, binary emulsion samples were of larger droplet size in comparison with simple emulsion systems, more profoundly for the EO:WM ratio of 1:2. This could be an indication of effective interactions between MD and WPC, which has also been reported by others [30].

The consistency (uniformity) of droplet size in emulsions can be deduced from the polydispersity index (PDI), as shown in Table 3. The PDI results unveiled that almost all emulsions were of acceptable uniformity, the best observed for samples at an EO:WM ratio of 1:3, which were of smaller droplet size. Furthermore, modification of whey protein using both methods resulted in more uniform emulsions than those prepared by intact whey protein. The emulsions prepared by MD alone had the highest PDI values (0.512–0.630). Due to the lower PDI and droplet size of samples at emulsions with an EO:WM ratio of 1:3 it was decided to produce spray-dried particles with this EO:WM ratio.

Based on creaming index results presented in Table 3, it was found that all emulsion samples were stable without any measurable phase separation, except those samples prepared by MD, which had phase separation above 60%. The high instability of the MD-prepared emulsion could probably be related to the lack of surface-active properties of

MD [31]. These results are consistent with the droplet size and polydispersity index (PDI) data obtained in this investigation. The stability of emulsions could be attributed to the enhanced emulsifying and emulsion stabilizing properties of HWPC and UWPC [17].

Table 3. Physical characteristics of emulsions with different concentrations of *Ferulago angulata* EO.

Sample Code	Droplet Size (nm)	PDI *	Zeta Potential (mV)	CI (%) **	EE (%) ***
MD2	294.2	0.512	−11.6	87	12.2
MD3	213.0	0.630	−20.9	64	15.8
WPC2	338.2	0.475	−30.6	nd	89.6
WPC3	252.7	0.282	−21.5	nd	60.5
HWPC2	289.4	0.368	−17.7	nd	90.1
HWPC3	242.7	0.345	−21.6	nd	78.7
UWPC2	583.4	0.572	−36.1	nd	90.0
UWPC3	203.2	0.238	−23.6	nd	76.3
MD-WPC2	848.8	0.601	−18.0	nd	91.6
MD-WPC3	253.4	0.365	−25.1	nd	91.3
MD-HWPC2	1185.0	0.374	−25.2	nd	93.5
MD-HWPC3	251.2	0.271	−27.4	nd	93.0
MD-UWPC2	655.4	0.518	−38.3	nd	92.4
MD-UWPC3	230.3	0.268	−29.2	nd	90.2

* PDI—polydispersity index; ** CI—creaming index; *** EE—encapsulation efficiency; nd—not determined.

The charge density on the surface of emulsions characterized by Zeta potential (Table 3) revealed that all samples were of moderately high charge, inferring the stability of droplets in the continuous phase, as observed by the results of CI. Zeta potential values of UWPC emulsions were higher than those for intact WPC, which in turn were higher than HWPC emulsions. The Zeta potential values of modified whey proteins found in the present work were higher than those reported by Sarabandi et al. [13]. Generally, it can be postulated from Zeta potential data that blending MD with WPC reduces the charge density of particles, an indication of successful interactions between the two components. The larger the emulsion droplets, the lower the Zeta potential value.

The lowest encapsulation efficiency (EE) of essential oil was observed for emulsions prepared by MD (the most unstable sample), whereas the EE of binary emulsions was generally higher than 90% (Table 3), showing the most successful systems for encapsulation of EO. The emulsions with the best EE were MD-HWPC at both EO:WM ratios used. Although simple emulsions of WPC (intact and modified) had acceptable EE values, the EE increased at higher EO concentrations. The high EE values of binary blend systems have been attributed to the combined emulsification properties of MD and WPC [32], especially for modified WPCs.

3.3. *Physical Characterization of Spray-Dried Microparticles*

According to the results in Table 4, the largest particle sizes corresponded to binary blends of untreated WPC with MD (30.21 μm). Untreated sole MD and WPC as wall materials gave particles with an average size of 19.52–24.61 μm. Application of hydrolyzed (HWPC) and sonicated WPC (UWPC) led to the production of microparticles with the smallest size with an average diameter of 6.23 and 12.95 μm, respectively. However, the application of binary blends of MD with HWPC or UWPC increased their mean particle size to 8.24 μm and 13.5 μm, respectively. It was concluded from dynamic particle size measurement that enzymatic hydrolysis of WPC led to the formation of the smallest particles.

Moreover, the application of binary blends of MD with each of the treated WPC (both HWPC and UWPC) had an improving effect on particle size. Different particle sizes of samples can be attributed to the viscosity of the produced feed and, thereby, an increase in the size of the atomized droplets inside the chamber [33]. Spray-drying yield varied in a range of 64.1–85.0% for different powder samples. Higher yields were obtained for binary blends of MD with UWPC samples. Apparently, large particles, especially those produced from binary blends of carriers, resulted in more recovery in the drying chamber. Contrary to powders obtained from ultrasound-treated WPC, enzymatic hydrolysis of

WPC led to the formation of small particles, resulting in less recovery from the drying chamber (yield = 64.1–67.0%).

Table 4. Production yield and physical-functional properties of spray-dried powders containing EO.

Sample Code *	Yield (%)	Mean Particle Size (μm)	PDI	Zeta Potential (mV)	EE (%)
MD	75.1 ± 2.5 [b]	19.52 ± 4.50 [b]	0.60 ± 0.05 [ab]	−25.7 ± 0.45 [c]	78.9 ± 1.1 [c]
WPC	79.0 ± 1.7 [a]	24.61 ± 3.19 [b]	0.65 ± 0.08 [a]	−24.9 ± 0.13 [c]	80.0 ± 1.5 [c]
HWPC	64.1 ± 2.8 [c]	6.23 ± 0.55 [e]	0.55 ± 0.05 [b]	−18.5 ± 0.12 [e]	58.2 ± 1.6 [e]
UWPC	79.2 ± 1.1 [b]	10.95 ± 0.51 [d]	0.33 ± 0.06 [c]	−26.3 ± 0.18 [b]	85.3 ± 1.9 [b]
MD-WPC	82.3 ± 2.1 [a]	30.21 ± 4.52 [a]	0.65 ± 0.03 [a]	−25.5 ± 0.12 [c]	80.2 ± 0.8 [c]
MD-HWPC	67.0 ± 1.2 [c]	7.24 ± 0.85 [e]	0.53 ± 0.05 [b]	−20.3 ± 0.21 [d]	61.3 ± 1.0 [d]
MD-UWPC	85.0 ± 2.7 [a]	12.92 ± 0.81 [c]	0.31 ± 0.07 [c]	−29.1 ± 0.37 [a]	89.0 ± 1.2 [a]
	Bulk density (g/mL)	Tapped density (g/mL)	Angle of repose (°)	Hausner ratio	Carr index
MD	0.55 ± 0.01 [a]	0.65 ± 0.02 [a]	31.3 ± 0.7 [c]	1.15 ± 0.07 [a]	0.13 ± 0.05 [a]
WPC	0.48 ± 0.02 [b]	0.58 ± 0.02 [b]	33.3 ± 0.5 [b]	1.19 ± 0.08 [a]	0.14 ± 0.06 [a]
HWPC	0.44 ± 0.01 [c]	0.50 ± 0.01 [c]	35.0 ± 0.4 [a]	1.19 ± 0.05 [a]	0.17 ± 0.07 [a]
UWPC	0.42 ± 0.02 [c]	0.51 ± 0.02 [c]	36.7 ± 0.3 [a]	1.21 ± 0.04 [a]	0.17 ± 0.08 [a]
MD-WPC	0.58 ± 0.02 [a]	0.62 ± 0.01 [a]	30.8 ± 1.4 [c]	1.14 ± 0.02 [a]	0.12 ± 0.04 [a]
MD-HWPC	0.55 ± 0.03 [a]	0.59 ± 0.02 [a]	31.3 ± 0.5 [c]	1.27 ± 0.10 [a]	0.21 ± 0.05 [a]
MD-UWPC	0.53 ± 0.02 [a]	0.63 ± 0.02 [a]	31.7 ± 0.7 [c]	1.18 ± 0.03 [a]	0.15 ± 0.07 [a]
	Moisture content (%)	a_w	Solubility (%)	Hygroscopicity (%)	
MD	3.11 ± 0.11 [a]	0.33 ± 0.03 [a]	74.90 ± 3.51 [c]	9.15 ± 0.31 [d]	
WPC	2.86 ± 0.05 [a]	0.34 ± 0.01 [a]	75.79 ± 2.19 [c]	9.17 ± 0.40 [d]	
HWPC	3.42 ± 0.05 [a]	0.31 ± 0.02 [a]	78.70 ± 1.92 [c]	10.21 ± 0.24 [c]	
UWPC	3.07 ± 0.08 [a]	0.36 ± 0.02 [a]	75.54 ± 2.23 [c]	9.66 ± 0.42 [d]	
MD-WPC	3.23 ± 0.06 [a]	0.33 ± 0.01 [a]	81.34 ± 1.58 [b]	10.78 ± 0.72 [b]	
MD-HWPC	3.04 ± 0.10 [a]	0.35 ± 0.01 [a]	86.35 ± 0.81 [a]	12.36 ± 0.16 [a]	
MD-UWPC	3.13 ± 0.05 [a]	0.32 ± 0.01 [a]	79.71 ± 0.82 [b]	11.94 ± 0.36 [b]	

* Sample codes are explained in Table 1. PDI—polydispersity index; EE—encapsulation efficiency. Data are represented as means of triplicate measurements ± SD. Different alphabetical letters show significant differences between means ($p < 0.05$).

The particle size distribution of spray-dried particles varied in the range of 0.31–0.65, which were of acceptable PDI values. Of course, there was a particular dependency of PDI values on the type of carriers used. Application of sonicated WPC (alone or in combination with MD) led to the production of microparticles with the least PDIs (0.31–0.33). This could be an indication of effective interactions between MD and modified WPCs as binary blends of wall materials, which has also been confirmed in other studies [30]. The results of Zeta potential (ZP) obtained for different spray-dried microparticles revealed that HWPC (alone or combined with MD) produced particles with the least surface charge (−18.5 to −20.3 mV). In contrast, sonication of WPC led to a higher surface charge with ZP values of −26.3 mV.

Interestingly, binary blends of UWPC with MD resulted in even more surface charge density with ZP values of −29.1 mV. This may be explained as an indication of successful interactions between sonicated WPC and MD in the atomization process in spray-drying. It has been reported that the difference in Zeta potential values might be attributed to the presence of depolymerized protein fractions with different solubility, composition of charged amino acids, surface activity, flexibility, and migration rate at the emulsion interface [34].

The lowest encapsulation efficiency of essential oil was observed for powders produced by hydrolyzed WPC alone or combined with MD (EE = 58.2–61.3%). The EE of binary systems obtained from UWPC (89.0%) was highest among SD samples, indicating the most successful covering systems for encapsulation of EO. The higher EE of binary blend systems has been attributed to the combined emulsification properties of MD and whey protein [32], especially with modified whey proteins [13].

Moisture is one of the effective indicators of the physical, oxidative, and microbial stability and maintaining the flow behavior of powders. The moisture value (2.86–3.42%)

and water activity (0.31–0.36) of the samples indicated the appropriate physical, microbial and oxidative stability of spray-dried powders. In this regard, the solubility, molecular weight, degree of hydrophilicity/hydrophobicity, and water-holding capacity of treated (hydrolyzed or sonicated) proteins are factors affecting the amount of moisture and water activity of powders [35,36]. Solubility and hygroscopicity are other essential indicators of powders, especially in applications, and they improve the miscibility and compatibility of lipophilic compounds in the food system [37]. The highest amount of solubility (86.4%) and hygroscopicity (12.4%) were related to the MD-HWPC sample (Table 4). The lowest amounts of these indices corresponded to SD powders produced from MD or WPC alone. The values of these indices are significantly affected by the nature, structural characteristics, and the amount of hydrophilic areas available to the carrier (proteins) [33]. Other morphological, structural, porosity, and particle size indicators are also influential on the solubility of powders [38]. Hygroscopicity is usually affected by the molecular weight of the peptide/protein, the amount of surface oil, the hydrophilic and moisture absorbing areas on the surface of the particles, the chemical and structural nature of the carriers and, finally, the amount of moisture and the size of the particles [38]. The solubility and hygroscopicity of spray-dried microparticles from the MD-HWPC carrier were greater than those of other carriers. This phenomenon may be attributed to structural modifications in the HWPC, resulting in increased accessibility and exposure of hydrophilic regions.

Density (bulk/tapped) and angle of repose were other indices that may affect the packaging and storage of powders. In general, there is an inverse relationship between the angle of repose and the bulk/tapped densities—a lower angle of repose means higher densities due to better particle packing. A lower angle of repose indicates the particles flow more freely over each other, allowing for better packing and fewer air spaces when piled. Conversely, a powder with a higher angle of repose (worse flow) will have lower bulk/tapped densities. The particles cannot pack as tightly due to greater friction and interparticle forces. As shown in Table 4, the highest values of these indices were related to the samples produced with modified WPC (HWPC and UWPC) with values of 35.0°–36.7°, respectively. Carriers with modified WPCs decreased bulk/tapped densities of the spray powders (0.50–0.51 g/mL). This variation can be ascribed to discrepancies in structural features, surface morphology, particle size distribution, adhesion, and agglomeration, which ultimately result in diverse porosities. However, the decrease in bulk/tapped densities leads to an increase in the amount of air trapped between the particles and may result in the risk of lipid oxidation. However, the ability of the carriers to create an insulating film/coating within the matrix is a key factor in the oxidative stability of the particles [37,39]. Based on flowability data for spray-dried powders calculated using the Carr index and Hausner Ratio (Table 4), the flowability of all powders was satisfactory and remained unaffected by both the carrier type and any modifications made.

3.4. Morphological Properties Using FE-SEM Observations

Morphological evaluation of powders with the FE-SEM technique is one of the crucial observation methods to determine the physical stability of microparticles (occurrence of mechanisms such as Ostwald ripening, agglomeration, and integration) during storage [40]. The surface properties of microparticles during spray-drying are impacted by factors such as the composition and concentration of components in the wall material, the rate of moisture evaporation, the transfer of solids to the wall layer, and the formation of a crust [41]. Figure 1A–C shows the appearance and morphological characteristics of spray-dried microparticles stabilized with different wall materials: (A) intact WPC or combination of MD-WPC, (B) UWPC or combination of MD-UWPC, and (C) HWPC or combination of MD-HWPC.

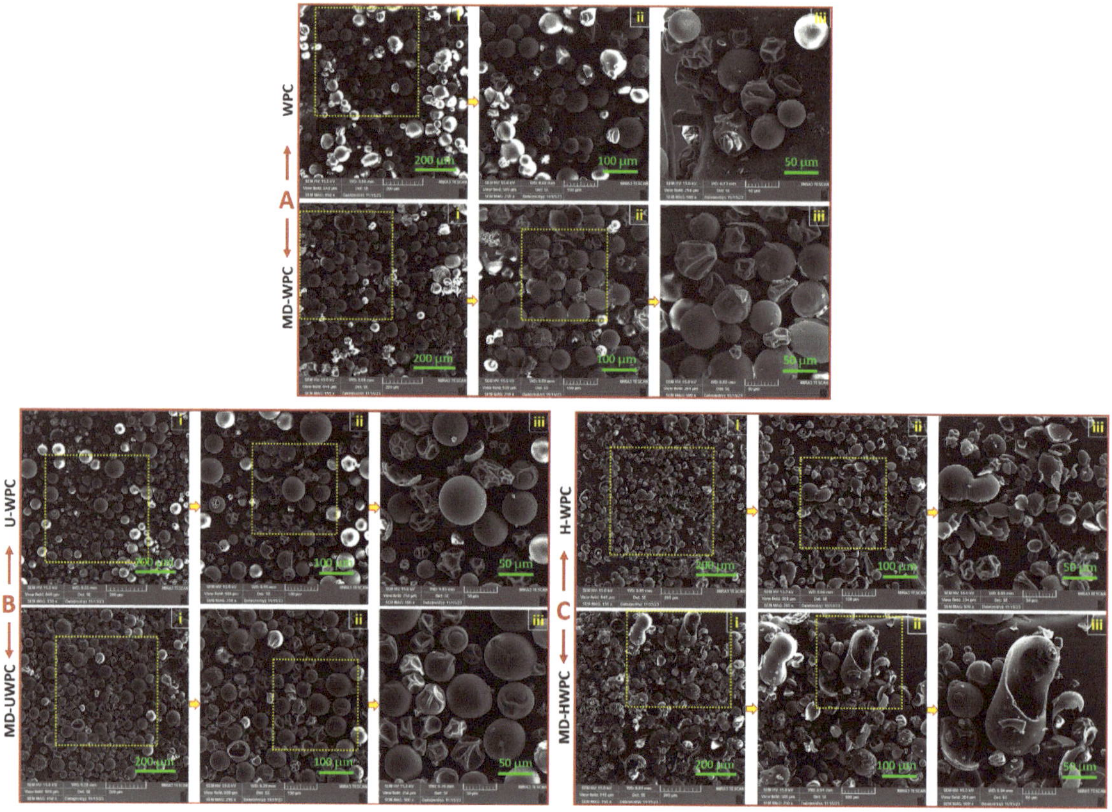

Figure 1. SEM images of spray-dried powders produced with (**A**) intact WPC and combination of MD-WPC, (**B**) ultrasound-treated WPC (UWPC) and combination of MD-UWPC, and (**C**) hydrolyzed WPC (HWPC) and combination of MD-HWPC. Images 'ii' and 'iii' are sequential magnifications of image 'i' in each row driven from yellow-dotted square areas.

The SEM study showed that the morphology of microparticles was influenced by the type and physical modification of the wall material. The powders produced with intact WPC were somehow irregular and wrinkled, showing the presence of indentations and some pores on the surface of particles (Figure 1A). At the same time, the powders obtained from a binary blend of MD and WPC had considerably improved surface morphology compared to that of intact WPC with more uniform shapes and sizes (with a diameter of 10–50 µm). These particles were relatively spherical with small surface grooves, sometimes showing tiny reservoir-type pores. Similar reports show that emulsion particles incorporating chia oil [42] and grape seed oil [13] stabilized by modified WPC exhibited spherical particles with uniform and regular surfaces. Applying UWPC (Figure 1B) as wall material in spray-drying produced microparticles with increased small cracks and holes on their surface. However, a combination of UWPC with MD as a feed solution led to the production of particles with a higher degree of homogeneity, relatively spherical structures with smooth surfaces, and no evidence of aggregation. This indicates that the formation of microcapsules was successful. Of all the wall material treatments tested in this study, the MD-UWPC system showed the most significant enhancement in the morphology of microparticles, exhibiting a consistent shape and size. This suggests that the essential oil as the core material was effectively encapsulated by the wall material system through the spray-drying process. Figure 1C shows that enzymatic hydrolysis of WPC negatively affected the integrity of microparticles. Most of the produced particles lost their spherical shapes. A higher magnification of the SEM images in Figure 1C(ii,iii) revealed the presence of broken microparticles with irregular shapes and surface cracks. These findings suggest that the process of microcapsule formation was inferior. Interestingly, the application of binary blends of MD with HWPC did not help in the recovery of spherical-shaped particles in the spray-drying process, and the corresponding SEM images clearly showed no improvement in the surface morphology of microparticles (Figure 1C). In other similar studies, the use of protein and polysaccharide carriers (alone or in combination) as emulsifiers and carriers in the microencapsulation of drumstick (*Moringa oleifera*) oil [38], European eel (*Anguilla anguilla*) oil [35], walnut oil [43], baltic herring (*Clupea harengus membras*) oil [44], fish oil [45], squalene [32], and chia oil [42] resulted in the generation of particles characterized by irregular, spherical structures with either smooth or indented surfaces, and walls that were either intact or broken, exhibiting numerous pores.

3.5. Color Analysis of Spray-Dried Powders

Color parameters determine the quality characteristics, conditions of use, and usability of a formula/or combination in food formulations. In this research, the effect of carrier modification and blends of MD with untreated or treated WPC on the color parameters of microparticles of encapsulated Chavir EO. Figure 2 demonstrates the color properties of microencapsulated EO powders obtained from different wall materials. Total color difference (ΔE) reflects the color differences of powders as compared to standard color plates (in this study, RAL 080 85 05 with the color name of white wheat flour). Treatments incorporating UWPC showed the highest ΔE values (19.1). In contrast, HWPC or MD-WPC corresponded to the lowest ΔEs (3.9–4.2), reflecting the lowest color difference concerning the standard RAL plate used in this study. Ultrasound-treated WPC created yellowish powders with yellowness indices (YI) of 40.2 compared to 17.3 for intact WPC. HWPC as wall material produced powders with a creamy to white appearance and a YI of 10.5. However, blending MD to those carriers before spray-drying significantly ($p < 0.05$) decreased YIs from 40.2 to 25.8, 17.3 to 11.2, and 10.45 to 5.0 for UWPC, WPC, and HWPC treatments, respectively, and the resulting powders appeared as creamy to white. Whiteness index (WI) data confirmed results as mentioned above in a way that WI values were increased with blends of MD to each of the other carriers. Also, the highest WI values corresponded to MD-HWPC and MD-WPC or MD-HWPC, with values of 91.6 and 82.9.

Chroma data can reflect the color of spray-dried powders by providing information about the intensity or purity of the color. Chroma, which represents the saturation or vividness of a color, can help in determining how vibrant or dull the color of the spray-dried powder is. By analyzing the chroma data of the powder samples, researchers can quantify the color attributes and assess factors such as color uniformity, stability, and quality. This information can be valuable in industries such as food, pharmaceuticals, and cosmetics, where color consistency is crucial for product quality and consumer acceptance. Chroma analysis showed the same trend observed and reported for ΔE values YI data. Microencapsulated particles based on HWPC or MD-HWPC showed the highest chroma values, and treatments with HWPC and MD-HWPC demonstrated the lowest values. Studies have indicated that the carrier type (whether chemically or physically modified) and other chemical processes, like pigment degradation and fatty acid oxidation, including browning reactions during spray-drying, may result in alterations to the color characteristics of the final powders [46].

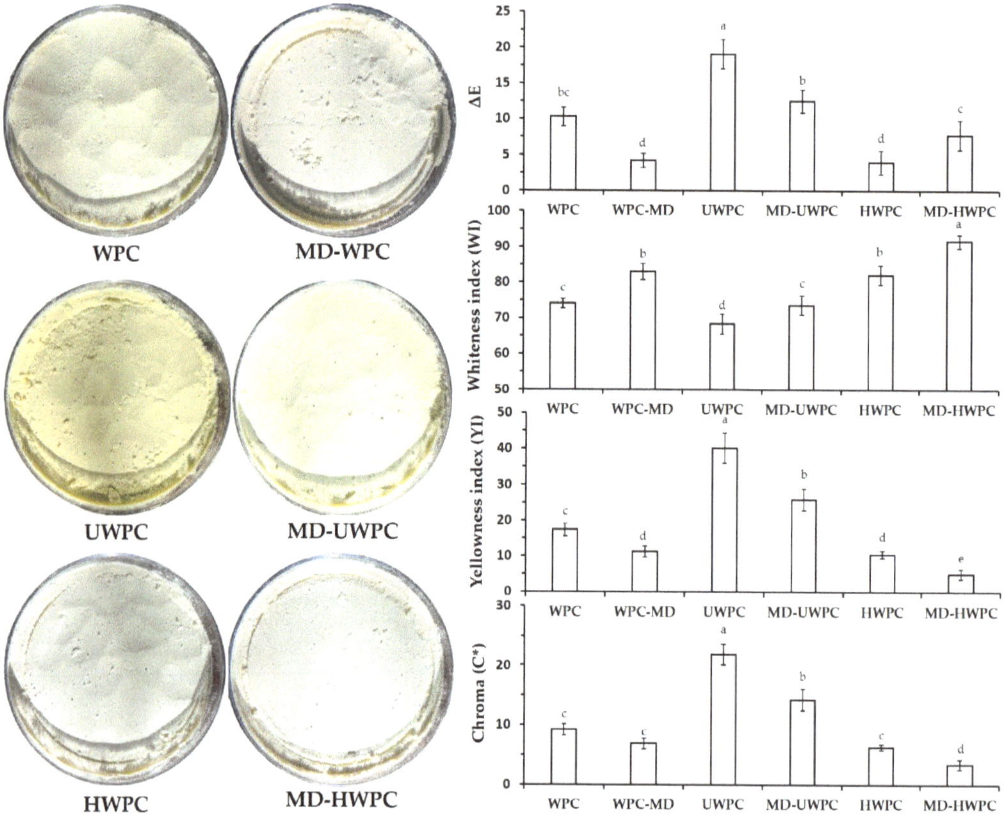

Figure 2. Digital images representing visual appearance (left-hand side) and different color parameters (right-hand side) of spray-dried powders produced using different carriers: WPC, MD-WPC, ultrasound-treated WPC (UWPC), MD-UWPC, hydrolyzed WPC (HWPC), and MD-HWPC blends. Data represent mean values of triplicate measurements. Error bars indicate SD values. Different alphabetical letters show significant differences between means ($p < 0.05$).

3.6. Analysis of Chemical Structure of Spray-Dried Microparticles by FTIR

FTIR analysis was employed to assess the chemical structure and interactions among the components of neat FA essential oil (EO) and various wall materials. The results offer valuable insights into the interaction of EO within the structure of loaded microparticles. In this study, the chemical structures of wall material components (in forms of intact and modified) as well as core material (FA essential oil) in preparation of encapsulated microparticles were investigated. Figure 3 demonstrates FTIR peaks of neat FA essential oil (Figure 3A) and different spray-dried powders (Figure 3B). In general, the main structural regions of whey proteins were identified as follows [47]: (1) amide-A region (N-H stretch) at frequencies 3300–3420 cm^{-1} for WPC; (2) amide-B region (C-H and O-H stretch) at 2926 cm^{-1} frequency (for all intact or modified proteins); (3) amide-I region (C=O stretch) at 1650 cm^{-1}; (4) amide-II (N-H deformation and C-N stretch at 1550 cm^{-1}; (5) amide-III (including secondary structures of α-helix, β-sheet, β-turn, and random coils) at 1243 cm^{-1}; and (6) other polysaccharide factors and C=O stretch in the frequency of 1030–1080 cm^{-1} and N-H bending at 600–650 cm^{-1}.

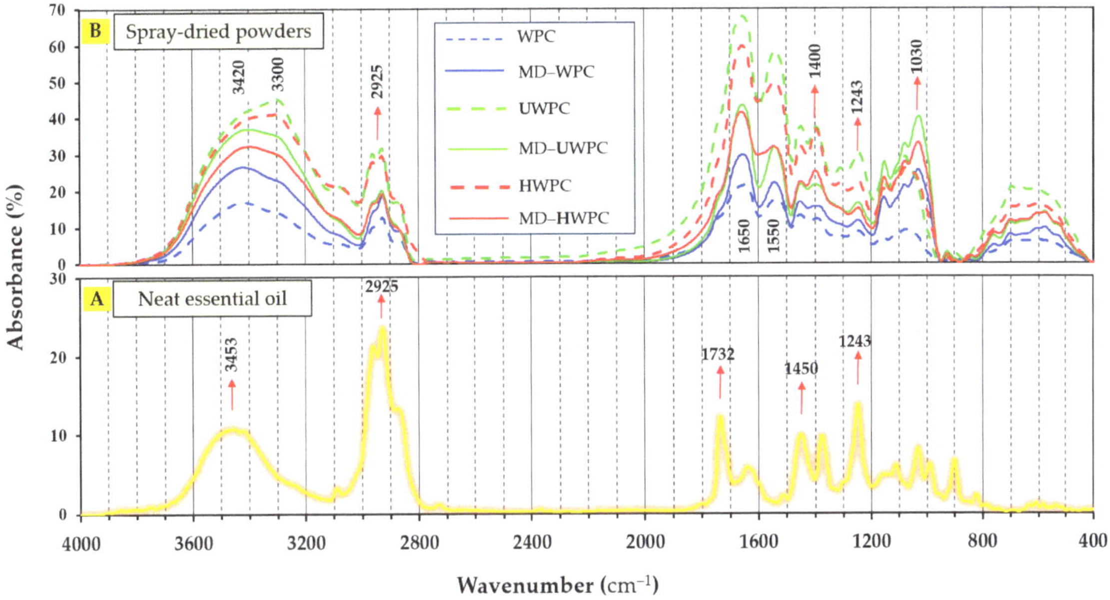

Figure 3. FTIR spectra of (**A**) neat *Ferulago angulata* essential oil, (**B**) spray-dried microcapsules produced by different carriers (WPC, MD-WPC, ultrasound-treated WPC (UWPC), MD-UWPC, hydrolyzed WPC (HWPC), and MD-HWPC blends.

FTIR spectrum of FA essential oil (Figure 3B) showed a broad absorption band observed at 3400–3500 cm^{-1} relating to C—H stretching of cis-alkene groups, stretching vibrations of cis double bands, —CH_2 stretching, and bending [48,49]. There are also prominent peaks at 2925 cm^{-1} (associated with O—H, N—H, and C—H stretching of aldehydes), 1732 cm^{-1} (indicative of C = O and C = C stretching of aldehyde/ketones), and 1242 cm^{-1} (attributed to —CH_3 and C—H in-plane bending). These peaks are characteristic of α-Pinene and β-Ocimene identified as the major constituents of FA essential oil, a finding supported by GC-MS analysis in this study (Table 2) and the previous literature [50,51]. Additionally, peaks at 1450 and 1242 cm^{-1} are linked to the methyl band and C—H vibration absorption of benzene rings, respectively [52].

Comparing FTIR spectra of neat EO (Figure 3A) with those of microencapsulated EO (Figure 3B) revealed the appearance of high-intensity peaks at 3300–3420 cm^{-1} (C—H stretching), suggesting the incorporation of EO oil within the carriers' structure. Previous findings have suggested that this phenomenon could be linked to the effective encapsulation of essential oil within the carriers' structure [53]. After the spray drying of essential oil with different carriers used in this study, several important changes occurred: (1) a sharp decrease in the intensity of EO peak in the regions of 2925 cm^{-1} after encapsulation as a result of potential formation of hydrogen bonds between the cis-alkene groups of fatty acids and the hydroxyl groups of the amide-A and B regions of proteins, along with the shielding of these regions within the protein matrix [35]; (2) a decrease in the intensity of the peak related to carbonyl ester groups (1732 cm^{-1}) was observed after encapsulation (especially in the samples produced with WPC or MD-WPC). These changes are the result of trapping and placing of oil in the carrier matrix without chemical reaction between them [49]; (3) a significant increase in the intensity of absorption peaks at 1030 cm^{-1} after spray drying with almost all combinations of wall materials relating to the involvement of amide regions (I, II, and III) of the WPC proteins, which indicates the complete distribution and placement of the functional groups.

3.7. Thermal Characteristics of Spray-Dried Powders

Thermogravimetric analysis (TGA) provides essential information about the thermal stability, decomposition temperature, and residual mass of spray-dried samples by measuring the weight change as a function of temperature. Differential scanning calorimetry (DSC) is another powerful tool that offers insights into the thermal transitional events occurring in the sample by measuring the heat flow associated with those transitions. In this study, simultaneous thermal analysis (STA) provided concurrent measurement of thermal events using both TGA and DSC techniques. Furthermore, derivative thermogravimetry (DTG) in STA experiments offers additional information by analyzing the rate of weight change as a function of applied temperature, offering a more detailed insight into the thermal decomposition processes. By integrating TGA, DSC, and DTG in STA experiments, it is possible to gain a thorough understanding of the thermal properties of the microcapsules and identify key thermal parameters critical for their characterization.

Figure 4 demonstrates STA thermal diagrams of neat FA EO (Figure 4A), spray-dried microcapsules produced using MD-WPC (Figure 4B), MD-UWPC (Figure 4C), and MD-HWPC (Figure 4D). The temperature of maximum degradation rate (T_d, °C), weight loss (ΔW, %), and weight residue of experimental samples are also shown in Table 5. In Figure 4A, a distinct and abrupt weight loss (WL) was noted, attributed to the thermal decomposition of pure essential oil within the temperature range of 55–110 °C, with a peak degradation temperature (Td) of 55 °C. This thermal process, resulting in a 95% reduction in the initial weight of the essential oil, was marked by an exothermic peak at around 155 °C, indicating a thermal decomposition event in the DSC thermogram of the essential oil. As reported in other studies, EO from different plants shows shallow thermal stability in TGA analysis [54,55]. In a general outcome from Figure 4, it can be seen that encapsulating EO using MD-WPC wall materials, regardless of carrier type, increased the thermal stability of EO from 55 °C to 180–250 °C. This indicates an enhancement of 2 to 2.5-fold in the degradation temperatures (T_d) of EO within spray-dried microcapsules.

There is a three-step thermal degradation event in all spray-dried powders (Figure 4B–D). In the MD-WPC sample (Figure 4B), the first weight loss was minimal (2.1% of initial weight) observed at T below 100 °C. This phenomenon can be associated with the evaporation of moisture and residual essential oils adhering to the microparticles. This occurrence was supported by an initial endothermic peak (at 120–150 °C) observed in the corresponding DSC thermogram. The second weight loss was at T_d of 250 °C, which led to a 10.5% weight reduction due to thermal decomposition of microparticles. The third thermal event, occurring between 320 °C and 500 °C, resulted in a final weight loss of 7.5%, likely attributed to the thermal decomposition of the encapsulated essential oil within the MD-WPC carrier. This stage did not show any additional exothermic peak in the DSC analysis. As shown in Figure 4B, the MD-WPC sample is more stable in terms of thermal degradation than other samples. This can be proven by the remaining weight of almost 80% after STA analysis. Meanwhile, for MD-UWPC and MD-HWPC treatments, more weight losses were observed (ca. 55%), with a remaining total material of ca. 45% after thermal analysis. This difference can be related to the modification of WPC by either ultrasonication or enzymatic hydrolysis, which in turn may have led to the susceptibility of the carriers to thermal degradation.

STA thermal characterization of the MD-UWPC sample showed similar three-stage degradation profiles but at different weight loss and thermal degradation temperatures at three analysis steps, as shown in Table 5. However, the MD-HWPC sample demonstrated weaker thermal stability with a two-stage weight loss, demonstrating the loss of essential oil at earlier stages (T_d of 182 °C) with concomitant endothermic DSC event at this stage, indicating destruction of core material due to evaporation. This thermal behavior was not seen for MD-WPC or MD-UWPC samples, indicating that the MD-HWPC sample is susceptible to thermal decomposition. As shown in SEM images (Figure 1C), the latter sample already exhibited more open-structured microparticles, which may explain the lower thermal stability of these microparticles.

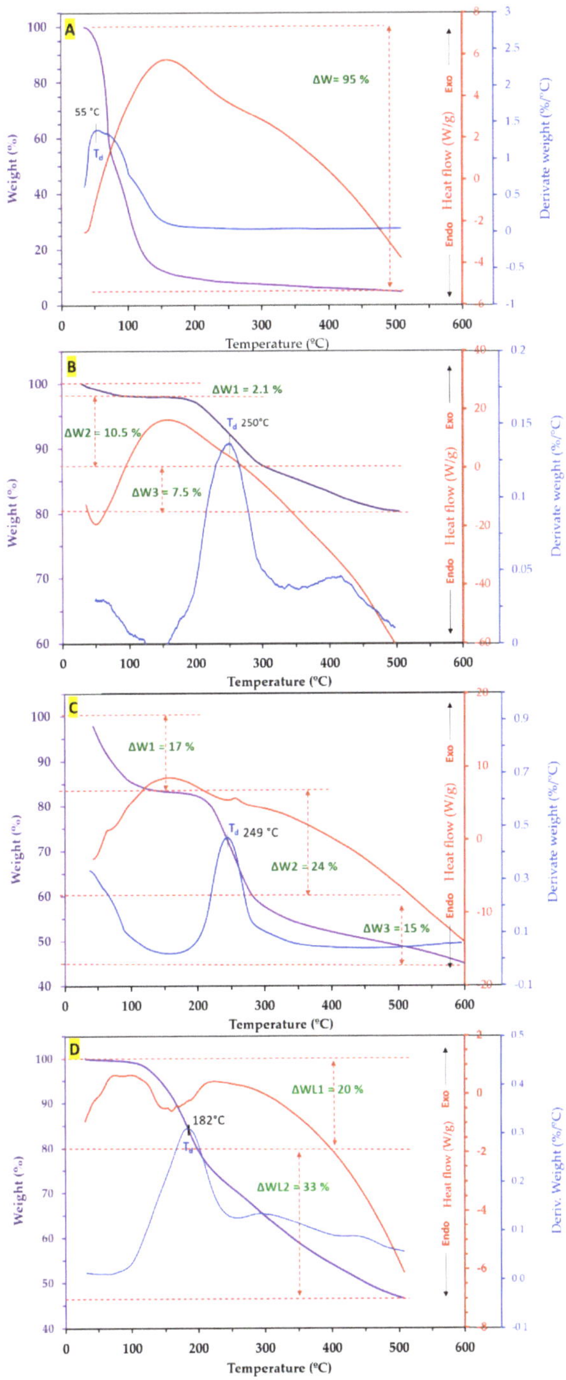

Figure 4. STA thermograms of neat *Ferulago angulate* EO (**A**) and encapsulated EO using different carriers: MD–WPC (**B**), MD–UWPC (**C**), and MD–HWPC (**D**).

Table 5. Thermal and weight transition phases (Δ), degradation temperature (T_d), and weight loss (ΔW) parameters of TGA-DTG-DSC thermograms for experimental materials subjected to thermal analysis.

Samples	Δ1		Δ2		Δ3		Weight Residue after Thermal Analysis (%)
	$T_{d1, onset}$ (°C)	ΔW_1 (%)	$T_{d2, onset}$ (°C)	ΔW_2 (%)	$T_{d3, onset}$ (°C)	ΔW_3 (%)	
Neat EO	55–100	95.0	-	-	-	-	5.0
MD–WPC	55–100	2.1	250	10.5	400	7.5	81.0
MD–UWPC	55–100	17.0	249	24.0	400	15.0	44.0
MD–HWPC	55–100	20.0	182	33.0	-	-	47.0

In summary, the thermal analysis results indicate that encapsulating unstable substances like essential oils into thermally stable microparticles produced through spray-drying using carriers such as MD-WPC (in either intact or modified forms) could enhance their thermal stability when compared to pure materials. Therefore, HWPC or UWPC, in combination with MD as a wall compound, could cover free EO as heat-sensitive compounds and boost their thermal endurance. Our results are in accordance with those reported by other researchers [56,57].

3.8. Total Phenolic Content and Antioxidant Properties

It is shown in Figure 5 that the neat EO extracted from *Ferulago angulate* (in its free form) exhibited the highest amount of TPC and $DPPH^-/ABTS^+$ radicals scavenging activity (RSA).

This is due to the presence of phenolic and antioxidant compounds such as ocimene (23.3%), α-pinene (14.5%), bornyl acetate (11.3%), and verbenol (7.2%) that were found in GC-MS analysis of FA EO shown in (Table 2). Stabilization of EO within spray-dried microcapsules using WPC (either intact or modified) significantly ($p < 0.05$) reduced its TPC and antioxidant activity. As reported in earlier studies, it is anticipated that there may be a decrease in the antioxidant activity of microencapsulated essential oil due to the stabilization process during spray-drying, which reduces its availability during antioxidant and TPC determination tests [13,58]. However, the type of WPC modification (enzyme hydrolysis vs. ultrasonication) and the use of binary blends of WPC-MD affected the available phenolic compounds and antioxidant activity of the produced microparticles. Among different spray-dried powders, those produced with HWPC or MD-HWPC as wall material showed significantly ($p < 0.05$) higher TPC and $DPPH^-/ABTS^+$ RSA values than UWPC and WPC carriers. This enormous difference in TPC and antioxidant activity of microparticles produced from HWPC could be related to their instability during and after the spray-drying process. As shown in SEM results (in Figure 1C), enzymatic hydrolysis of WPC negatively affected the integrity of spray-dried microparticles with the occurrence of broken particles with irregular shape and surface cracks, suggesting that the process of microcapsule formation was inferior. This could lead to the release of EO to the surface of microparticles, which, in turn, could increase the TPC and antioxidant activity of these particles. Figure 5. showed that the application of intact WPC (alone or in combination with MD) as wall materials produced particles with the lowest antioxidant properties because the EO cannot migrate to the surface of particles. In a previous study, it was shown that the application of power ultrasound and partial enzymatic hydrolysis of WPC increased $DPPH^-/ABTS^+$ radicals scavenging activities of spray-dried microcapsules incorporating grape seed oil compared to the situation where intact WPC were used as wall material [13]. These authors also showed that the hydrolyzed WPC had an even more significant effect in increasing the antioxidant properties of obtained microcapsules compared to sonicated WPC. These findings support our results in Figure 5, which show the enhanced antioxidant properties of HWPC samples compared to intact WPC or UWPC carriers.

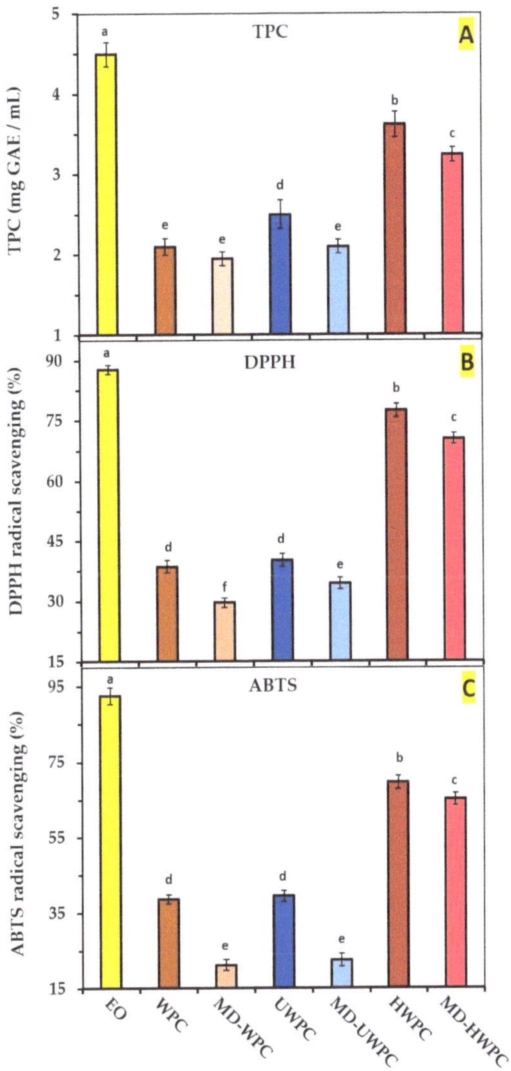

Figure 5. TPC (**A**), DPPH (**B**), and ABTS (**C**) radicals scavenging activities of neat *Ferulago angulate* EO (**A**) and its encapsulated form using different wall materials during the spray-drying process. Data represent mean values of triplicate measurements. Error bars indicate SD values. Different alphabetical letters show significant differences between means ($p < 0.05$).

3.9. Release Study of EO from Spray-Dried Microparticles

The cumulative release of EO content from spray-dried microparticles produced with different wall materials (WPC, UWPC, and HWPC) in phosphate buffer (pH = 7.3) was measured during 300 min, and the results are shown in Figure 6. This figure illustrates a biphasic release pattern for EO, wherein the release rate was rapid during the initial 50 min, followed by a gradual slowdown for the remainder of the measurement period. This pattern has been documented in other studies involving chitosan nanoparticles and microspheres as well [6,20]. The initial phase in the release profile is likely due to the swelling mechanism of the microparticles in the phosphate buffer [59], leading to a rapid release rate. The subsequent phase of the release profile may be associated with surface erosion and/or diffusion mechanisms, representing a slower process that results in a reduced release of the encapsulated essential oil. In this study, the lowest release content was observed for microparticles prepared with WPC and UWPC as wall material. This low-release content is probably attributed to the formation of stable microparticles (with denser encapsulating layer) by those carriers, which could retain the EO from escaping into release media [60]. However, the release content of spray-dried powders prepared by HWPC was significantly ($p < 0.05$) higher than other wall material treatments. This distinct difference in WPC is related to instability, such as microparticles. As shown in SEM results (in Figure 1C), enzymatic hydrolysis of WPC negatively affected the integrity of spray-dried microparticles with the occurrence of broken particles with irregular shape and surface cracks, indicating that the process of microcapsule formation was inferior. This could lead to increased release of EO to the surface of these microparticles. Moreover, successful encapsulation occurs by using wall materials with strong film-forming properties. Since the modifications applied in the current study disrupted the molecular structure of whey protein, it could be expected that the film-forming ability has probably been lost, resulting in a disrupted molecular structure and, thus, less integrity of the capsule surface, as observed in SEM micrographs. Overall, the low integrity of the microparticles' surface led to higher and faster release in an aqueous medium, as was evident in the release kinetics results.

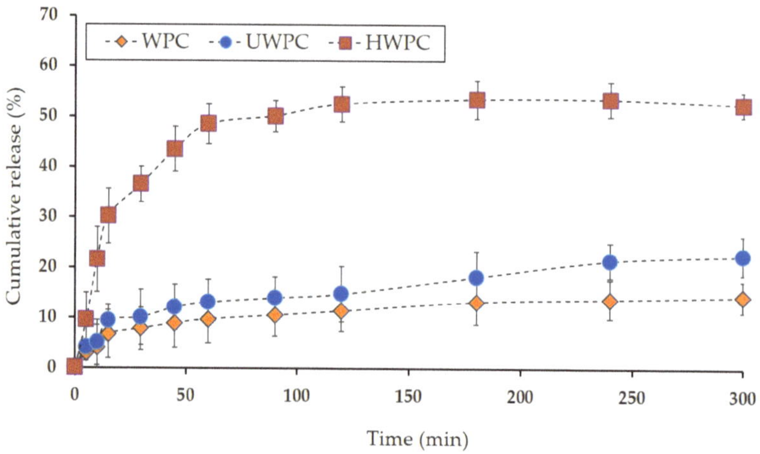

Figure 6. Release profile of *Ferulago angulate* EO from spray-dried microparticles produced with different wall materials (WPC, UWPC, and HWPC) in phosphate buffer (pH = 7.3). Data represent mean values of triplicate measurements, and error bars indicate SD values.

4. Conclusions

This study focused on investigating the efficiency of binary blends of modified WPC and MD in spray-drying encapsulation of *Ferulago angulata* essential oil. The choice of wall materials and their modifications (via power ultrasound or enzymatic hydrolysis) significantly influenced the physical characteristics and encapsulation efficiency of emulsions containing essential oils. The physical characterization of spray-dried microparticles revealed essential insights into particle size, yield, distribution, surface charge, encapsulation efficiency, moisture content, solubility, hygroscopicity, and flowability, emphasizing the impact of carrier type and modification on the physical properties of microparticles. Binary blends of modified WPC-MD showed improvements in particle sizes and yields, particularly with UWPC. Zeta potential measurements highlighted differences in surface charge density, with UWPC-MD blends exhibiting superior values. The SEM images revealed distinct differences in the surface characteristics of the particles based on the type and modification of the wall materials used. Thermal analysis elucidated the thermal stability of microcapsules, emphasizing the role of wall materials in enhancing thermal endurance. Evaluation of total phenolic content and antioxidant properties showcased the influence of carrier selection on the availability of phenolic compounds and antioxidant activity. The release study unveiled a biphasic release pattern of essential oil, with wall materials playing a crucial role in release kinetics. Microparticles prepared with HWPC exhibited higher release content, attributed to their instability, while enzymatic hydrolysis of WPC negatively impacted microparticle integrity, potentially increasing EO release. These findings collectively emphasize the pivotal role of wall materials in shaping the physical, morphological, thermal, antioxidant, and release properties of spray-dried microparticles, offering valuable insights for the development of microencapsulation techniques for essential oils.

Author Contributions: Conceptualization, S.H.P.; Methodology, N.B. and K.A.; Validation, A.M.A.; Formal analysis, N.B.; Investigation, A.M.A.; Data curation, N.B., A.M.A. and K.A.; Writing—review & editing, S.H.P.; Project administration, S.H.P. All authors have read and agreed to the published version of the manuscript.

Funding: This research received no external funding.

Institutional Review Board Statement: Not applicable.

Informed Consent Statement: Not applicable.

Data Availability Statement: The original contributions presented in the study are included in the article, further inquiries can be directed to the corresponding author.

Conflicts of Interest: The authors declare no conflict of interest.

References

1. Moghaddam, M.; Mehdizadeh, L.; Mirzaei Najafgholi, H.; Ghasemi Pirbalouti, A. Chemical Composition, Antibacterial and Antifungal Activities of Seed Essential Oil of *Ferulago angulata*. *Int. J. Food Prop.* **2018**, *21*, 158–170. [CrossRef]
2. Rustaiyan, A.; Sedaghat, S.; Larijani, K.; Khossravi, M.; Masoudi, S. Composition of the Essential Oil of *Ferulago angulata* (Schlecht.) Boiss. from Iran. *J. Essent. Oil Res.* **2002**, *14*, 447–448. [CrossRef]
3. Ghasemi Pirbalouti, A.; Izadi, A.; Malek Poor, F.; Hamedi, B. Chemical Composition, Antioxidant and Antibacterial Activities of Essential Oils from *Ferulago angulata*. *Pharm. Biol.* **2016**, *54*, 2515–2520. [CrossRef] [PubMed]
4. Bagherifar, S.; Sourestani, M.M.; Zolfaghari, M.; Mottaghipisheh, J.; Zomborszki, Z.P.; Csupor, D. Variation of Chemical Constituents and Antiradical Capacity of Nine *Ferulago Angulata* Populations from Iran. *Chem. Biodivers.* **2019**, *16*, e1900302. [CrossRef] [PubMed]
5. Salehi, M.; Moslehishad, M.; Hosseini, S.M. Physicochemical Properties and Stability of Encapsulated *Ferulago angulata* Subsp. Carduchorum Essential Oil Using Polymer Coating. *J. Med. Plants Prod.* **2021**, *10*, 1–10.
6. Kaboudi, Z.; Peighambardoust, S.H.; Nourbakhsh, H.; Soltanzadeh, M. Nanoencapsulation of Chavir (*Ferulago angulata*) Essential Oil in Chitosan Carrier: Investigating Physicochemical, Morphological, Thermal, Antimicrobial and Release Profile of Obtained Nanoparticles. *Int. J. Biol. Macromol.* **2023**, *237*, 123963. [CrossRef] [PubMed]
7. Akhlaghi, H. Volatile Constituents from the Aerial Parts of *Ferulago angulata* (Schlecht.) Boiss. Growing Wild Northeast Iran. *Anal. Chem. Lett.* **2012**, *2*, 133–138. [CrossRef]
8. Soltanzadeh, M.; Peighambardoust, S.H.; Ghanbarzadeh, B.; Amjadi, S.; Mohammadi, M.; Lorenzo, J.M.; Hamishehkar, H. Active Gelatin/Cress Seed Gum-Based Films Reinforced with Chitosan Nanoparticles Encapsulating Pomegranate Peel Extract: Preparation and Characterization. *Food Hydrocoll.* **2022**, *129*, 107620. [CrossRef]
9. Aghamirzaei, M.; Peighambardoust, S.H.; Azadmard-Damirchi, S.; Majzoobi, M. Effects of Grape Seed Powder as a Functional Ingredient on Flour Physicochemical Characteristics and Dough Rheological Properties. *J. Agric. Sci. Technol.* **2015**, *17*, 365–373.
10. Nah, J.-W.; Jeong, G.-W. Preparation and Encapsulation Techniques of Chitosan Microsphere for Enhanced Bioavailability of Natural Antioxidants. *Carbohydr. Res.* **2021**, *500*, 108218. [CrossRef] [PubMed]
11. Golshan Tafti, A.; Peighambardoust, S.H.; Behnam, F.; Bahrami, A.; Aghagholizadeh, R.; Ghamari, M.; Rafat, S.A. Effects of Spray-Dried Sourdough on Flour Characteristics and Rheological Properties of Dough. *Czech J. Food Sci.* **2013**, *31*, 361–367. [CrossRef]
12. Jafari, S.; Jafari, S.M.; Ebrahimi, M.; Kijpatanasilp, I.; Assatarakul, K. A Decade Overview and Prospect of Spray Drying Encapsulation of Bioactives from Fruit Products: Characterization, Food Application and in Vitro Gastrointestinal Digestion. *Food Hydrocoll.* **2023**, *134*, 108068. [CrossRef]
13. Sarabandi, K.; Tamjidi, F.; Akbarbaglu, Z.; Samborska, K.; Gharehbeglou, P.; Kharazmi, M.S.; Jafari, S.M. Modification of Whey Proteins by Sonication and Hydrolysis for the Emulsification and Spray Drying Encapsulation of Grape Seed Oil. *Pharmaceutics* **2022**, *14*, 2434. [CrossRef] [PubMed]
14. Ravash, N.; Peighambardoust, S.H.; Soltanzadeh, M.; Pateiro, M.; Lorenzo, J.M. Impact of High-Pressure Treatment on Casein Micelles, Whey Proteins, Fat Globules and Enzymes Activity in Dairy Products: A Review. *Crit. Rev. Food Sci. Nutr.* **2022**, *62*, 2888–2908. [CrossRef] [PubMed]

15. Soltanzadeh, M.; Peighambardoust, S.H.; Gullon, P.; Hesari, J.; Gullón, B.; Alirezalu, K.; Lorenzo, J. Quality Aspects and Safety of Pulsed Electric Field (PEF) Processing on Dairy Products: A Comprehensive Review. *Food Rev. Int.* **2020**, *38*, 96–117. [CrossRef]
16. Meng, Y.; Liang, Z.; Zhang, C.; Hao, S.; Han, H.; Du, P.; Li, A.; Shao, H.; Li, C.; Liu, L. Ultrasonic Modification of Whey Protein Isolate: Implications for the Structural and Functional Properties. *LWT* **2021**, *152*, 112272. [CrossRef]
17. Akbarmehr, A.; Peighambardoust, S.H.; Ghanbarzadeh, B.; Sarabandi, K. Physicochemical, Antioxidant, Antimicrobial, and in Vitro Cytotoxic Activities of Corn Pollen Protein Hydrolysates Obtained by Different Peptidases. *Food Sci. Nutr.* **2023**, *11*, 2403–2417. [CrossRef]
18. Moghadam, M.; Salami, M.; Mohammadian, M.; Emam-Djomeh, Z.; Jahanbani, R.; Moosavi-Movahedi, A.A. Physicochemical and Bio-Functional Properties of Walnut Proteins as Affected by Trypsin-Mediated Hydrolysis. *Food Biosci.* **2020**, *36*, 100611. [CrossRef]
19. Dehghan-Manshadi, A.; Peighambardoust, S.H.; Azadmard-Damirchi, S.; Niakousari, M. Effect of Infrared-Assisted Spouted Bed Drying of Flaxseed on the Quality Characteristics of Its Oil Extracted by Different Methods. *J. Sci. Food Agric.* **2020**, *100*, 74–80. [CrossRef] [PubMed]
20. Duman, F.; Kaya, M. Crayfish Chitosan for Microencapsulation of Coriander (*Coriandrum sativum* L.) Essential Oil. *Int. J. Biol. Macromol.* **2016**, *92*, 125–133. [CrossRef] [PubMed]
21. Bahrampour, Z.; Peighambardoust, S.H.; Amini, A.M.; Soltanzadeh, M. Application of Low-, and Medium-Molecular Weight Chitosan for Preparation of Spray-Dried Microparticles Loaded with *Ferulago Angulata* Essential Oil: Physicochemical, Antioxidant, Antibacterial and in-Vitro Release Properties. *Int. J. Biol. Macromol.* **2023**, *253*, 126554. [CrossRef] [PubMed]
22. Frascareli, E.C.; Silva, V.M.; Tonon, R.V.; Hubinger, M.D. Effect of Process Conditions on the Microencapsulation of Coffee Oil by Spray Drying. *Food Bioprod. Process.* **2012**, *90*, 413–424. [CrossRef]
23. Marques, G.R.; de Fernandes, R.V.B.; Lago, A.M.T.; Borges, S.V.; Bertolucci, S.K.V.; de Jesus Boari Lima, A.; Botrel, D.A. Spray-Dried Thyme Essential Oil Microparticles Using Different Polymeric Matrices. *Dry. Technol.* **2021**, *39*, 1883–1894. [CrossRef]
24. Ozdemir, N.; Bayrak, A.; Tat, T.; Altay, F.; Kiralan, M.; Kurt, A. Microencapsulation of Basil Essential Oil: Utilization of Gum Arabic/Whey Protein Isolate/Maltodextrin Combinations for Encapsulation Efficiency and in Vitro Release. *J. Food Meas. Charact.* **2021**, *15*, 1865–1876. [CrossRef]
25. Sakooei-Vayghan, R.; Peighambardoust, S.H.; Hesari, J.; Soltanzadeh, M.; Peressini, D. Properties of Dried Apricots Pretreated by Ultrasound-Assisted Osmotic Dehydration and Application of Active Coatings. *Food Technol. Biotechnol.* **2020**, *58*, 249–259. [CrossRef] [PubMed]
26. Peighambardoust, S.H.; Yaghoubi, M.; Hosseinpour, A.; Alirezalu, K.; Soltanzadeh, M.; Dadpour, M. Development and Application of Dual-Sensors Label in Combination with Active Chitosan-Based Coating Incorporating Yarrow Essential Oil for Freshness Monitoring and Shelf-Life Extension of Chicken Fillet. *Foods* **2022**, *11*, 3533. [CrossRef]
27. Shahi, Z.; Sayyed-Alangi, S.Z.; Najafian, L. Effects of Enzyme Type and Process Time on Hydrolysis Degree, Electrophoresis Bands and Antioxidant Properties of Hydrolyzed Proteins Derived from Defatted Bunium Persicum Bioss. Press Cake. *Heliyon* **2020**, *6*, e03365. [CrossRef] [PubMed]
28. de Moura, S.C.S.R.; Berling, C.L.; Garcia, A.O.; Queiroz, M.B.; Alvim, I.D.; Hubinger, M.D. Release of Anthocyanins from the Hibiscus Extract Encapsulated by Ionic Gelation and Application of Microparticles in Jelly Candy. *Food Res. Int.* **2019**, *121*, 542–552. [CrossRef] [PubMed]
29. Sefidkon, F.; Omidbaigi, R. Chemical Composition of the Essential Oil of *Ferulago angulata* from Iran. *J. Essent. Oil Bear. Plants* **2004**, *7*, 60–63. [CrossRef]
30. Campelo, P.H.; do Carmo, E.L.; Zacarias, R.D.; Yoshida, M.I.; Ferraz, V.P.; de Barros Fernandes, R.V.; Botrel, D.A.; Borges, S.V. Effect of Dextrose Equivalent on Physical and Chemical Properties of Lime Essential Oil Microparticles. *Ind. Crops Prod.* **2017**, *102*, 105–114. [CrossRef]
31. Turasan, H.; Sahin, S.; Sumnu, G. Encapsulation of Rosemary Essential Oil. *LWT-Food Sci. Technol.* **2015**, *64*, 112–119. [CrossRef]
32. Lekshmi, R.G.K.; Tejpal, C.S.; Anas, K.K.; Chatterjee, N.S.; Mathew, S.; Ravishankar, C.N. Binary Blend of Maltodextrin and Whey Protein Outperforms Gum Arabic as Superior Wall Material for Squalene Encapsulation. *Food Hydrocoll.* **2021**, *121*, 106976. [CrossRef]
33. Gomes, M.H.G.; Kurozawa, L.E. Improvement of the Functional and Antioxidant Properties of Rice Protein by Enzymatic Hydrolysis for the Microencapsulation of Linseed Oil. *J. Food Eng.* **2020**, *267*, 109761. [CrossRef]
34. Zia, Q.; Tabassum, M.; Meng, J.; Xin, Z.; Gong, H.; Li, J. Polydopamine-Assisted Grafting of Chitosan on Porous Poly (L-Lactic Acid) Electrospun Membranes for Adsorption of Heavy Metal Ions. *Int. J. Biol. Macromol.* **2021**, *167*, 1479–1490. [CrossRef] [PubMed]
35. Taktak, W.; Nasri, R.; Lopez-Rubio, A.; Hamdi, M.; Gomez-Mascaraque, L.G.; Ben Amor, N.; Kabadou, A.; Li, S.; Nasri, M.; Karra-Chaâbouni, M. Improved Antioxidant Activity and Oxidative Stability of Spray Dried European Eel (*Anguilla anguilla*) Oil Microcapsules: Effect of Emulsification Process and Eel Protein Isolate Concentration. *Mater. Sci. Eng. C* **2019**, *104*, 109867. [CrossRef] [PubMed]
36. Rezvani, M.; Hesari, J.; Peighambardoust, S.H.; Manconi, M.; Hamishehkar, H.; Escribano-Ferrer, E. Potential Application of Nanovesicles (Niosomes and Liposomes) for Fortification of Functional Beverages with Isoleucine-Proline-Proline: A Comparative Study with Central Composite Design Approach. *Food Chem.* **2019**, *293*, 368–377. [CrossRef]

37. Elik, A.; Yanık, D.K.; Göğüş, F. A Comparative Study of Encapsulation of Carotenoid Enriched-Flaxseed Oil and Flaxseed Oil by Spray Freeze-Drying and Spray Drying Techniques. *LWT* **2021**, *143*, 111153. [CrossRef]
38. Premi, M.; Sharma, H.K. Effect of Different Combinations of Maltodextrin, Gum Arabic and Whey Protein Concentrate on the Encapsulation Behavior and Oxidative Stability of Spray Dried Drumstick (*Moringa oleifera*) Oil. *Int. J. Biol. Macromol.* **2017**, *105*, 1232–1240. [CrossRef]
39. Nottagh, S.; Hesari, J.; Peighambardoust, S.H.; Rezaei-Mokarram, R.; Jafarizadeh-Malmiri, H. Development of a Biodegradable Coating Formulation Based on the Biological Characteristics of the Iranian Ultra-Filtrated Cheese. *Biology* **2018**, *73*, 403–413. [CrossRef]
40. dos Santos, P.P.; de Aguiar Andrade, L.; Flôres, S.H.; de Oliveira Rios, A. Nanoencapsulation of Carotenoids: A Focus on Different Delivery Systems and Evaluation Parameters. *J. Food Sci. Technol.* **2018**, *55*, 3851–3860. [CrossRef] [PubMed]
41. Botrel, D.A.; de Barros Fernandes, R.V.; Borges, S.V.; Yoshida, M.I. Influence of Wall Matrix Systems on the Properties of Spray-Dried Microparticles Containing Fish Oil. *Food Res. Int.* **2014**, *62*, 344–352. [CrossRef]
42. Noello, C.; Carvalho, A.G.S.; da Silva, V.M.; Hubinger, M.D. Spray Dried Microparticles of Chia Oil Using Emulsion Stabilized by Whey Protein Concentrate and Pectin by Electrostatic Deposition. *Food Res. Int.* **2016**, *89*, 549–557. [CrossRef] [PubMed]
43. Zhou, D.; Pan, Y.; Ye, J.; Jia, J.; Ma, J.; Ge, F. Preparation of Walnut Oil Microcapsules Employing Soybean Protein Isolate and Maltodextrin with Enhanced Oxidation Stability of Walnut Oil. *LWT-Food Sci. Technol.* **2017**, *83*, 292–297. [CrossRef]
44. Damerau, A.; Ogrodowska, D.; Banaszczyk, P.; Dajnowiec, F.; Tańska, M.; Linderborg, K.M. Baltic Herring (*Clupea harengus* Membras) Oil Encapsulation by Spray Drying Using a Rice and Whey Protein Blend as a Coating Material. *J. Food Eng.* **2022**, *314*, 110769. [CrossRef]
45. Gómez-Guillén, M.C.; Pérez-García, S.; Alemán, A.; Vázquez, J.A.; Montero, M.P. The Role of the Drying Method on Fish Oil Entrapment in a Fish Muscle Protein—κ-Carrageenan—Fish Protein Hydrolysate Wall Matrix and the Properties of Colloidal Dispersions. *Food Hydrocoll.* **2022**, *131*, 107799. [CrossRef]
46. Carneiro, H.C.F.; Tonon, R.V.; Grosso, C.R.F.; Hubinger, M.D. Encapsulation Efficiency and Oxidative Stability of Flaxseed Oil Microencapsulated by Spray Drying Using Different Combinations of Wall Materials. *J. Food Eng.* **2013**, *115*, 443–451. [CrossRef]
47. He, K.; Zeng, G.; Chen, A.; Huang, Z.; Peng, M.; Huang, T.; Chen, G. Graphene Hybridized Polydopamine-Kaolin Composite as Effective Adsorbent for Methylene Blue Removal. *Compos. Part B Eng.* **2019**, *161*, 141–149. [CrossRef]
48. Jamwal, R.; Kumari, S.; Sharma, S.; Kelly, S.; Cannavan, A.; Singh, D.K. Recent Trends in the Use of FTIR Spectroscopy Integrated with Chemometrics for the Detection of Edible Oil Adulteration. *Vib. Spectrosc.* **2021**, *113*, 103222. [CrossRef]
49. Karaaslan, M.; Şengün, F.; Cansu, Ü.; Başyiğit, B.; Sağlam, H.; Karaaslan, A. Gum Arabic/Maltodextrin Microencapsulation Confers Peroxidation Stability and Antimicrobial Ability to Pepper Seed Oil. *Food Chem.* **2021**, *337*, 127748. [CrossRef] [PubMed]
50. Xiang, F.; Zhao, Q.; Zhao, K.; Pei, H.; Tao, F. The Efficacy of Composite Essential Oils against Aflatoxigenic Fungus Aspergillus Flavus in Maize. *Toxins* **2020**, *12*, 562. [CrossRef] [PubMed]
51. Shahbazi, Y.; Shavisi, N.; Karami, N.; Kakaei, S. Chemical Composition and in Vitro Antibacterial Activity of *Ferulago angulata* (Schlecht.) Boiss Essential Oil. *Pharm. Sci.* **2015**, *21*, 6–11. [CrossRef]
52. Natrajan, D.; Srinivasan, S.; Sundar, K.; Ravindran, A. Formulation of Essential Oil-Loaded Chitosan–Alginate Nanocapsules. *J. Food Drug Anal.* **2015**, *23*, 560–568. [CrossRef] [PubMed]
53. van der Goot, A.J.; Peighambardoust, S.H.; Akkermans, C.; Van Oosten-Manski, J. Creating Novel Structures in Food Materials: The Role of Well-Defined Shear Flow. *Food Biophys.* **2008**, *3*, 120–125. [CrossRef]
54. Himed, L.; Merniz, S.; Monteagudo-Olivan, R.; Barkat, M.; Coronas, J. Antioxidant Activity of the Essential Oil of Citrus Limon before and after Its Encapsulation in Amorphous SiO_2. *Sci. Afr.* **2019**, *6*, e00181. [CrossRef]
55. Cazón, P.; Antoniewska, A.; Rutkowska, J.; Vázquez, M. Evaluation of Easy-Removing Antioxidant Films of Chitosan with Melaleuca Alternifolia Essential Oil. *Int. J. Biol. Macromol.* **2021**, *186*, 365–376. [CrossRef] [PubMed]
56. Gharanjig, H.; Gharanjig, K.; Farzi, G.; Hosseinnezhad, M.; Jafari, S.M. Novel Complex Coacervates Based on Zedo Gum, Cress Seed Gum and Gelatin for Loading of Natural Anthocyanins. *Int. J. Biol. Macromol.* **2020**, *164*, 3349–3360. [CrossRef] [PubMed]
57. Khoshakhlagh, K.; Koocheki, A.; Mohebbi, M.; Allafchian, A. Development and Characterization of Electrosprayed Alyssum Homolocarpum Seed Gum Nanoparticles for Encapsulation of D-Limonene. *J. Colloid Interface Sci.* **2017**, *490*, 562–575. [CrossRef] [PubMed]
58. Cabral, B.R.P.; de Oliveira, P.M.; Gelfuso, G.M.; de Quintão, T.S.C.; Chaker, J.A.; de Oliveira Karnikowski, M.G.; Gris, E.F. Improving Stability of Antioxidant Compounds from Plinia Cauliflora (Jabuticaba) Fruit Peel Extract by Encapsulation in Chitosan Microparticles. *J. Food Eng.* **2018**, *238*, 195–201. [CrossRef]
59. Desai, K.G.; Park, H.J. Preparation and Characterization of Drug-loaded Chitosan–Tripolyphosphate Microspheres by Spray Drying. *Drug Dev. Res.* **2005**, *64*, 114–128. [CrossRef]
60. Ogunjimi, A.T.; Fiegel, J.; Brogden, N.K. Design and Characterization of Spray-Dried Chitosan-Naltrexone Microspheres for Microneedle-Assisted Transdermal Delivery. *Pharmaceutics* **2020**, *12*, 496. [CrossRef] [PubMed]

foods

MDPI

Article

A Simple High-Throughput Technology for Microorganism Detection and Quantitative Analysis

Liping Wang and Ziyun Wu *

School of Agriculture and Biology, Shanghai Jiao Tong University, Shanghai 200030, China; lorraine2014@sjtu.edu.cn
* Correspondence: wuziyun@sjtu.edu.cn

Abstract: Normal and damaged microorganisms are related to food safety. The colony-forming unit (CFU) assay and viability of microorganisms have broad applications in food. Traditionally, the CFU assay has been the benchmark for assessing microbial viability across various fields. However, the normal and damaged microorganisms cannot be distinguished. Here, we introduce an improved technology for foods that uses a visible absorbance microplate reader platform for high-throughput quantitative analysis of microbial lag time, doubling time, and CFU. This platform utilizes a 96-well plate and a microplate reader to accurately determine the viable cell number from a five-microliter sample. It boasts the capability to measure a dynamic range spanning from five to seven orders of magnitude, significantly reducing the time required by over 20-fold in comparison to traditional spread plate methods. Additionally, it demonstrates a remarkable ability to detect a single cell within a well. A mild temperature treatment for cell viability detection was implemented and was able to reflect the real microbial quality. Consequently, the high-throughput method as an improved technology provides essential technical support for microbial detection.

Keywords: colorimetric; high throughput; microbial counts; trace sample; microplate reader

Citation: Wang, L.; Wu, Z. A Simple High-Throughput Technology for Microorganism Detection and Quantitative Analysis. *Foods* **2024**, *13*, 2954. https://doi.org/10.3390/foods13182954

Academic Editor: Alejandro Castillo

Received: 15 July 2024
Revised: 11 August 2024
Accepted: 13 September 2024
Published: 18 September 2024

1. Introduction

Measuring microbial viability has diverse applications across a broad range of fields where microbiology plays a crucial role, including the food industry, agriculture, public health, environmental science, pharmaceuticals, and healthcare [1,2]. Traditionally, the dilution and plating method has been the standard procedure for quantifying live bacteria in terms of colony-forming units (CFUs). Among these techniques, the spread plate method is widely recognized for its utility in microbial enumeration. This process involves dispersing a bacterial sample across an agar plate, followed by incubation to allow colony development, after which the colonies are counted [3]. To facilitate accurate colony counts, samples often require dilution. However, the manual enumeration of bacterial CFUs on agar plates not only demands considerable time but also subjects the process to potential inaccuracies [4].

Image analysis for CFU counts is a useful approach in microbiological processing. A new segmentation algorithm alongside specialized colony counter hardware, a phase-contrast microscope, and geometric viability assay (GVA) were developed for counting bacterial CFUs [4–6]. The methodologies outlined for bacterial CFU enumeration significantly streamline labor, time, and resource expenditure to varying extents. Despite the adaptability of imaging methods to high-throughput detection, distinguishing living from dead cells necessitates staining [7]. The microbial biosensors, including an ultra-sensitive surface-enhanced Raman scattering biosensor utilizing gold nanostars [8], a colorimetric pattern and sensor array composed of metal complexes (C1–C11) based on a single azophenol dye-based receptor [9], and phosphorescent oxygen sensors [10] have gained recognition for their exceptional sensitivity and specificity [11]. The mentioned microbial sensors demonstrate a notable level of specificity in their applications.

In this study, we introduced a high-throughput microbial detection method utilizing quantitative colorimetric technology, centered around the viability assay [12,13]. Three typical strains of BY4742, OP50, and LGG were used in the experiments. Baker's yeast (*Saccharomyces cerevisiae*) has been referred to as the *Escherichia coli* of the eukaryotic world [14]. *Escherichia coli* of OP50 and *Lactobacillus rhamnosus* are both procaryotic organisms [15]. Meanwhile, *Escherichia coli* is a Gram-negative bacterium, and *Lactobacillus rhamnosus* is a Gram-positive bacterium. They have a wide application in studies relevant to human aging and age-related diseases [16], as food for laboratory maintenance of *C. elegans*, and in the manufacturing of fermented dairy products and other fermented foods [17].

2. Materials and Methods

2.1. Materials

The wild-type yeast strain *Saccharomyces cerevisiae* BY4742 (MATα *his3Δ1 leu2Δ0 lys2Δ0 ura3Δ0*) was obtained from Horizon Discovery (Cambridge, UK). The *Escherichia coli* strain of OP50 was obtained from the Caenorhabditis Genetics Center (Minneapolis, MN, USA). The *Lactobacillus rhamnosus* strain of LGG was a gift from Professor Xianming Shi at Shanghai Jiao Tong University. The culture of each strain was kept in storage at −80 °C. The yeast extract peptone dextrose (YPD) medium was from Solarbio life science (Beijing, China), LB Lennox broth was from Sangon biotech (Shanghai, China), and MRS broth was from Qingdao Hope Bio-Technology Co., Ltd. (Qingdao, China).

2.2. Optimization of Medium for Three Strains

The cells of BY4742, OP50, and LGG were prepared by streaking a strain from frozen stocks onto YPD, LB, and MRS agar plates, respectively. After incubating the cells until colonies appeared, a single colony was picked and inoculated into a 1.0 mL liquid medium in a 5 mL tube and cultured at 30 °C for 2 days in a flat incubator with shaking at 200 rpm. The 2-day cultures were diluted with the corresponding medium. The 5 μL dilutions were pipetted into each well of 96-well microplate. A total of 100 μL of medium was then added to each well. The growth curves were first analyzed in individual media, namely, YPD, LB, and MRS. Subsequently, binary mixtures were explored, comprising 90% YPD + 10% LB, 70% YPD + 30% LB, 50% YPD + 50% LB, 90% YPD + 10% MRS, 70% YPD + 30% MRS, and 50% YPD + 50% MRS. Finally, ternary combinations were examined, including 80% YPD + 10% LB + 10% MRS, 60% YPD + 20% LB + 20% MRS, and 40% YPD + 30% LB + 30% MRS. The monitoring time was 20 h, and each set of experiments was repeated at least 6 times.

2.3. Reproducibility, Precision, and Accuracy of the High-Throughput Platform

The 2-day cultures of the three strains were diluted with medium (10^0, 10^{-1}, 10^{-2}, 10^{-3}, 10^{-4}, 10^{-5}, 10^{-6}, 10^{-7}). The 5 μL dilutions were pipetted into each well of a 96-well microplate. A total of 100 μL of optimized combined media was then added to each well. Each set of experiments was repeated at least 6 times. The traditional method was employed for CFU enumeration by spread plating [18]. According to the type of bacteria, an agar medium was selected and prepared. A ten-fold dilution series was diluted with the corresponding medium. The inoculum volume was 0.1 mL in terms of spreading. After the inoculum had been absorbed into the agar for 5–10 min, the plates were inverted and incubated to the desired condition. When enumerating CFUs, plates with between 20 and 300 CFUs can be used to calculate the number of CFUs/mL in the original sample. The detailed procedures are shown in Figure 1.

2.4. Trace Cell Detection

The 2-day cultures of OP50 were counted using the spread plate method. Subsequently, the cultures were diluted using LB medium to yield 1 and 2 cells per 5 μL. The 5 μL dilution was pipetted into each well of the 96-well microplate. A total of 100 μL of optimized medium was then added to each well. The 5 μL dilution was carefully pipetted onto an agar plate that had been divided into grids. Each set of experiments was repeated at least 12 times.

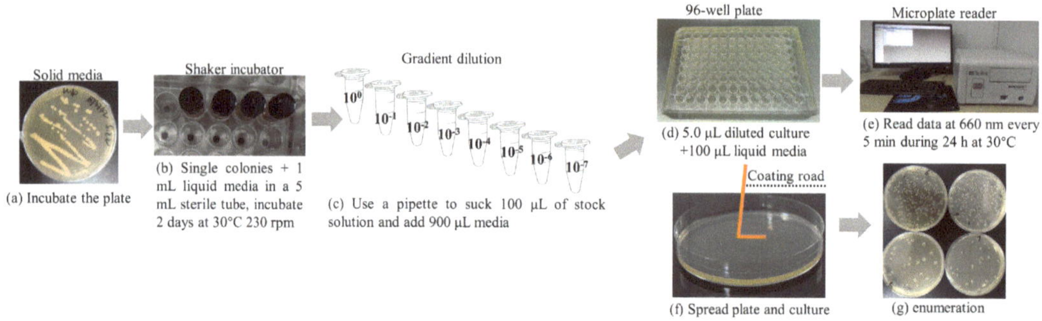

Figure 1. A roadmap of the platform-based colorimetric method to detect microbial counts using the high-throughput method. (**a**) Strains were activated in solid plates; (**b**) the strains were cultured in the corresponding liquid media; (**c**) the bacterial solution was gradient diluted using liquid media; (**d**) 5.0 μL of diluted culture was injected into a 96-well plate, followed the addition of 100 μL of combined media; (**e**) a commonly used microplate reader with shaking and temperature was used to read data at 660 nm every 5 min for 24 h at 30 °C; (**f**) the culturable cells were counted using the spread plate method; (**g**) enumeration.

The 2-day cultures of BY4742 were counted using the spread plate method. Subsequently, the BY4742 cultures were diluted using YPD medium to yield 100, 200, 300, 400, 500, 600, and 1000 cells per 5 μL. The 5 μL of each dilution were pipetted into a 96-well microplate. A total of 100 μL of optimized medium was then added to each well. Each set of experiments was repeated at least 12 times.

2.5. Cell Viability Detection

The OP50 cultures were diluted to 1 cell per 5 μL with LB medium. A total of 1mL of the dilutions was added to Eppendorf tubes and heated in a metal bath. The heating temperatures were 50 °C and 60 °C, and the heating times were 15, 30, and 45 min. When the heating was finished, the dilutions of 5 μL were pipetted into a 96-well microplate. A total of 100 μL of optimized medium was then added to each well. The 5 μL dilutions were carefully pipetted onto an agar plate that had been divided into grids. Each set of experiments was repeated at least 24 times.

2.6. Detection Procedure of the Microplate Reader

The microorganisms were cultured using an Epoch2 microplate spectrophotometer (BioTek, Winooski, VT, USA) with a constant agitation frequency of 567 cpm at 30 °C, and the intensity of the cultured cells was monitored by recording the optical density (OD) value every 5 min over a 24 h period at 660 nm.

2.7. Data Analysis

The raw data were exported to Microsoft Excel by GEN5, and the OD curve was plotted. The biomass, doubling time, lag time, and viability can be calculated according to our method described previously [12,13]. It is important to note that the OD value is 0.3 for both BY4742 and OP50, and it is 0.15 for the LGG strain.

3. Results

3.1. Detection Principle of High-Throughput Method for Microbial Detection

The combined medium (100 μL) was inoculated into a 96-well plate containing a 5 μL sample, and a microplate reader was applied to monitor the growth curve. Specifically, the microplate reader can incubate a maximum of 96 wells/samples per assay at 30 °C and a constant agitation frequency of 567 cpm. The OD value of each well at 660 nm was recorded every 5 min for a duration of 24 h. The quantification method is shown in Figure 2. Sample serial dilution was used to establish the upper and lower detection limits. Based on the CFUs on the agar plate, the dilution ratio of CFUs can be computed with great precision. The biomass, lag time, and viability can be calculated according to our method described previously [12,13]. The minimal dilution in the linear range is defined as 100% viability. Relative survival can be obtained by measuring its viability. The CFU/mL in sample (Q) can be calculated as follows:

$$Q = N_{\text{min dilution}} \times V_n \times 200$$

where $N_{\text{min dilution}}$ is the CFU in the minimal dilution in the linear range, Q is the total number of CFU/mL in a sample, and V_n is the relative survival.

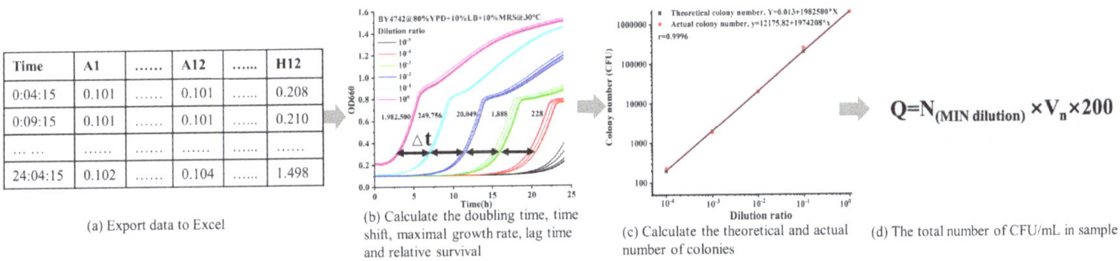

Figure 2. A roadmap of the platform-based colorimetric method to quantify microbial counts using the high-throughput method. (**a**) The data were exported to excel; (**b**) the formula for doubling time, lag time, and survival was calculated; (**c**) the diluted ratio and actual number of colonies were evaluated; (**d**) survival was used to determine the total CFU/mL in each sample.

3.2. Effect of Different Media on the Growth of Microorganisms

The majority of studies published have been concerned with the relationship between the growth of an organism and the amount of some essential nutrient in the medium [19]. Every strain typically has a preferred medium. The development of a combined medium is a key area of research for this approach because a broadly applicable medium can detect a variety of strains. The results of the study, which examined the effects of three media (YPD, LB, and MRS) and two temperatures (37 °C and 30 °C), are displayed in Figure 3a–f. The growth curves of BY4742, OP50, and LGG were significantly influenced by the effects of YPD, LB, and MRS media. In YPD medium, BY4742 and OP50 grew more effectively, and LGG could also grow. The strains' growth tendency in all three media remained unchanged by the incubation temperature. The results indicated that YPD medium might be used in more microorganisms. Subsequently, the different proportions of two media were studied (Supporting Figure S1), with YPD accounting for 50%, 70%, and 90%, respectively. The growth curves of BY4742 and LGG could be affected when the proportion of LB was set at 50%.

Furthermore, as shown in Figure 3g–l, the different proportions of the three media were studied. Among them, the percentages of YPD were 40%, 60%, and 80%, and the percentages of LB and MRS were 30%, 20%, and 10%. Considering the growth of three strains, these results indicated that the most widely applied combination among strains of BY4742, OP50, and LGG was 80% YPD + 10% LB + 10% MRS. The doubling time and lag time were further calculated, as shown in Supporting Figure S2. For BY4742, the optimal medium (80% YPD + 10% LB + 10% MRS) had little effect on the doubling time and lag time. Even though the doubling time for OP50 was noticeably longer, the additional time was roughly 5 min. For LGG, the doubling time and lag time had no significant change with only a slight reduction.

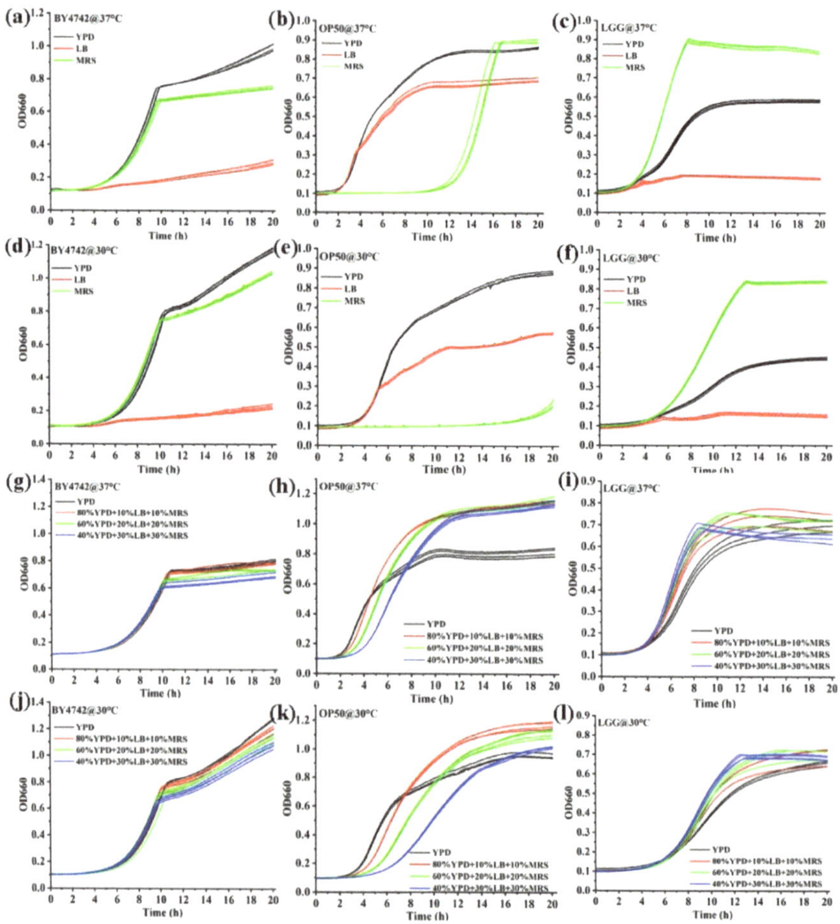

Figure 3. The YPD, LB, and MRS media significantly influenced the growth curves of BY4742, OP50, and LGG. (**a**) BY4742 grew better in YPD and MRS than LB at 37 °C; (**b**) OP50 grew better in YPD and LB than MRS at 37 °C; (**c**) LGG grew better in MRS and YPD than LB at 37 °C; (**d**) BY4742 grew better in YPD and MRS than LB at 30 °C; (**e**) OP50 grew better in YPD and LB than MRS at 30 °C; (**f**) LGG grew better in MRS and YPD than LB at 30 °C; (**g**) the growth curves of BY4742 at 37 °C in three combined media; (**h**) the growth curves of OP50 at 37 °C in three combined media; (**i**) the growth curves of LGG at 37 °C in three combined media; (**j**) the growth curves of BY4742 at 30 °C in three combined media; (**k**) the growth curves of OP50 at 30 °C in three combined media; (**l**) the growth curves of LGG at 30 °C in three combined media.

3.3. Evaluation of Reproducibility, Precision, and Accuracy of the High-Throughput Method

To validate the platform-based colorimetric method to quantify microbial counts using a high-throughput method, the growth curves of different dilutions (10^0–10^{-7}) were investigated. For three strains (BY4742, OP50, and LGG), a 5 μL culture was inoculated into a 96-well plate, and each well was supplemented with 100 μL of optimal medium (80% YPD + 10% LB + 10% MRS). The cell growth kinetics were monitored by recording the OD value every 5 min for 24 h at 660 nm. Based on the growth curves, the viability of cells was obtained. The relative survivals were computed according to the viability. In a linear range between the dilution ratio and relative survival, the cell viability of the minimal dilution is defined as 100%, and the CFU of the minimal dilution was recorded using the spread plate. The actual CFU was calculated by the survival, and the diluted ratio CFU was calculated by the dilution times. The actual CFUs are shown in Figure 4a,d,g. The actual and diluted ratio CFUs were fitted linearly, as shown in Figure 4b,e,h. The correlations between the actual and theoretical CFUs were quite high in the three different strains: BY4742 (r = 0.9996), OP50 (r = 0.9991), and LGG (r = 0.9995). The accuracy of the high-throughput method was evaluated using the standard deviation, as shown in Figure 4c,f,i. According to the difference in strains' doubling times, the orders of magnitude that could be detected by the high-throughput method were five, seven, and seven for BY4742, OP50, and LGG, respectively. The CFU/mL could be calculated as follows: $Q = N_{MIN\ dilution} \times V_n \times 200$, and the $N_{MIN\ dilution}$ values were as follows: BY4742 (N_0 = 1,982,500), OP50 (N_{-1} = 1,315,000), and LGG (N_{-1} = 2,935,000). The platform-based colorimetric method provides high precision detection for quantifying microbial counts. In the meantime, the CFU measurements of our protocol ranged from five to seven orders of magnitude.

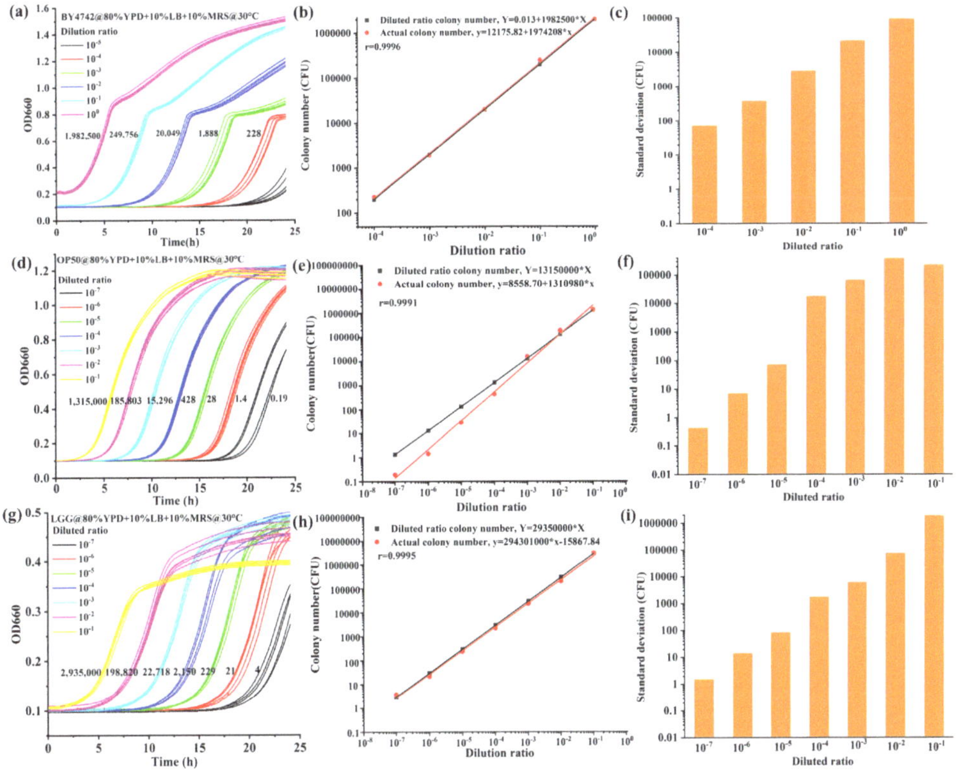

Figure 4. The growth curves of the three strains varied in differently diluted ratios. (**a**) The growth curves of BY4742 were different according to dilutions of 100 and 10-5; (**b**) the r value of correlation between the actual and diluted ratio CFU for different dilutions was 0.9996 in BY4742; (**c**) the standard deviation of the actual CFU for different dilutions of BY4742; (**d**) the growth curves of OP50 were different according to dilutions from 10^{-1} to 10^{-7}; (**e**) The r value of correlation between the actual and theoretical CFU for different dilutions was 0.9991 in OP50; (**f**) the standard deviation of the actual CFU for different dilutions of OP50; (**g**) the growth curves of LGG were different according to dilutions from 10^{-1} to 10^{-7}; (**h**) the r value of correlation between the actual and theoretical CFU for different dilutions was 0.9995 in LGG; (**i**) the standard deviation of the actual CFU for different dilutions of LGG.

3.4. The High-Throughput Method for Microorganism Detection at Trace Cell Level

Trace samples of OP50 and BY4742 were detected by the high-throughput platform. The representative results of single cell detection based on OP50, including growth curves, biomass, and lag time, are shown in Figure 5d–f. The OD value of two cells was greater than that of one cell in 24 h. The biomass of the two cells group was twice that of the one cell group. The two cells group had a shorter lag time than that of the one cell group. In order to further observe the actual cell number of OP50, the dilutions of 5 µL were dropped onto the solid plates, and the results are shown in Figure 5a, b. When the dilution of one cell per 5 µL was dropped onto the solid plate, 57 CFUs were observed in a 52-well solid plate. A total of 57 CFUs were recorded on a solid plate with 52 grids. Among all the wells, one cell per 5 µL was acquired at 34.6% probability (Figure 5c). Using the same method, the dilution of two cells per 5 µL were dropped one by one. There were 95 CFUs observed in a 50-well solid plate, and two cells per 5 µL were acquired at 26.0% probability. The representative growth curves of BY4742 (cell numbers: 0, 100, 200, 300, 400, 500, 600, and 1000) are shown in Figure 5g, and the OD value can reach 0.3 with 100 cells. The correlation between the actual CFU and the diluted ratio CFU in the trace sample was calculated, and the correlation coefficient of r was 0.9931(Figure 5h).

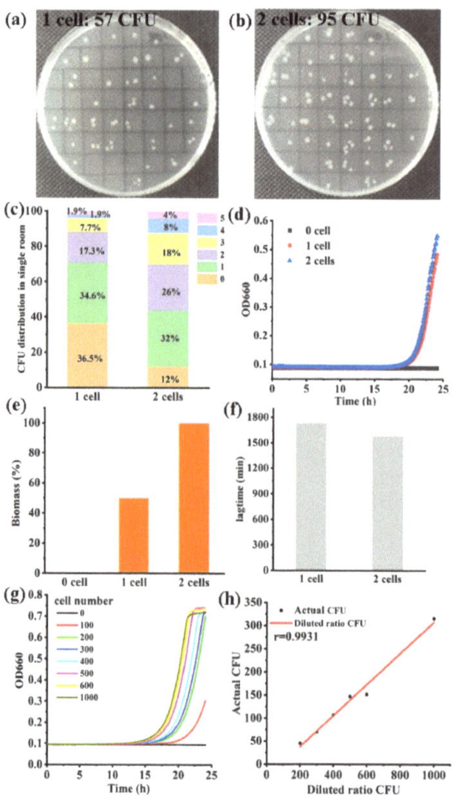

Figure 5. The high-throughput platform realized a trace sample for cell number detection. (**a**) The dilution of 1 cell per 5 µL was dropped onto the solid plate; (**b**) the dilution of 2 cells per 5 µL was dropped onto the solid plate; (**c**) the cell number distribution in the solid plates was counted; (**d**) the representative growth curves of single cell detection based on OP50; (**e**) the biomass result of a single cell was detected based on OP50; (**f**) the lag time of a single cell was detected based on OP50; (**g**) the representative growth curves of the trace cell were detected based on BY4742; (**h**) the correlation between the actual CFU and the diluted ratio CFU in the BY4742 sample was calculated.

3.5. The High-Throughput Method for Cell Viability Detection

The thermal damage degree of OP50 was further studied. The OP50 was diluted to one cell per 5 µL, and the dilutions were heated for 15 min, 30 min, and 45 min at 50 °C in a metal bath, respectively. The actual cell number of OP50 was dropped onto the solid plate, and the result is shown in Figure 6c, and the CFU number slightly decreased after heat treatment at 50 °C at different times. The CFU distributions were not significantly different on the solid plate, as shown in Figure 6e. The growth curves are shown in Figure 6a, and the results indicate that the growth curve changed with different degrees of thermal damage. But the spread plate method can only observe whether the cells are growing or not. The OP50 dilutions were heated for 15 min, 30 min, and 45 min at 60 °C in a metal bath, respectively. The actual cell number of OP50 was dropped onto the solid plate, and the CFU number changed significantly after heat treatment at 60 °C at different times, as shown in Figure 6d. The growth curves are shown in Figure 6b. The colonies did not grow on the solid plate when the cells were treated at 60 °C for 45 min. Although colonies grew on the solid plate at 60 °C for 15 min or 30 min, whether the cells were damaged or not could not be observed on the solid plate. Through the growth curve, the high-throughput method achieved damaged cell detection.

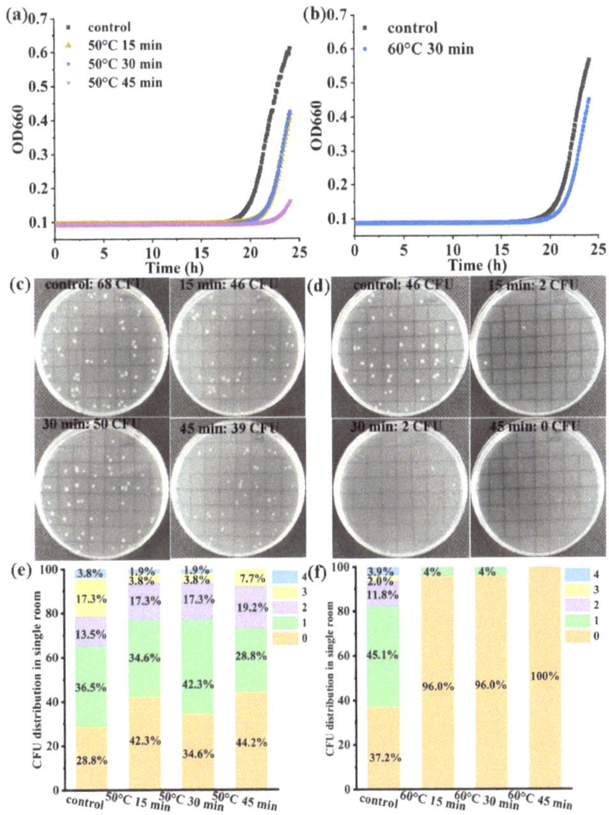

Figure 6. The high-throughput method achieved damaged cell detection. (**a**) The representative growth curve of OP50 was detected after 50 °C for 15 min, 30 min, and 45 min treatment; (**b**) the representative growth curve of OP50 was detected after 60 °C for 15 min, 30 min, and 45 min; (**c**) the dilution of 1 cell per 5 µL treated at 50 °C was dropped onto the solid plate; (**d**) the dilution of 1 cell per 5 µL treated at 60 °C was dropped onto the solid plate; (**e**) the cell number distribution on the solid plate was counted after 50 °C for 15 min, 30 min, and 45 min; (**f**) the cell number distribution was counted in solid medium after 60 °C for 15 min, 30 min, and 45 min.

4. Discussion

The main advantage of the high-throughput method is a more then 20-fold reduction in time, reagent use and plastic cost as compared with the spread plate method (Table 1). To measure the time saved by the high-throughput method, we compared two main steps including the preparation time and the experimental time. More time was saved in the plating and counting step. The spread plate method required 26 h to manually plate and count 288 conditions. The high-throughput method required 1.25 h to add samples into a 96-well plate and calculate the cells. In terms of preparation, the high-throughput method was also quicker than the spread plate method. Using a microplate reader, one researcher required 1.25 h to measure the CFUs of 96 samples. The resource savings and plastic waste reduction of the two methods were then compared. Seven pipette tips per sample were required as standard in the spread plate method as each sample must be diluted and then separately transferred to a solid plate. In the high-throughput method, two pipette tips were used per sample. Compared with the spread plate method, the plastic required was reduced from 288 culture dishes to one 96-well plate in the high-throughput method and the volume of medium required was reduced from 5760 mL to 9.6 mL. Calculating the cost of pipette tips, medium, and culture plates for 96 samples, we discovered that the spread plate method was more expensive in terms of consumables, costing an average of US$0.9 per sample compared with the high-throughput method which cost US$0.08 per sample. The most time- and resource-intensive step of the classic spread plate method is the dilution series [20], and many of the plastic items are used in this process. One of the objectives of the United Nations' sustainable development goals is responsible consumption and production [21]. The plastic items used in a microbiology laboratory with the platform-based colorimetric method to quantify microbial counts could be reduced by using high-throughput method.

Table 1. Comparison of spread plate method and high-throughput method.

	Spread Plate Method	High-Throughput Method
		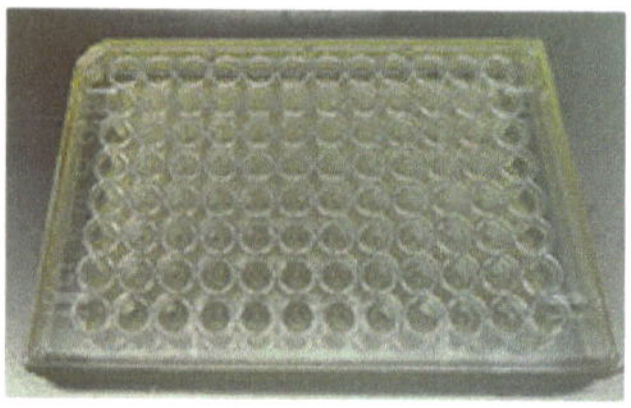
dish	288 culture dishes	one 96-well plate
volume of medium	5760 mL	9.6 mL
preparation time	5 h	2 h
incubation	48 h	24 h
count method	Manual count	Plate reader
experimental time	26 h	1.25 h
US$ per sample	0.9	0.08
result	CFU	lag time, doubling time, survival, CFU

Note: The preparation time of the spread plate method includes the medium preparation and time taken to pour, and the preparation time of high-throughput method includes medium preparation; the experimental time of the spread plate method includes dilution, spread, and counting, and the experimental time of high-throughput method includes sample addition to a 96-well plate and calculation.

The high-throughput method was established for the detection of BY4742, OP50, and LGG. A combined medium (80%YPD + 10%LB + 10%MRS) was utilized for culturing which can be easily applied to these three microorganisms. The high-throughput microbial detection assay offered significant advancements by (1) measuring viability across five–seven orders of magnitude, (2) minimizing the use of consumables, and (3) substantially reducing both the experimental duration and the cost per sample by over 20-fold and

11.25-fold, respectively, in comparison to the traditional spread plate method. According to Codex [22], microbiological risk assessment should explicitly consider the dynamics of microbiological growth, survival, and death in foods [23]. The most common imaging methods used to distinguish between dead and living cells generally include cell staining, segmentation and microscopy, which is costly and technically challenging [24]. However, the high-throughput method not only has the capability to distinguish between living and dead cells, but also to assess whether the cell is damaged or not. Flow cytometry can also determine the absolute bacterial number via staining, and the range of detection is greater than 5000 [7]. Compared to flow cytometry, the high-throughput method is capable of detecting cell numbers as low as one cell. A multiplex RT-qPCR can significantly reduce the time of analysis of culture-based methods and can specifically detect live *L. monocytogenes* [25]. In contrast to RT-PCR, the high-throughput method lacks specificity and has a wide range of applications. The platform-based colorimetric method to quantify microbial counts can be easily applied to different microbial detection in a microbiology laboratory without technical barriers. The colorimetric method used on the platform quantifies microbial counts and provides measurements of the biomass, doubling time, lag time, and survival detections. Furthermore, a trace sample can be realized by this method. The doubling time of the strains significantly influenced the detection limitation, as observed from the growth curves. Although the high-throughput method allowed for trace cell detection, it is important to note that strains with a long doubling time are not suitable for trace cell detection. Additionally, any precipitation, colored media, or turbid samples produced by bacteria in a 96-well plate are not suitable for detection.

5. Conclusions

In total, the high-throughput method is a reproducible, accurate, and simple way to estimate the number of living bacteria cells. It is not laborious and can be performed without any specialized training or equipment beyond a basic microplate reader. Our results showed that the high-throughput method substantially reduced operator time and reagent and plastics costs. Furthermore, the high-throughput method obtained measurements of biomass, lag time, doubling time, and viability from the growth curves. Subsequently, the number of living cells can be calculated based on the viability and survival rates. The CFU number of bacteria was easy to obtain from the survival rates and one-time spread plate count. Finally, not only can living cells be detected, but also the state of the cells can be demonstrated at a single-cell level.

Supplementary Materials: The following supporting information can be downloaded at: https://www.mdpi.com/article/10.3390/foods13182954/s1, Figure S1: Two combined media had different effects on the growth curves of BY4742, OP50, and LGG. (a) The growth curves of BY4742 at 37 °C; (b) The growth curves of BY4742 at 30 °C; (c) The growth curves of OP50 at 37 °C; (d) The growth curves of OP50 at 30 °C; (e) The growth curves of LGG at 37 °C; (f) The growth curves of LGG at 30 °C. Figure S2: The doubling time and lag time of the strains (BY4742, OP50, and LGG) were analyzed in both combined media and YPD medium. (a) The doubling time of BY4742; (b) The lag time of BY4742; (c) The doubling time of OP50; (d) The lag time of OP50; (e) The doubling time of LGG; (f) The lag time of LGG.

Author Contributions: Conceptualization, Z.W.; methodology, Z.W. and L.W.; software, Z.W. and L.W.; validation, Z.W.; formal analysis, L.W.; investigation, Z.W. and L.W.; resources, Z.W.; data curation, L.W.; writing—original draft preparation, L.W.; writing—review and editing, Z.W. and L.W.; visualization, L.W.; supervision, Z.W.; project administration, Z.W.; funding acquisition, Z.W. All authors have read and agreed to the published version of the manuscript.

Funding: This study was supported by the National Natural Science Foundation of China (82171550) and National Key Research and Development Program of China (2023YFF1104204). This study was also supported by the Shanghai Jiao Tong University Medical Engineering Cross Research Fund (23X010302236) and the startup funding from Shanghai Jiao Tong University (23X010502198).

Institutional Review Board Statement: Not applicable.

Informed Consent Statement: Not applicable.

Data Availability Statement: The original contributions presented in the study are included in the article/Supplementary Materials, further inquiries can be directed to the corresponding authors.

Conflicts of Interest: The authors declare no conflicts of interest.

References

1. Ayrapetyan, M.; Oliver, J.D. The viable but non-culturable state and its relevance in food safety. *Curr. Opin. Food Sci.* **2016**, *8*, 127–133. [CrossRef]
2. Khan, Z.A.; Warden, A.R.; Jiang, H.; Abdullah, A.; Ahmad, M.; Jiang, L.; Ding, X. Time-lapse proteomics unveil constant high exposure of non-antibiotic drug induces synthetic susceptibility towards regular antibiotics. *Microbiol. Res.* **2023**, *269*, 127320. [CrossRef] [PubMed]
3. Sanders, E.R. Aseptic laboratory techniques: Plating methods. *J. Vis. Exp.* **2012**, *63*, e3064.
4. Brugger, S.D.; Baumberger, C.; Jost, M.; Jenni, W.; Brugger, U.; Mühlemann, K. Automated counting of bacterial colony forming units on agar plates. *PLoS ONE* **2012**, *7*, e33695. [CrossRef]
5. Bhuyan, S.; Yadav, M.; Giri, S.J.; Begum, S.; Das, S.; Phukan, A.; Priyadarshani, P.; Sarkar, S.; Jayswal, A.; Kabyashree, K.; et al. Microliter spotting and micro-colony observation: A rapid and simple approach for counting bacterial colony forming units. *J. Microbiol. Methods* **2023**, *207*, 106707. [CrossRef]
6. Meyer, C.T.; Lynch, G.K.; Stamo, D.F.; Miller, E.J.; Chatterjee, A.; Kralj, J.M. A high-throughput and low-waste viability assay for microbes. *Nat. Microbiol.* **2023**, *8*, 2304–2314. [CrossRef]
7. Hazan, R.; Que, Y.-A.; Maura, D.; Rahme, L.G. A method for high throughput determination of viable bacteria cell counts in 96-well plates. *BMC Microbiol.* **2012**, *12*, 259. [CrossRef]
8. Tao, Y.; Liu, Q.; Cheng, N. Sea hedgehog-inspired surface-enhanced Raman scattering biosensor probe for ultrasensitive determination of Staphylococcus aureus in food supplements. *Biosens. Bioelectron.* **2024**, *252*, 116146. [CrossRef]
9. Singh, H.; Singh, G.; Kaur, N.; Singh, N. Pattern-based colorimetric sensor array to monitor food spoilage using automated high-throughput analysis. *Biosens. Bioelectron.* **2022**, *196*, 113687. [CrossRef]
10. Santovito, E.; Elisseeva, S.; Bukulin, A.; Kerry, J.P.; Papkovsky, D.B. Facile biosensor-based system for on-site quantification of total viable counts in food and environmental swabs. *Biosens. Bioelectron.* **2021**, *176*, 112938. [CrossRef]
11. Zhang, J.; Huang, H.; Song, G.; Huang, K.; Luo, Y.; Liu, Q.; He, X.; Cheng, N. Intelligent biosensing strategies for rapid detection in food safety: A review. *Biosens. Bioelectron.* **2022**, *202*, 114003. [CrossRef] [PubMed]
12. Murakami, C.J.; Burtner, C.R.; Kennedy, B.K.; Kaeberlein, M. A method for high-throughput quantitative analysis of yeast chronological life span. *J. Gerontol. Ser. A: Biol. Sci. Med. Sci.* **2008**, *63*, 113–121. [CrossRef] [PubMed]
13. Wu, Z.; Song, L.; Liu, S.Q.; Huang, D. A high throughput screening assay for determination of chronological lifespan of yeast. *Exp. Gerontol.* **2011**, *46*, 915–922. [CrossRef]
14. Bergman, L.W. Growth and maintenance of yeast. In *Two-Hybrid Systems: Methods and Protocols*; MacDonald, P.N., Ed.; Humana Press: Totowa, NJ, USA, 2001; pp. 9–14.
15. Freyre-González, J.A.; Treviño-Quintanilla, L.G.; Valtierra-Gutiérrez, I.A.; Gutiérrez-Ríos, R.M.; Alonso-Pavón, J.A. Prokaryotic regulatory systems biology: Common principles governing the functional architectures of *Bacillus subtilis* and *Escherichia coli* unveiled by the natural decomposition approach. *J. Biotechnol.* **2012**, *161*, 278–286. [CrossRef] [PubMed]
16. Guo, C.; Zhang, H.; Guan, X.; Zhou, Z. The anti-aging potential of neohesperidin and its synergistic effects with other citrus flavonoids in extending chronological lifespan of *Saccharomyces Cerevisiae* BY4742. *Molecules* **2019**, *24*, 4093. [CrossRef] [PubMed]
17. Sun, J.; Chen, H.; Qiao, Y.; Liu, G.; Leng, C.; Zhang, Y.; Lv, X.; Feng, Z. The nutrient requirements of *Lactobacillus rhamnosus* GG and their application to fermented milk. *J. Dairy Sci.* **2019**, *102*, 5971–5978. [CrossRef]
18. Wise, K. Preparing spread plates protocols. *Am. Soc. Microbiol.* **2006**, 1–8.
19. Bauchop, T.; Elsden, S.R. The growth of micro-organisms in relation to their energy supply. *Microbiology* **1960**, *23*, 457–469. [CrossRef]
20. Mountcastle, S.E.; Vyas, N.; Villapun, V.M.; Cox, S.C.; Jabbari, S.; Sammons, R.L.; Shelton, R.M.; Walmsley, A.D.; Kuehne, S.A. Biofilm viability checker: An open-source tool for automated biofilm viability analysis from confocal microscopy images. *npj Biofilms Microbiomes* **2021**, *7*, 44. [CrossRef]
21. Alves, J.; Sargison, F.A.; Stawarz, H.; Fox, W.B.; Huete, S.G.; Hassan, A.; McTeir, B.; Pickering, A.C. A case report: Insights into reducing plastic waste in a microbiology laboratory. *Access Microbiol.* **2021**, 3, 000173. [CrossRef]
22. CODEX Alimentarius Commission. *Draft Principles and Guidelines for the Conduct of Microbiological Risk Assessment*; Alinorm 99/13A; FAO: Rome, Italy, 1998.
23. Ross, T.; McMeekin, T.A. Modeling microbial growth within food safety risk assessments. *Risk Anal.* **2003**, *23*, 179–197. [CrossRef] [PubMed]

24. Davey Hazel, M. Life, death, and in-between: Meanings and methods in microbiology. *Appl. Environ. Microb.* **2011**, *77*, 5571–5576. [CrossRef] [PubMed]
25. Azinheiro, S.; Ghimire, D.; Carvalho, J.; Prado, M.; Garrido-Maestu, A. Next-day detection of viable Listeria monocytogenes by multiplex reverse transcriptase real-time PCR. *Food Control* **2022**, *133*, 108593. [CrossRef]

MDPI AG
Grosspeteranlage 5
4052 Basel
Switzerland
Tel.: +41 61 683 77 34

Foods Editorial Office
E-mail: foods@mdpi.com
www.mdpi.com/journal/foods

www.ingramcontent.com/pod-product-compliance
Lightning Source LLC
Chambersburg PA
CBHW041345210926
51850CB00013B/2991
* 9 7 8 3 7 2 5 8 2 4 3 1 1 *